Ecological Studies, Vol. 144

Analysis and Synthesis

Edited by

M.M. Caldwell, Logan, USA
G. Heldmaier, Marburg, Germany
O.L. Lange, Würzburg, Germany
H.A. Mooney, Stanford, USA
E.-D. Schulze, Jena, Germany
U. Sommer, Kiel, Germany

Ecological Studies

Volumes published since 1994 are listed at the end of this book.

Springer

Berlin
Heidelberg
New York
Barcelona
Hong Kong
London
Milan
Paris
Singapore
Tokyo

U. Seeliger   B. Kjerfve (Eds.)

Coastal Marine Ecosystems of Latin America

With 69 Figures and 6 Tables

Springer

Dr. Ulrich Seeliger
Universidade do Rio Grande
Depto. de Oceanografia
Av. Italia km 8
96201-900 Rio Grande, RS
Brazil

Dr. Björn Kjerfve
University of South Carolina
Marine Science Program
Department of Geological Sciences
Earth and Water Sciences Building 508 a
Columbia, SC 29208
USA

ISSN 0070-8356
ISBN 978-3-642-08657-1

Library of Congress Cataloging-in-Publication Data.

Coastal marine ecosystems of Latin America / U. Seeliger, B. Kjerfve (eds.)
   p. cm. - (Ecological studies ; v. 144)
   Includes bibliographical references (p.).

   1. Coastal ecology–Latin America. I. Seeliger, U. (Ulrich), 1944- II. Kjerfve, Björn, 1944- III. Series

QH106.5 .C597 2000
577.5'1'098–dc21   00-044015

This work is subject to copyright. All rights are reserved, whether the whole or part of the material is concerned, spcifically the rights of translation, reprinting, reuse of illustrations, recitation, broadcasting, reproduction on microfilm or in any other way, and storage in data banks. Duplication of this publication or parts thereof is permitted only under the provisions of the German Copyright Law of September 9, 1965, in its current version, and permissions for use must always be obtained from Springer-Verlag. Violations are liable for prosecution under the German Copyright Law.

Springer-Verlag Berlin Heidelberg New York
a member of BertelsmannSpringer Science+Business Media GmbH
© Springer-Verlag Berlin Heidelberg 2001
Softcover reprint of the hardcover 1st edition 2001

The use of general descriptive names, registered names, trademarks, etc. in this publication does not imply, even in the absence of a specific statement, that such names are exempt from the relevant protective laws and regulations and therefore free for general use.

Cover design: *design & production* GmbH, Heidelberg

Preface

The Pacific and Atlantic margins of Central and South America extend for more than 50,000 km between cold polar and tropical climatic zones, encompassing a great diversity of geographically extensive and functionally significant coastal and marine ecosystems. These ecosystems decisively control the rich resources of inshore environments and coastal seas, often far beyond their boundaries. Early European colonizers tapped natural resources initially without harm to the environment, however, a seemingly limitless resource potential stimulated ever increasing exploitation. By the end of the 19th century, long stretches of coastline with major settlements had already undergone significant changes. Owing to unreasonable demands of society, this situation intensified during the 20th century and led to a general decline of remaining resources and irreversible changes in the nature of many ecosystems. Sad examples of this phenomenon are the reduction of over 90 % of Brazil's Atlantic Tropical Forest due to logging and agricultural activities, and the loss of nearly 60 % of Ecuadorian mangroves due to shrimp farming. Lack of proper urban sanitation and unplanned industrial development has led to further destruction. Yet other, still pristine coastal regions have remained relatively unaffected by human intervention mainly due to their relative inaccessibility, but their basic biological components and ecological processes are also poorly known.

The rate of coastal resource development, the expansion of maritime jurisdiction into the 200-mile zone, and the potential threat of sea level rise present a tremendous challenge for coastal area management. Coastal communities now have a vital stake in nearshore ocean uses. Conflicts between growing socio-economic needs but decreasing revenues and expanding ecotourism but mounting environmental degradation provoked a growing awareness of ecological problems in coastal environments by populations and motivated considerable financial support for restoration and management by governments of most Latin American coastal nations.

However, our comprehension of the organizational patterns, the behavior and function, and the rapidly changing nature of coastal marine

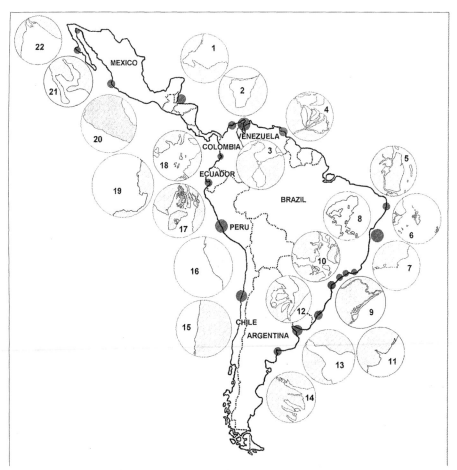

Fig. 1. *1* Gulf of Honduras, *2* Ciénaga Grande de Santa Marta, *3* Maracaibo system, *4* Orinoco River Delta, *5* Itamaracá estuary ecosystem, *6* Abrolhos reefs, *7* Cabo Frio upwelling, *8* Baía de Guanabara, *9* Cananéia lagunar and estuary ecosystem, *10* Paranaguá Bay, *11* Southwest Atlantic Convergence ecosystem, *12* Patos Lagoon estuary, *13* Río de la Plata estuary, *14* Bahia Blanca, *15* Chilean sand beach ecosystem, *16* Peruvian coastal upwelling system, *17* Gulf of Guayaquil and Guayas River estuary, *18* Buenaventura Bay, *19* Eastern Pacific coral reefs, *20* tropical Pacific coast of Mexico, *21* upwelling and lagoonal ecosystems of Baja California, *22* Upper Gulf of California and Colorado River estuary

ecosystems in Latin America is still lagging behind and, without the pertinent knowledge, sustainable management of ecosystem resources is in jeopardy. Perhaps as a first step, a competent revision of the existing, albeit sparse information based on individual and isolated studies often dispersed in local sources outside the international mainstream is now timely and important.

This book, therefore, attempts to present the principal system components that constitute the main control structures of the flows and processes in 22 important ecosystems of Latin America's Pacific and Atlantic coasts (Fig. 1). To facilitate comparisons among systems, each review describes the environmental settings, biotic components and structure of the system, considers trophic processes and energy flow, traces interfaces with adjacent systems, evaluates the modifying influence of natural and/or human perturbations, and suggests management needs. Although the focus of the book is on basic research, by examining many of the unresolved management issues the results have application for coastal managers. The book highlights advances of ecological research in major Latin American coastal marine ecosystems, however, it also addresses gaps in our understanding, and thus should alert scientists, managers, and decision-makers to the lingering danger of neglecting basic research.

We gratefully acknowledge the enthusiasm of our contributing colleagues for distilling the pertinent information from the body of published literature and for making available their many years of research experience in each ecosystem. The Brazilian National Research Council (CNPq) provided supportive funding for this book. Special thanks go to Dr. Luiz Drude de Lacerda for the critical review of several contributions and his helpful suggestions during the editing process of this book.

August 2000 *Ulrich Seeliger*

Contents

The Tectonic and Geological Environment of Coastal South America 1
J.N. Kellog and W.U. Mohriak

1	Introduction	1
2	The Atlantic Continental Margin	1
2.1	Physiographic Features	1
2.2	Structural and Stratigraphic Evolution	4
3	The Andean Margin	6
3.1	Geomorphology and Active Tectonics	6
3.2	Tectonic History	9
4	The North Andean and Caribbean Margins	10
4.1	Geomorphology and Active Tectonics	10
4.2	Tectonic History	11
References	....	13

| 1 | **The Gulf of Honduras** .... | 17 |
| | W.D. Heyman and B. Kjerfve | |

1.1	Introduction	17
1.2	Hydrometeorology and Oceanography	18
1.3	Habitats, Diversity, and Productivity	21
1.3.1	Coastal Lagoons	22
1.3.2	Coastal Embayments	23
1.3.3	Inner Cays	24
1.3.4	Mid Lagoon Cays	25
1.3.5	Barrier Reef	25
1.3.6	Open Ocean Environment	26
1.4	Land Use and Economic Activities	26
1.5	Marine-Protected Areas	28
1.6	Integrated Coastal Zone Management	29
References	....	31

2	The Coastal Lagoon Ciénaga Grande De Santa Marta, Colombia	33
	J. Polanía, A. Santos-Martínez, J.E. Mancera-Pineda, and L. Botero Arboleda	
2.1	Introduction	33
2.2	Environmental Setting	33
2.3	Biotic Components	36
2.3.1	Plankton	36
2.3.2	Invertebrates	36
2.3.3	Fishes	37
2.3.4	Birds, Reptiles, and Mammals	38
2.3.5	Mangroves	38
2.4	Trophic Structure	39
2.5	Human Impact and Fisheries	40
2.6	Management	43
References		44
3	The Maracaibo System, Venezuela	47
	G. Rodríguez	
3.1	Introduction	47
3.2	Environmental Factors	47
3.2.1	Tides and Circulation	49
3.2.2	Physico-chemical Characteristics	50
3.2.3	Nutrients	51
3.3	Biological Components	51
3.3.1	Benthic Communities	51
3.3.2	Plankton Communities	53
3.3.3	Vertebrates	54
3.4	Production Processes	55
3.5	Environmental Issues	56
3.6	Management Considerations	58
References		59
4	The Orinoco River Delta, Venezuela	61
	J.E. Conde	
4.1	Introduction	61
4.2	Environmental Setting	61

4.3	Biological Components	64
4.3.1	Phytoplankton Community	64
4.3.2	Vegetation	64
4.3.3	Vertebrate Fauna	65
4.4	Human Impact and Exploitation	66
4.5	Management Considerations	68
References		69

5 The Itamaracá Estuarine Ecosystem, Brazil — 71
C. Medeiros, B. Kjerfve, M. Araujo, and S. Neumann-Leitão

5.1	Introduction	71
5.2	Environmental Setting	71
5.3	Hydrodynamics	73
5.4	Hydrology	74
5.5	Biota	76
5.6	Trophic Relations	78
5.7	Environmental Problems	78
References		80

6 The Abrolhos Reefs of Brazil — 83
Z.M.A.N. Leão and R.K.P. Kikuchi

6.1	Introduction	83
6.2	Environmental Setting	85
6.3	Reef Organisms	86
6.3.1	Corals	86
6.3.2	Algae	87
6.3.3	Other Biota	88
6.4	Reef Types	90
6.4.1	Coastal Arc	90
6.4.2	Fringing Reefs of the Abrolhos Archipelago	92
6.4.3	Outer Arc	93
6.5	Environmental Impacts and Management	93
References		95

7	The Cabo Frio Upwelling System, Brazil	97
	J.L. Valentin	

7.1	Introduction	97
7.2	Climate and Hydrology	98
7.3	Biological Community	99
7.3.1	Plankton	99
7.3.2	Benthos	101
7.3.3	Nekton	103
References		104

8	Baía de Guanabara, Rio De Janeiro, Brazil	107
	B. Kjerfve, L.D. de Lacerda, and G.T.M. Dias	

8.1	Introduction	107
8.2	Geological Setting	108
8.3	Bathymetry and Bottom Sediment	109
8.4	Climate and Weather	110
8.5	Runoff	111
8.6	Tidal Variability	112
8.7	Salinity, Circulation and Flushing	112
8.8	Coastal Ecosystems	113
8.9	Anthropogenic Impacts	114
References		115

9	The Lagoon Region and Estuary Ecosystem of Cananéia, Brazil	119
	J.G. Tundisi and T. Matsumura-Tundisi	

9.1	Introduction	119
9.2	Environmental Setting	119
9.3	Biotic Components	123
9.3.1	Mangroves	123
9.3.2	Plankton Community	124
9.3.3	Benthic Community	125
9.3.4	Fish Fauna	126
9.4	Nutrient Cycles, Energy Flow, and Food Chains	126
9.5	Sustainable Development and Management Needs	127
References		127

10	**The Subtropical Estuarine Complex of Paranaguá Bay, Brazil**	131
	P. C. Lana, E. Marone, R. M. Lopes, and E. C. Machado	
10.1	Introduction	131
10.2	Environmental Settings	131
10.2.1	Geomorphologic Processes	133
10.2.2	Physical Characteristics	133
10.2.3	Chemical Characteristics	135
10.3	Biotic Components	136
10.4	Trophic Structure and Energy Flow	140
10.5	Human Impacts and Management Needs	141
References		143
11	**The Convergence Ecosystem in the Southwest Atlantic**	147
	C. Odebrecht and J. P. Castello	
11.1	Introduction	147
11.2	Environmental Setting	148
11.3	Fertilization Processes	149
11.4	The Organisms	152
11.4.1	Plankton	152
11.4.2	Macrobenthic Invertebrates	153
11.4.3	Fishes and Cephalopods	155
11.5	Biological Production and Trophic Structure	157
11.6	Human Impacts	158
11.6.1	Fisheries	158
11.6.2	Pollution and Blooms	160
11.7	Management Considerations	161
References		162
12	**The Patos Lagoon Estuary, Brazil**	167
	U. Seeliger	
12.1	Introduction	167
12.2	Environmental Setting	167
12.3	Estuarine Habitats	169
12.3.1	The Water Column	169
12.3.2	Unvegetated Subtidal Soft-Bottoms and Intertidal Flats	170

12.3.3	Sea Grass Beds	171
12.3.4	Marginal Marshes	172
12.3.5	Artificial Hard Substrates	173
12.4	Energy Flow	174
12.4.1	Primary Production Cycles	174
12.4.2	Trophic Relations	176
12.5	Estuary-Coast Interactions	179
12.6	Impact and Management	180
References		182

13	**The Río de la Plata Estuary, Argentina-Uruguay**	185
	H. Mianzan, C. Lasta, E. Acha, R. Guerrero, G. Macchi, and C. Bremec	
13.1	Introduction	185
13.2	Environmental Setting	186
13.2.1	Climate	186
13.2.2	Estuarine Dynamics	187
13.3	Biotic Components	188
13.3.1	Freshwater Environment	188
13.3.2	Mixohaline Environment	190
13.3.2.1	Plankton	190
13.3.2.2	Benthos	191
13.3.2.3	Nekton	192
13.3.2.4	Mammals and Birds	193
13.3.3	Continental Shelf Environment	193
13.3.3.1	Plankton	193
13.3.3.2	Benthos	194
13.3.3.3	Nekton	195
13.3.3.4	Mammals	195
13.4	Biological Significance of the Salt Wedge Regime	196
13.4.1	Head of the Salt Wedge	196
13.4.2	Halocline and Surface Salinity Front	198
13.5	Human Impacts and Management Needs	199
References		200

14	**The Bahia Blanca Estuary, Argentina**	205
	G.M.E. Perillo, M.C. Piccolo, E. Parodi, and R.H. Freije	
14.1	Introduction	205
14.2	Geomorphology	205

14.3	Physical Processes	207
14.3.1	Freshwater Input	207
14.3.2	Tides	208
14.3.3	Winds	209
14.3.4	Salinity and Temperature	209
14.4	Biological Communities	211
14.4.1	Benthos	211
14.4.2	Plankton	212
14.4.3	Nekton	213
14.5	Impact and Management	214
References		215

15 The Sand Beach Ecosystem of Chile ... 219
E. Jaramillo

15.1	Introduction	219
15.2	Environmental Setting	219
15.3	Biological Components	220
15.4	Human Impacts	225
References		226

16 The Peruvian Coastal Upwelling System ... 229
J. Tarazona and W. Arntz

16.1	Introduction	229
16.2	Environmental Characteristics	229
16.2.1	Currents and Winds	229
16.2.2	Coastal Upwelling	231
16.2.3	Oceanographic Features	232
16.3	Community Structure and Dynamics	233
16.3.1	Phytoplankton	233
16.3.2	Zooplankton	234
16.3.3	Benthic Organisms	235
16.3.4	Fishes and Other Vertebrates	236
16.4	El Niño Impact on the Ecosystem	239
16.5	Need for Cautious Management	242
References		243

17	The Gulf of Guayaquil and the Guayas River Estuary, Ecuador	245
	R.R. Twilley, W. Cárdenas, V.H. Rivera-Monroy, J. Espinoza, R. Suescum, M. M. Armijos, and L. Solórzano	
17.1	Introduction	245
17.2	Environmental Setting	246
17.3	Biogeochemistry	249
17.4	Estuarine Habitats and Communities	251
17.4.1	Mangroves	251
17.4.2	Plankton	253
17.4.3	Benthos	253
17.5	Productivity and Trophic Structure	254
17.6	Coupling of Coastal and Estuarine Ecosystems	257
17.7	Human Impacts	258
17.8	Preliminary Models and Ecosystem Management	259
References		260

18	The Estuary Ecosystem of Buenaventura Bay, Colombia	265
	J.R. Cantera and J.F. Blanco	
18.1	Introduction	265
18.2	Environmental Settings	265
18.3	Coastal Habitats and Communities	268
18.3.1	Sandy Beaches	268
18.3.2	Cliffs and Rocky Shores	269
18.3.3	Mangrove Swamps	270
18.3.4	Mud Flats	272
18.3.5	Pelagic Estuarine Environment	273
18.4	Trophic Relations	273
18.4.1	Primary Production	273
18.4.2	Food Webs	274
18.5	Human Impact	276
18.5.1	Pollution	277
18.5.2	Exploitation	277
18.6	Management Needs	278
References		279

19	**Eastern Pacific Coral Reef Ecosystems**	281
	P.W. Glynn	

19.1	Introduction	281
19.2	Environmental Setting	282
19.3	The Eastern Pacific Coral Reef Region	285
19.4	Ecological Processes	291
19.5	Nutrient Cycling, Carbon Production and Trophic Relationships	295
19.6	Functional Interfaces with Adjacent Biotopes	297
19.7	Natural and Anthropogenic Impacts	297
19.8	Management Needs	299
References		303

20	**The Tropical Pacific Coast of Mexico**	307
	F.J. Flores-Verdugo, G. de la Lanza Espino, F.C. Espinosa, and C.M. Argraz-Hernández	

20.1	Introduction	307
20.2	Climate	307
20.3	Coastal Ecosystems and Biota	308
20.3.1	Lagoons	308
20.3.2	Mangroves and freshwater wetlands	309
20.4	Human Impacts and Management Needs	311
20.4.1	Fisheries	311
20.4.2	Environmental Problems	311
References		312

21	**Upwelling and Lagoonal Ecosystems of the Dry Pacific Coast of Baja California**	315
	S.E. Ibarra-Obando, V.F. Camacho-Ibar, J.D. Carriquiry, and S.V. Smith	

21.1	Introduction	315
21.2	Oceanography and Climate	316
21.3	The Upwelling Ecosystem of Baja California	317
21.3.1	Biotic Components and Trophic Relations	318
21.3.2	El Niño Effect on the California Current Ecosystem	321
21.4	Lagoonal Ecosystems	322
21.4.1	Bahía San Quintín	323

21.4.1.1	Environmental Setting	323
21.4.1.2	Biotic Components	324
21.4.1.3	Ecosystem Metabolism	325
21.5	Natural Resources and Human Impact	327
References		329

22 The Colorado River Estuary and Upper Gulf of California, Baja, Mexico 331
S. Alvarez-Borrego

22.1	Introduction	331
22.2	Meteorology	331
22.3	Environmental Settings	333
22.4	Biological Communities	336
22.5	Human Impacts	338
References		339

A Summary of Natural and Human-Induced Variables in Coastal Marine Ecosystems of Latin America 341
B. Kjerfve, U. Seeliger, and L. Drude De Lacerda

1	Climate and Hydrology	341
2	El Niño–La Niña Cycle	344
3	Ocean Currents	345
4	Tides, Waves, and Relative Sea Level	347
5	Human Impacts	349
References		352

Subject Index 355

Contributors

Eduardo Acha

INIDEP, P.O. Box 175, 7600 Mar del Plata, Argentina

C.M. Agraz-Hernández

Instituto de Ciencias del Mar y Limnologia, Universidad Nacional Autonoma de México, Estacion Mazatlan, Joel Montes Camarena s/n, CP 811, Mazatlán 82000, Sinaloa, México

Saúl Alvarez-Borrego

División de Oceanología, Centro de Investigación Científica y de Educación, Superior e Ensenada, B. C., Carretera Tijuana-Ensenada Km 105, Ensenada, Baja California, México

Moacyr Araujo

Departamento de Oceanografia, Universidade Federal de Pernambuco, Av. Arquitetura s/n, 50739-540, Recife, PE, Brazil
e-mail: tritton@elogica.com.br

Leonor Botero Arboleda

COLCIENCIAS, Santafé de Bogotá, Colombia

Mariano Montaño Armijos

Escuela Superior Politécnica del Litoral, Instituto de Química, Guayaquil, Ecuador

Wolf Arntz

Alfred Wegener Institute for Polar and Marine Research, P.O. Box 120161, D-27515, Bremerhaven, Germany

Juan F. Blanco

    Centro de Investigaciones Marinas y Estuarinas, Facultad de Ciencias, Universidad del Valle, A.A.25360, Cali, Colombia
e-mail: jfblanco73@hotmail.com

Claudia Bremec

    CONICET, P.O. Box 175, 7600 Mar del Plata, Argentina

Victor F. Camacho-Ibar

    Instituto de Investigaciones Oceanológicas, Universidad Autónoma de Baja California (IIO-UABC), km 103 Carretera Tijuana-Ensenada, Ensenada, 22860, Baja California, México
e-mail: vcamacho@bahia.ens.uabc.mx

Jaime R. Cantera

    Centro de Investigaciones Marinas y Estuarinas, Facultad de Ciencias, Universidad del Valle, A.A.25360, Cali, Colombia
e-mail: jcantera@biologia.univalle.edu.co

Washington Cárdenas

    Department of Biology, P.O. Box 42451, University of Southwestern Louisiana, Lafayette, LA 70504, USA

José D. Carriquiry

    Instituto de Investigaciones Oceanológicas, Universidad Autónoma de Baja California (IIO-UABC), km 103 Carretera Tijuana-Ensenada, Ensenada, 22860, Baja California, México
e-mail: jdcarriq@bahia.ens.uabc.mx

Jorge P. Castello

    Depto. de Oceanografia, Universidade do Rio Grande, Av. Italia km 8, 96201-900 Rio Grande, RS, Brazil

Jesús E. Conde

    Centro de Ecología, Instituto Venezolano de Investigaciones Científicas (IVIC), Apartado 21827, Caracas 1020-A, Venezuela
e-mail: jconde@oikos.ivic.ve

Contributors

Eunice da Costa Machado

  Centro de Estudos do Mar, Universidade Federal do Paraná,
  Av. Beira-Mar s/n, 83255-000 Pontal do Sul, Paraná, Brazil

Paulo da Cunha Lana

  Centro de Estudos do Mar, Universidade Federal do Paraná,
  Av. Beira-Mar s/n, 83255-000 Pontal do Sul, Paraná, Brazil
  e-mail: lana@aica.cem.ufpr.br

Gilberto T.M. Dias

  Departamento de Geologia – LAGEMAR,
  Universidade Federal Fluminense, 24251-970 Niterói, RJ, Brazil

Francisco Contreras Espinosa

  Universidad Autonoma Metropolitana Iztapalapa,
  Av. Michoacán y la Purísima, Iztapalapa, 09340 México DF

Jorge Espinoza

  Laboratorio de Contaminación, Instituto Nacional de Pesca, Guayaquil, Ecuador

Francisco J. Flores-Verdugo

  Instituto de Ciencias del Mar y Limnologia, Universidad Nacional Autonoma de México, Estacion Mazatlan, Joel Montes Camarena s/n, CP 811, Mazatlán 82000, Sinaloa, México
  e-mail: verduz@mar.icmyl.unam.mx

Rubén H. Freije

  Departamento de Química e Ingeniería Química, Universidad Nacional del Sur, Av. Alem 1253, 8000 Bahía Blanca, Argentina

Peter W. Glynn

  Division of Marine Biology and Fisheries, Rosenstiel School of Marine and Atmospheric Science, University of Miami, 4600 Rickenbacker Causeway, Miami, FL 33149, USA
  e-mail: pglynn@rsmas.miami.edu

Raul Guerrero

    INIDEP, P.O. Box 175, 7600 Mar del Plata, Argentina

William D. Heyman

    The Nature Conservancy, 62 Front Street, Punta Gorda, Belize
    e-mail: will@btl.net

Silvia E. Ibarra-Obando

    Centro de Investigación Científica y Educación Superior de Ensenada
    (CICESE), km 107 carretera Tijuana-Ensenada,
    Ensenada, 22860, Baja California, México
    e-mail: sibarra@cicese.mx

Eduardo Jaramillo

    Instituto de Zoologia, Universidad Austral de Chile, Casilla 567,
    Valdivia, Chile
    e-mail: ejaramil@valdivia.uca.uach.cl

James N. Kellogg

    701 Sumter St., EWSC 203, Department of Geological Sciences,
    University of South Carolina, Columbia, SC 29208 USA
    e-mail: kellogg@sc.edu

Ruy K.P. Kikuchi

    Departamento de Ciências Exatas, Universidade Estadual de Feira de
    Santana, BR116 km 3 s/n, Campus Universitário,
    44031-460 Feira de Santana, Bahia, Brazil
    e-mail: kikuchi@ufba.br

Björn Kjerfve

    Marine Science Program, Department of Geological Sciences, and the
    Belle W. Baruch Institute for Marine Biology and Coastal Research,
    University of South Carolina, Columbia, SC 29208, USA
    e-mail: bjorn@sc.edu

Contributors

Luiz Drude de Lacerda

> Depto. de Geoquimíca, Universidade Federal Fluminense,
> 24020-007 Niterio, RJ, Brazil
> e-mail: geodrud@vm.uff.br

G. de la Lanza Espino

> Instituto de Biologia, Universidad Nacional Autonoma de México,
> Circuito de la Investigación Cientifica s/n,
> C.P. 04510, Cdad. Universitaria, Coyoacan, México, DF

Carlos Lasta

> INIDEP, P.O. Box 175, 7600 Mar del Plata, Argentina

Zelinda M.A.N. Leão

> Laboratório de Estudos Costeiros, Centro de Pesquisa em Geofísica e Geologia, Universidade Federal da Bahia, Rua Caetano Moura 123, 40210-340, Salvador, Bahia, Brazil
> e-mail: zelinda@ufba.br

Rubens Mendes Lopes

> Centro de Estudos do Mar, Universidade Federal do Paraná,
> Av. Beira-Mar s/n, 83255-000 Pontal do Sul, Paraná, Brazil

Gustavo Macchi

> CONICET, P.O. Box 175, 7600 Mar del Plata, Argentina

José Ernesto Mancera-Pineda

> Instituto de Investigaciones Marinas y Costeras de Punta Betín "José Benito Vives de Andréis", Apartado 1016, Santa Marta, Colombia

Eduardo Marone

> Centro de Estudos do Mar, Universidade Federal do Paraná,
> Av. Beira-Mar s/n, 83255-000 Pontal do Sul, Paraná, Brazil

Takako Matsumura-Tundisi

> International Institute of Ecology, Rua Alfredo Lopes 1717,
> 13560-460 São Carlos, SP, Brazil
> e-mail: iie@parqtec.com.br

Carmen Medeiros

> Departamento de Oceanografia, Universidade Federal de Pernambuco,
> Av. Arquitetura s/n, 50739-540, Recife, PE, Brazil
> e-mail: tritton@elogica.com.br

Hermes Mianzan

> CONICET, P.O. Box 175, 7600 Mar del Plata, Argentina
> e-mail: hermes@mdp.edu.ar

Webster U. Mohriak

> Petrobras – Exploration and Production, GEREX/GESIP, Avda. Chile,
> 65 S 1301 E, 20.031-900 Rio de Janeiro, RJ, Brazil
> e-mail: webmohr@ep.petrobras.com.br

Sigrid Neumann-Leitão

> Departamento de Oceanografia, Universidade Federal de Pernambuco,
> Av. Arquitetura s/n, 50739-540, Recife, PE, Brazil
> e-mail: tritton@elogica.com.br

Clarisse Odebrecht

> Depto. de Oceanografia, Universidade do Rio Grande, Av. Italia km 8,
> 96201-900 Rio Grande, RS, Brazil
> e-mail: doclar@super.furg.br

Elisa Parodi

> Instituto Argentino de Oceanografía, CC 107, 8000 Bahía Blanca,
> Argentina

Gerardo M.E. Perillo

> Instituto Argentino de Oceanografía, CC 107, 8000 Bahía Blanca,
> Argentina
> e-mail: perillo@criba.edu.ar

Contributors

M. Cintia Piccolo

    Instituto Argentino de Oceanografía, CC 107, 8000 Bahía Blanca, Argentina

Jaime Polanía

    Instituto de Estudios Caribeños, Universidad Nacional de Colombia, Sede San Andrés, Apartado 438, San Andrés Isla, Colombia
    e-mail: jhpolanv@bacata.usc.unal.edu.co

Victor H. Rivera-Monroy

    Department of Biology, P.O. Box 42451, University of Southwestern Louisiana, Lafayette, LA 70504, USA

Gilberto Rodríguez

    Centro de Ecología, Instituto Venezolano de Investigaciones Científicas, Apartado 21827, Caracas 1020 A, Venezuela
    e-mail: grodrigu@oikos.ivic.ve

Adriana Santos-Martínez

    Instituto de Estudios Caribeños, Universidad Nacional de Colombia Sede San Andrés, Apartado 438, San Andrés Isla, Colombia
    e-mail: jhpolanv@bacata.usc.unal.edu.co

Ulrich Seeliger

    Depto. de Oceanografia, Universidade do Rio Grande, Av. Italia km 8, 96201-900 Rio Grande, RS, Brazil
    e-mail: uli@ecoscientia.com.br

Stephen V. Smith

    School of Ocean and Earth Science and Technology, University of Hawaii, 1000 Pope Road, Honolulu, Hawaii 96822, USA
    e-mail: svsmith@soest.hawaii.edu

Lucía Solórzano

    Escuela Superior Politécnica del Litoral, Instituto de Química, Guayaquil, Ecuador

Rocío Suescum

    Laboratorio de Contaminación, Instituto Nacional de Pesca, Guayaquil, Ecuador

Juan Tarazona

    DePSEA Group, Antonio Raimondi Institute, San Marcos University, Box 1898, Lima 100, Peru

José Galizia Tundisi

    International Institute of Ecology, Rua Alfredo Lopes 1717, 13560-460 São Carlos, SP, Brazil
    e-mail: iie@parqtec.com.br

Robert R. Twilley

    Department of Biology, P.O. Box 42451, University of Southwestern Louisiana, Lafayette, LA 70504, USA
    e-mail: rtwilley@usl.edu

Jean L. Valentin

    Departamento de Biologia Marinha, Instituto de Biologia, Universidade Federal do Rio de Janeiro, 21949-900 Rio de Janeiro, Brazil
    e-mail: jlv@plugue.com.br

The Tectonic and Geological Environment of Coastal South America

J.N. KELLOGG and W.U. MOHRIAK

1 Introduction

The tectonic history and geological factors, such as the present-day geomorphology and vertical motions of the coastline, influence the coastal and marine ecosystems of South America. The continent's Phanerozoic tectonic history is dominated by its separation from Africa and the Mid-Miocene uplift of the Andes. Tectonically, South America is divided into two parts, the Andean chain to the west and a vast stable platform to the east, consisting of exposed Precambrian rocks and shallow sedimentary cover rocks. The Pacific Andean coastline is characterized by high relief, a relatively narrow shelf bordering a deep trench, small drainage basins, and rapid vertical motions of the coast. Low relief, a broad shelf, and extremely large drainage basins and alluvial fans characterize the Atlantic coastline. Today, approximately 93 % of South America's drainage is to the Caribbean and the Atlantic away from the Andes and provides the world's best example of present-day continent-scale drainage control by plate tectonics (Inman and Nordstrom 1971; Hoorn et al. 1995; Potter 1997).

2 The Atlantic Continental Margin

2.1 Physiographic Features

The breakup of Western Gondwana in the Mesozoic (Rabinowitz and LaBrecque 1979) is characterized by aborted rifts in the onshore northeastern region of Brazil (e.g., Reconcavo-Tucano-Jatobá Rift System) and several rifts that evolved to form one of the world's largest series of passive basins (e.g., Pelotas, Santos, Campos, Espírito Santo, Mucuri, Cumuruxatiba, Jequitinhonha, Camamu-Almada, Jacuípe, and Sergipe/Alagoas) along continental margins. A shallow platform with Tertiary sediments onlapping the Precambrian basement to the west, a deep rift trough filled with Neocomian to Aptian sediments, and thinner rift sequences from the slope towards the boundary between continental and oceanic crust characterize sedimentation along the continental margin (Mohriak et al. 1998). Salt tectonics is expressed by thin subhorizontal layers in the proximal regions near the boundary faults or hinge lines that correspond to the western limits of the rift troughs. Towards the slope and deep waters, salt and extensional tectonics affecting the overburden dominate the basin architecture (Mohriak et al. 1995). Gravity-driven compressional tectonics may be found towards the boundary between the continental and oceanic crust (Demercian et al. 1993; Cobbold et al. 1995). The thermal phase of subsidence may result in thick depocenters in some basins (e.g., Santos and Espírito Santo Basin), whereas other basins are characterized by an abrupt shelf-edge and by thin sequences overlying tilted rift blocks or oceanic crust in deep water regions (e.g., Jacuípe Basin).

Physiographic features related to the tectonic framework of the South Atlantic Ocean are (1) the spreading ridge between the South American and the African continents, located closer to the coastline in the northern basins of the South American Margin, and (2) segments of the South Atlantic Margin that are about perpendicular to the ridge (Fig. 1). Linear tectonic features are the Walvis Ridge in Africa, the Vitória-Trindade Ridge in the Eastern Brazilian Margin, and the Florianópolis Fracture Zone, northwest of the Rio Grande Rise. The continental platform is wide in the southern basins and in the northernmost provinces (in front of the Amazon Cone), but is rather narrow in the northeastern margin.

Some major tectonic features in the Eastern Brazilian Margin include the Rio Grande Rise south of the Florianópolis Lineament and the São Paulo Plateau, which is characterized by a large salt diapir province. In the deep water region, salt tectonics is responsible for mini-basins and evacuation troughs, expressed in the sea bottom as concave irregularities. Basinwards from the outer limit of the salt diapir province, several structures with a circular outline correspond to volcanic plugs, such as the Almirante Saldanha Seamount (Fig. 1). Bathymetric and potential field data are helpful to characterize volcanic features, such as the Rio Grande Rise and the Rio Grande Fracture Zone, in the oceanic domain of the Pelotas Basin (Gamboa and Rabinowitz 1981). The Florianópolis lineament (north of the Rio Grande Rise and south of the São Paulo Plateau) is the western prolongation of the Rio Grande Fracture Zone. It is aligned in an E-W direction, and is associated with several igneous plugs (São Paulo or Florianópolis Ridge). The abyssal plain, the continental rise and the platform of the San-

The Tectonic and Geological Environment of Coastal South America

Fig. 1. Main physiographic features of the Brazilian margin. (Modified from Asmus and Baisch 1983)

tos and Campos basins are crossed by a major NW-trending lineament that extends along the Jean Charcot seamounts, and advances onland towards the Cabo Frio region (Souza et al. 1993).

The southeastern segment of the Brazilian margin (Fig. 1) is characterized by several E–W inflections of the coastline along Rio de Janeiro State. This province (Cabo Frio region between the Campos and Santos basins) is characterized by the deflection of the pre-Aptian hinge line from the more general NE trend, and by widespread post-rift volcanic activity (Mohriak et al. 1995). The offshore and onland alkaline plugs are aligned in an E–W direction, from Poços de Caldas to Cabo Frio, and they have been dated as Late Cretaceous to Early Tertiary, with a peak of magmatic activity in the Eocene (Misuzaki and Mohriak 1992).

A major tectonic feature in the eastern Brazilian Margin is the Vitória-Trindade Ridge, an E–W lineament probably associated with a hot-spot (Fig. 1). There are several submarine seamounts adjacent to the Abrolhos Volcanic Complex, and towards the abyssal plain, volcanic islands may reach the seafloor (e.g, Trindade and Martin Vaz Islands). Other volcanic seamounts are aligned along E–W and NW-SE directions, off Bahia, Sergipe, and Alagoas states. The southern portions of the Espírito Santo and the Cumuruxatiba basins correspond to reentrants of the bathymetry (concavities from sea to land), approximately westward from the prolongation of the volcanic features in the oceanic crust. Dating (K-Ar) of the Abrolhos and Royal Charlotte volcanic complexes indicates that they contain mainly Tertiary basaltic rocks, which intruded into the continental platform and extruded above previously deposited sedimentary rocks, masking the possible occurrence of salt layers above the rift.

Northeastern Brazil is characterized by one large onshore rift (Recôncavo-Tucano-Jatobá Rift System), which failed to develop a thermal phase of subsidence, and by the elongated Jacuípe and Sergipe-Alagoas basins, which correspond to rifts that evolved as continental margin sedimentary basins (Ojeda 1982; Matos 1992). The platform in the northeastern margin is characterized by an abrupt shelf-break and by a continental-oceanic crust boundary very close to the shelf-edge (Mohriak et al. 1998).

2.2 Structural and Stratigraphic Evolution

The tectonic evolution of the basins can be divided into pre-rift, rift, protoceanic and continental margin phases (Asmus and Ponte 1973; Ponte and Asmus 1978). Stratigraphic divisions, which take tectonic phases into account, establish four megasequences (pre-rift, continental, transitional, and marine) that are normally separated by erosional unconformities. The pre-rift megasequence occurs only in the northeastern margin (both onshore and offshore) and is subdivided into Paleozoic and Jurassic supersequences. The marine megasequence may be divided into restricted and open marine supersequences. The transitional megasequence is characterized by salt tectonics, which imparts one of the most important controls on the evolution of all the sedimentary basins along the Eastern Brazilian Margin, with the exception of the Pelotas Basin (Chang et al. 1992). The sedimentary fill and structural styles of the basins along the Brazilian Margin are intrinsically related to basement-involved rift phases and basement-detached drift phases that were created during the separation of the South American and African tectonic plates.

The rift phase in most Eastern Atlantic sedimentary basins is characterized by a mosaic of N–S or NE/SW down-stepping synthetic faults, sometimes interrupted by antithetic faults creating a network of half-grabens with internal highs. E-W or NW-SE transfer fault systems accommodate the different stretching rates between the basins. The drift phase started when the stretching and rifting of the continental crust ceased and accretion of oceanic crust began. Salt movements, affecting the overlying rocks, created a series of listric growth faults in the evacuation zones, intraslope sub-basins surrounded by piercing salt domes, salt walls, and thrust faults. The distribution of the salt along the Brazilian and African margins is very irregular. Huge diapirs and salt walls (e.g., Santos and Campos offshore Brazil, Kwanza and Gabon offshore Africa) characterize some basins whereas others have much smaller quantities of evaporites (e.g., Sergipe-Alagoas in Brazil, Rio Muni and Douala in Africa). Lithospheric stretching and rifting ceased in the Eastern Brazilian Margin with the onset of seafloor spreading in the early to middle Cretaceous (probably by late Aptian – early Albian), but there are some indications (e.g., offsets at the base of the Aptian salt and other younger reflectors) of localized reactivations of basement-involved normal faults in the Sergipe-Alagoas and Jacuípe basins up to late Cretaceous time. In these basins, the transition between the continental and marine environments is characterized by recurrent tectonic activity, with concomitant magmatic activity.

The drift phase is characterized by a shallow-water Albian to Cenomanian carbonate platform in most basins, and the Upper Cretaceous is characterized by a major transgression which culminates with drowning of these platforms and predominance of bathyal environments (Chang et al. 1992; Rangel et al. 1994). As a consequence of thermal subsidence and the initiation of oceanic spreading in the South Atlantic, bathymetric depths increased progressively. The late Cretaceous/early Tertiary deepwater sedimentary environments predominated in a number of basins (e.g., Campos Basin). This time interval (Cenomanian to Paleocene) is characterized by predominantly transgressive marine siliciclastic sequences in most basins, with the exception of the Santos Basin where phenomenal episodes of massive clastic progradation resulted in overall regression in the Late Cretaceous, with deposition of continental red beds on the present-day platform. The prograding sequences filled salt-evacuated troughs, particularly in the Cabo Frio region of the Santos Basin (Mohriak et al. 1995). The massive clastic progradation resulted in huge offlapping sequences, associated with a peculiar style of salt tectonics characterized by antithetic faults. The Tertiary is characterized by progradation from the platform towards the deep-water regions, in offlapping sequences, and also by local post-rift volcanic episodes that may occur both onland and in the offshore regions.

3 The Andean Margin

3.1 Geomorphology and Active Tectonics

The Cocos, Nazca, and Antarctic oceanic plates are converging with the South American continental plate along the western (Andean) margin of Central and South America at rates of 70 to 90 mm year^{-1} (Fig. 2). The convergence has produced an unstable coastline with the strongest recorded earthquakes of this century and the greatest relief anywhere in the world, formed by the Andean mountain range (maximum height over 7,000 m), which extends for over 8,000 km along the Pacific coast of South America, and the Perú-Chile Trench (as deep as 8,000 m). East of the Trench, the continental slope rises steeply to a narrow 5- to 120-km-wide shelf. For the first 15 km east of the trench, the slope consists of accretionary complex sediments stacked against the seaward edge of the continental crust. In northern Perú the upper slope and shelf are underlain by up to 5,000 m of Eocene and younger sediments unconformably overlying Mesozoic sedimentary and Paleozoic continental basement rocks. Off northern Chile (23°S to 33°S) the continental slope steepens, with only 100 m of Tertiary sediments overlying consolidated Mesozoic and Paleozoic rocks. The continental shelf is narrow or absent, and sediment along the trench axis is sparse (von Huene 1989). In Perú and northern Chile, the coastal region is a narrow desert belt with coastal plains, low hills, and marine terraces. The coastal belt receives almost no rain, but is watered in the subsurface along river courses by runoff from melting snows of the high Andes. As a result, the continental shelf, slope, and trench are receiving relatively little sediment input at present. Off south-central Chile, sediment floods the trench

The Tectonic and Geological Environment of Coastal South America

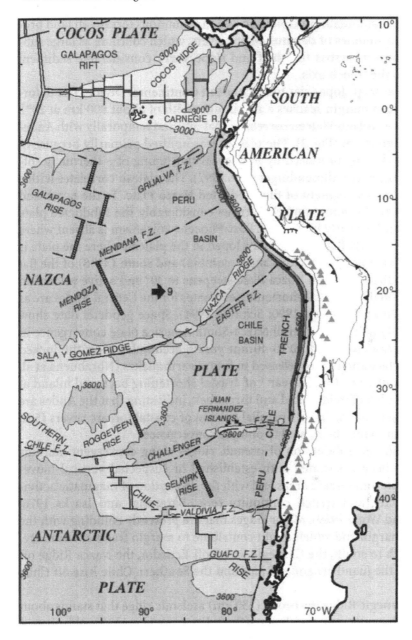

Fig. 2. Regional tectonic features of the Nazca plate and major neotectonic structures of the Andean margin (modified from Greene and Wong 1989; Dewey and Lamb 1992; Kellogg and Vega 1995). *Solid triangles* locate active volcanoes. *Plus signs* indicate active uplift of coastal areas; *minus signs* indicate active subsidence of coastline (Gonzalez-Ferran 1985). *Arrow* denotes relative motion of the Nazca and South American plates (cm year^{-1}). *White areas* denote high ground in the Andes (over 2000 m a.s.l.)

axis because abundant precipitation and mountainous terrain on land provided large volumes of detritus. Major rivers, which continue as large submarine canyons across the shelf and slope, carry considerable sediment directly to the trench axis.

Onshore steep slopes rise to the Andean Continental Divide. The deforming Andean margin reaches a maximum width of about 800 km at 20° S where the Altiplano Plateau correlates spatially and temporally with Andean arc magmatism (Fig. 2). The plateau was uplifted primarily because of crustal thickening produced by horizontal shortening of a thermally softened lithosphere (Allmendinger et al. 1997). The plateau correlates with a 30° east-dipping segment of the subducted Nazca Plate. To the north and south, where the mountain belt narrows considerably, the subducted plate shallows and is nearly horizontal. Post-Pliocene volcanism is absent where the plate is nearly flat, and well developed in the plateau where the plate is steeper. To the north (Ecuador and Colombia) and south (33° S) of the flat segments, the subducted Nazca Plate steepens to 30° and active volcanism resumes. Total horizontal shortening estimates for the Central Andes are as high as 320 km (Sheffels 1990; Schmitz 1994). Space geodetic data show that roughly half of the overall Nazca-South America plate convergence in Perú and Bolivia, about 30 to 40mm year^{-1}, accumulated on the locked plate interface and can be released in future earthquakes (Norabuena et al. 1998). About 10 to 15 mm year^{-1} of crustal shortening occurred inland at the sub-Andean foreland fold and thrust belt, indicating that the Andes are continuing to build. Trench-parallel motion of coastal forearc slivers (5 to 10mm year^{-1}) may be related to oblique convergence.

Collision and subduction of oceanic ridges along convergent margins have long been associated with established or suspected vertical movements of arc and forearc areas and with the disruption of magmatic activity, seismicity, and trench continuity (e.g., Barazangi and Isacks 1976; Greene and Wong 1989). Major ridges that are presently colliding with the Andean margin and which might contribute to margin fragmentation are, from north to south, the Carnegie Ridge off Ecuador, the Nazca Ridge off Perú, and the Juan Fernandez Ridge and the Southern Chile Rise off Chile (Fig. 2).

The Carnegie Ridge is a broad (250 km) aseismic ridge that stands about 1,500 m above the surrounding seafloor. It was produced by volcanism at the Galapagos hot spot. The Carnegie Ridge has been colliding with the Andean margin since at least 2 Ma based on examination of the basement uplift signal along trench parallel transects (Gutscher et al. 1999a). Increased seismic coupling due to the colliding Carnegie Ridge may be correlated with five great (MW >7.8) earthquakes this century and the northeastward displacement of the North Andes block or microplate. The Nazca

Ridge is a broad (250-km-wide) continuous aseismic ridge that rises to as much as 4000 m above the adjacent seafloor. Gutscher et al. (1999b) proposed that a lost oceanic plateau (Inca Plateau, a mirror image of the Marquesas Plateau) has subducted beneath northern Peru. The combined buoyancy of the Inca Plateau and the Nazca Ridge supports a 1,500-km-long segment of flat-dipping subducting Nazca Plate and shuts off arc volcanism. The coastal region of south-central Peru displays raised marine terraces indicating strong uplift since the Late Pliocene. Southward progressing uplift rates during the Late Pleistocene reached 0.7 mm year^{-1} (Macharé and Ortlieb 1992). Two major areas of ridge collisions occur along the Pacific margin of Chile, the aseismic Juan Fernandez Ridge and the Southern Chile Rise. The ridge is discontinuous, averaging 50 km in width, and is composed of irregular small seamounts (1,000 to 1,500 m above the seafloor). The slab dip is shallow, about 15°.

The Southern Chile Rise, a spreading-ridge system, has been colliding with the Andean margin since 14 Ma (Cande and Leslie 1986). The intersection point of the Chile Rise and the Andean margin is migrating northward with time. The Southern Chile Rise collision has caused uplift and plutonism on the continental margin adjacent to the impact point in late Neogene and Quaternary times.

3.2 Tectonic History

Throughout most of its Phanerozoic history, South America's western margin was bordered by a paleo-Pacific Ocean, which connected to the Paleozoic and Mesozoic basins of the ancestral Andean chain and also to pericratonic basins along the western side of South America. Twice during the Paleozoic Era this southwestern side of Gondwana was a convergent margin with proto-Andean chains that largely separated the paleo-Pacific from interior epicontinental seas (Gohrbandt 1992). Four major orogenies are associated with these convergent intervals and correspond to widespread unconformities on the platform (Zalán 1991). To the east of the ancestral Andean chain, four large basins developed on the South American platform: the Paraná, Solimões, Parnaíba and the Amazonas. In the late Paleozoic there was major accretion and subduction along the southwest side of Gondwana, and in the Permian and Triassic periods the La Ventana-Cape Fold Belt developed in north-central Argentina, one that correlates with southern Africa (Milani 1992).

Lateral growth of the continental margin ended in the Paleozoic, and in the early Mesozoic a magmatic arc and associated back-arc basins were developed on the late Paleozoic basement (Mpodozis and Ramos 1989).

Subsequent subduction erosion or strike-slip faulting eliminated large pieces of the Paleozoic forearc assemblages. The central and southern Andean region experienced severe deformation during the Middle to Late Cretaceous, including arc-continent collision and over 100 km of convergence (Vicente 1989). In pre-Miocene time, the western half of the present Amazon watershed is inferred to have had at least one large, western-flowing river called "Sanozama" – Amazonas spelled backward (de Almeida 1974). Some of these sediments were deposited in sedimentary basins up to 5,000 m thick on the subsiding Pacific continental shelf near the Guayaquil Gap.

Uplift in the region of the Altiplano began around 25 Ma, coincident with an increased plate convergence rate (Allmendinger et al. 1997). Shortening slowed in the Altiplano and shifted eastward beginning between 12 and 6 Ma. The base of the upper slope off northern Perú subsided to lower bathyal water depths by middle Miocene time. Thus, the seaward edge of the continent subsided at least 2 km during the early stages of the uplift of the modern Andes. This subsidence suggests a subduction zone dominated by subcrustal tectonic erosion rather than accretion. Sedimentation in the forearc basins of Perú records the detailed oceanographic history of one of the best-developed coastal upwelling regimes. Tectonic subsidence of the Lima Basin since late Miocene has resulted in a 100-km-eastward movement of coastal upwelling.

4 The North Andean and Caribbean Margins

4.1 Geomorphology and Active Tectonics

The Cordillera Occidental (Western Cordillera) and coastal range (Serrania de Baudo) of Colombia are part of the Basic Igneous Complex, which is one of the world's largest ophiolitic complexes, extending from Costa Rica through Panama and Colombia to Ecuador.

The Caribbean margins of Panama, Colombia, and Venezuela consist of extensive fold belts (Case 1974; Bowin 1976) which are part of a deformed zone extending from Costa Rica to eastern Venezuela. Up to 10 km of Tertiary turbidites, carbonates, and fluvial and lacustrine sediments have been preserved in the deformed belt in northern Colombia (Duque-Caro 1979). The North Panama and North Colombia fold belts are characterized by numerous volcanoes that are formed by overpressured Miocene muds and occasionally capped with reefs.

Seismic sections across the southern margin of the Caribbean reveal structures related to the oblique right-lateral convergence of the Caribbean plate and the northern margin of South America (Lu and McMillen 1982; Ladd et al. 1984). Undeformed Caribbean acoustic basement dips landward beneath folded sediments of the Panama and Colombia deformed belts with the youngest sediments involved in the folding.

Global Positioning System (GPS) measurements suggest the existence of a rigid Panama-Costa Rica microplate that is moving northward relative to the stable Caribbean plate (Kellogg and Vega 1995). Northward motion of Panama and Costa Rica relative to the Caribbean plate is independently suggested by the April 1991 Costa Rica earthquake, active folding in the North Panama deformed belt, and a south-dipping Wadati-Benioff zone beneath Panama. The 1991 earthquake produced up to 20–30 cm of uplift along the coast, measured by uplift of coral reefs and other shoreline features and by repeated geodetic leveling. Panama is also continuing to collide eastward with the northern Andes. The Panama-Colombia border area is one of the most seismically active areas in northwestern South America.

Rapid subduction of the Cocos and Nazca oceanic plates is occurring at the Middle America (76 mm year^{-1}), Ecuador (71 mm year^{-1}), and Colombia (54 mm year^{-1}) trenches. The subduction of the buoyant Cocos Ridge, the mirror image of the Carnegie Ridge on the Nazca Plate, is raising the coastline, and may be slowing down subduction at the southern end of the Middle America Trench (Pennington 1981) and detaching the Panama microplate from the Caribbean plate. The oblique subduction of the Nazca Plate and the Carnegie Ridge has resulted in about 8 mm year^{-1} of northeastward "escape" of the northern Andes relative to stable South America. GPS results and seismic imaging also reveal Caribbean-North Andean slow (about 17 mm year^{-1}) oblique amagmatic subduction.

4.2 Tectonic History

We propose a simplified sequence of events for the accretionary development of southern Central America, the northern Andes, and the southern Caribbean margin. Until the middle of the Middle Jurassic, the northern margin of South America was joined with North America on the north. By the end of the Middle Jurassic, South America began to rift apart from North America. Paleomagnetic reconstruction of the Pacific plate and the Pacific-Farallon spreading ridge suggests that an oceanic plateau may have formed the core of the present Caribbean plate (Duncan and Hargraves 1984). The plateau was postulated to have been erupted onto late

Jurassic to early Cretaceous oceanic lithosphere as the Farallon plate passed over the Galapagos hot-spot, initiated in mid to late Cretaceous time (100–75 Ma). The thickened Caribbean volcanic plateau collided with the Greater Antilles Arc, filling the gap between South America and nuclear Central America, in late Cretaceous time (80–70 Ma). Subduction of the Farallon plate commenced behind the plateau. Late Cretaceous marine terrigenous deposits bordered the Guyana Shield and only at the Cretaceous-Tertiary boundary did a collisional belt develop on the northwest margin of the South American plate when the Western Cordillera oceanic arc terrane collided with and overthrust the South American continental margin (Bourgois et al. 1982; Kellogg and Vega 1995). The collisional belt reversed paleoslopes along the North Andean chain to the east, produced incipient foreland basins with deposits that, with continued uplift, spread eastward and onlapped the western limits of the Guyana Shield.

The buoyant, indigestible Caribbean oceanic lithosphere drove the Greater Antilles Arc northeastward, accompanied by subduction of proto-Caribbean crust, until it collided with the Bahama platform in late Eocene time. This collision produced a major eastward change in Caribbean plate motion that resulted in the accretion of the Sinu-San Jacinto sedimentary wedge or South Caribbean deformed belt (Duque-Caro 1979; Toto and Kellogg 1992). Atlantic oceanic lithosphere of the North and South American plates commenced to subduct westward beneath the Lesser Antilles. This produced oblique convergence along the Caribbean-South American margin as the Lesser Antilles volcanic arc and accretionary wedge, the leading edge of the Caribbean plate, moved eastward along the South American margin. The southern end of the buoyant volcanic arc collided obliquely with the South American margin, resulting in a series of nappes and foredeep basins. The foredeeps were controlled by relative Caribbean advance, and thus young to the east: Maracaibo/Barinas Basin, latest Paleocene-early Late Eocene; Guarumen Basin, Early Oligocene; Guárico Basin, Oligocene-Early Miocene; Maturín Basin, Late Oligocene-Middle Miocene; South Trinidad Basin, Early and Middle Miocene. These foredeeps controlled clastic reservoir deposition, and the rapid sedimentation caused the onset of hydrocarbon generation (Pindell et al. 1998). In late Oligocene time, uplift of the Central Cordillera of Colombia (unpublished apatite fission-track ages), the Santa Marta massif, and the Sierra de Perijá (Kellogg 1984) began.

About 6–12 Ma the Panama-Choco island arc arrived on the Caribbean plate at the northwestern margin of South America (Keigwin 1982; Keller et al. 1989; Duque-Caro 1990). The arc-continent collision eventually formed a land bridge between the Americas (3.5 Ma), allowing the migration of mammals, closed the Pacific-Caribbean seaway, drastically changing ocean circulation patterns and perhaps the world's climate, and uplifted the Eastern Cordillera of Colombia. The coastal range (Serrania de Baudo) was accreted to the South American continental margin.

In mid-Miocene time most of the present Central and North Andes developed (e.g., Kroonenberg et al. 1990; Hoorn et al. 1995). Late Miocene deposits in the Magdalena Valley between the Eastern and Central Cordilleras of Colombia demonstrate a Late Miocene age for the Magdalena River (Dengo and Covey 1993). The paleo-Magdalena, and its principal tributary, the paleo-Cauca, both developed in tectonic lows as the Eastern and Central Cordilleras were uplifted, interrupting and severing the earlier foreland basin. This Mid-Miocene event completely disrupted earlier westward and northwestward drainage from the Guyana and Brazilian Shields. North of Colombia, sediments eroded from the uplifted Central and Eastern Cordilleras were deposited in the Cauca and Magdalena fans.

References

Allmendinger RW, Jordan TE, Kay SM, Isacks BL (1997) The evolution of the Altiplano-Puna Plateau of the Central Andes. Annu Rev Earth Planet Sci 25:139–174
Asmus HE, Baisch PR (1983) Geological evolution of the Brazilian continental margin. Episodes 4:3–9
Asmus HE, Ponte FC (1973) The Brazilian marginal basins. In: Nairn AEM, Stehili FG (eds) The ocean basins and margins, vol 1. The South Atlantic. Plenum Press, New York, pp 87–133
Barazangi M, Isacks B (1976) Spatial distribution of earthquakes and subduction of the Nazca plate beneath South America. Geology 4:686–692
Bourgois J, Calle B, Tournon J, Toussaint J-F (1982) The Andean ophiolitic megastructures on the Buga-Buenaventura transverse (Western Cordillera-Valle Colombia). Tectonophysics 82:207–299
Bowin C (1976) Caribbean gravity field and plate tectonics. Geol Soc Am Spec Pap 169
Cande SC, Leslie RB (1986) Late Cenozoic tectonics of the southern Chile trench. J Geophys Res 91(B1):471–496
Case JE (1974) Oceanic crust forms basement of eastern Panama. Geol Soc Am Bull 85:645–652
Chang HK, Kowsmann RO, Figueiredo AMF, Bender A (1992) Tectonics and stratigraphy of the East Brazil Rift System: an overview. Tectonophysics 213:97–138
Cobbold PR, Szatmari P, Demercian LS, Coelho D, Rossello EA (1995) Seismic experimental evidence for thin-skinned horizontal shortening by convergent radial gliding on evaporites, deep-water Santos Basin. In: Jackson MPA, Roberts RG, Snelson S (eds) Salt tectonics: a global perspective. AAPG Mem 6:305–321
de Almeida LFG (1974) A drenagem festonada e seu significado fotogeologico. Sociedade Brasileiro de Geologia. Anais de 38th Congresso, Porto Alegre, RS, 7:175–197
Demercian LS, Szatmari P, Cobbold PR (1993) Style and pattern of salt diapirs due to thin-skinned gravitational gliding, Campos and Santos basins, offshore Brazil. Tectonophysics 228:393–433

Dengo C, Covey M (1993) Structure of the Eastern Cordillera of Colombia: implications for trap styles and regional tectonics. Bull Am Assoc Petrol Geol 77(8):135-137

Dewey JF, Lamb SH (1992) Active tectonics of the Andes. Tectonophysics 205:79-95

Duncan RA, Hargraves RB (1984) Plate tectonic evolution of the Caribbean region in the mantle reference frame. In: Bonini WE, Hargraves RB, Shagam R (eds) Caribbean-South American Plate boundary and regional tectonics. Geol Soc Am Mem 162:81-93

Duque-Caro H (1979) Major structural elements of northern Colombia. AAPG Mem 29:329-351

Duque-Caro H (1990) The Choco Block in the northwestern corner of South America: structural, tectonostratigraphic, and paleogeographic implications. J South Am Earth Sci 3(1):71-84

Gamboa LAP, Rabinowitz PD (1981) The Rio Grande fracture zone in the western South Atlantic and its tectonic implications. Earth Planet Sci Lett 52:410-418

Gohrbandt KHA (1992) Paleozoic paleogeographic depositional developments on the central proto-Pacific margin of Gondwana: their importance for hydrocarbon accumulation. J South Am Earth Sci 6:267-287

Gonzalez-Ferran O (1985) Preliminary neotectonic map of South America. Prepared by Regional Seismological Center for South America (Ceresis), scale 1:5,000,000

Greene HG, Wong FL (1989) Ridge collisions along the plate margins of South America compared with those in the Southwest Pacific. In: Ericksen GE, Cañas Pinochet MT, Reinemund JA (eds) Geology of the Andes and its relation to hydrocarbon and mineral resources, Circum-Pacific Council for Energy and Mineral Resources Earth Sciences Series, vol 11. Houston, Texas, pp 39-57

Gutscher MA, Malavieille J, Lallemand S, Collot JY (1999a) Tectonic segmentation of the North Andean margin: impact of the Carnegie Ridge collision. Earth Planet Sci Lett 168:255-270

Gutscher MA, Olivet JL, Aslanian D, Eissen JP, Maury R (1999b) The "lost Inca Plateau": cause of flat subduction beneath Peru? Earth Planet Sci Lett 171:335-341

Hoorn C, Guerrero J, Sarmiento GA, Lorente MA (1995) Andean tectonics as a cause for changing drainage patterns in Miocene South America. Geology 23:237-240

Inman DL, Nordstrom CE (1971) The tectonic and morphologic classification of coasts. J Geol 79:1-21

Keigwin L (1982) Isotopic paleoceanography of the Caribbean and east Pacific; role of Panama uplift in late Neogene time. Science 217:350-353

Keller G, Zenker CE, Stone SM (1989) Late Neogene history of the Pacific-Caribbean gateway. J South Am Earth Sci 2(1):73-108

Kellogg JN (1984) Cenozoic tectonic history of the Sierra de Perijá and adjacent basins. In: Bonini WE, Hargraves RB, Shagam R (eds) Caribbean-South American Plate boundary and regional tectonics. Geol Soc Am Mem 162:239-261

Kellogg JN, Vega V (1995) Tectonic development of Panama, Costa Rica, and the Colombian Andes: constraints from global positioning system geodetic studies and gravity. In: Mann P (ed) Geologic and tectonic development of the Caribbean Plate Boundary in southern Central America. Geol Soc Am Spec Pap 295, Boulder, Colorado, pp 75-90

Kroonenberg SB, Bakker JGM, van der Wiel AM (1990) Late Cenozoic uplift and paleogeography of the Colombian Andes: constraints on the development of high Andean biotas. Geol Mijnbouw, Spec Issue, 69:279-290

Ladd JW, Truchan M, Talwani M, Stoffa PL, Buhl P, Houtz R, Mauffret A, Westbrook G (1984) Seismic reflection profiles across the southern margin of the Caribbean. In: Bonini WE, Hargraves RB, Shagam R (eds) Caribbean-South American Plate boundary and regional tectonics. Geol Soc Am Mem 162:153-159

Lu RS, McMillen KJ (1982) Multichannel seismic survey of the Colombia Basin and adjacent margins. In: Watkins JS, Drake CL (eds) Studies in continental margin geology. AAPG Mem 34:95-410

Macharé J, Ortlieb L (1992) Plio-Quaternary vertical motions and the subduction of the Nazca Ridge, central coast of Peru. Tectonophysics 205:97-108

Matos RMD (1992) The Northeastern Brazilian Rift System. Tectonics 11(4):766-791

Milani EJ (1992) Intraplate tectonics and the evolution of the Parana Basin, Brazil. In: de Wit MJ, Ransome IGD (eds) Inversion tectonics of the Cape Fold Belt, Karoo and Cretaceous basins of southern Africa. Proceedings of the Conference on Inversion tectonics of the Cape Fold Belt Capetown/South Africa. Balkema, Rotterdam, pp 101-108

Mizusaki AMP, Mohriak WU (1992) Sequências vulcano-sedimentares na região da plataforma continental de Cabo Frio, RJ. Anais do XXXVII Congresso Brasileiro de Geologia - Resumos Expandidos, São Paulo, SP, vol 2, pp 468-469

Mohriak WU, Macedo JM, Castellani RT, Rangel HD, Barros AZN, Latgé MAL, Ricci JA, Misuzaki AMP, Szatmari P, Demercian LS, Rizzo JG, Aires JR (1995) Salt tectonics and structural styles in the deep-water province of the Cabo Frio region, Rio de Janeiro, Brazil. In: Jackson MPA, Roberts DG, Snelson S (eds) Salt tectonics: a global perspective. AAPG Mem 65:273-304

Mohriak WU, Bassetto M, Vieira IS (1998) Crustal architecture and tectonic evolution of the Sergipe - Alagoas and Jacuípe Basins, Offshore Northeastern Brazil. Tectonophysics 288:199-220

Mpodozis C, Ramos V (1989) The Andes of Chile and Argentina. In: Ericksen GE, Cañas Pinochet MT, Reinemund JA (eds) Geology of the Andes and its relation to hydrocarbon and mineral resources. Circum-Pacific Council for Energy and Mineral Resources Earth Sciences Series, vol 11. Houston, Texas, pp 59-90

Norabuena E, Leffler-Griffin L, Mao A, Dixon T, Stein S, Sacks IS, Ocola L, Ellis M (1998) Space geodetic observations of Nazca-South America convergence across the central Andes. Science 279:358-362

Ojeda HAO (1982) Structural framework, stratigraphy, and evolution of Brazilian marginal basins. AAPG Bull 66:732-749

Pennington WD (1981) Subduction of the eastern Panama basin and the seismotectonics of northwestern South America. J Geophys Res 86:10753-10770

Pindell JL, Higgs R, Dewey JF (1998) Cenozoic Palinspastic reconstruction, paleogeographic evolution and hydrocarbon setting of the Northern Margin of South America. In: Paleogeographic evolution and non-glacial eustasy, Northern South America. SEPM Spec Publ 58:45-85

Ponte FC, Asmus HE (1978) Geological framework of the Brazilian continental margin. Geol Rundsch 67:201-235

Potter PE (1997) The Mesozoic and Cenozoic paleodrainage of South America: a natural history. J South Am Earth Sci 10(5/6):331-344

Rabinowitz PD, LaBrecque J (1979) The Mesozoic South Atlantic Ocean and evolution of its continental margins. J Geophys Res 84:5973-6002

Rangel HD, Martins FAL, Esteves FR, Feijó FJ (1994) Bacia de Campos. Bol Geocienc Petrobras 8:203-218

Schmitz M (1994) A balanced model of the southern central Andes. Tectonics 13:484-492

Sheffels BM (1990) Lower bound on the amount of crustal shortening in the central Bolivian Andes. Geology 18(9):812-815

Souza KG, Fontana RL, Mascle J, Macedo JM, Mohriak WU, Hinz K (1993) The southern Brazilian margin: an example of a South Atlantic volcanic margin. 3rd international congress of the Brazilian Geophysical Society, Rio de Janeiro 2:1336-1341

Toto EA, Kellogg JN (1992) Structure of the Sinu-San Jacinto fold belt – an active accretionary prism in northern Colombia. J South Am Earth Sci 5(1):211–222

Vicente JC (1989) Early Late Cretaceous overthrusting in the Western Cordillera of Southern Peru. In: Ericksen GE, Cañas Pinochet MT, Reinemund JA (eds) Geology of the Andes and its relation to hydrocarbon and mineral resources. Circum-Pacific Council for Energy and Mineral Resources Earth Sciences Series, vol 11. Houston, Texas, pp 91–117

von Huene R (1989) Structure of the Andean convergent margin and some implications for hydrocarbon resources. In: Ericksen GE, Cañas Pinochet MT, Reinemund JA (eds) Geology of the Andes and its relation to hydrocarbon and mineral resources. Circum-Pacific Council for Energy and Mineral Resources Earth Sciences Series, vol 11. Houston, Texas, pp 119–129

Zalán PV (1991) Influence of pre-Andean orogenies on the Paleozoic intracratonic basins of South America. IV Simposio Bolivariano, Exploracion Petrolera en Cuencas Subandinas, Memorias, Bogota, Colombia 7

1 The Gulf of Honduras

W.D. Heyman and B. Kjerfve

1.1 Introduction

The Gulf of Honduras is a 10,000 km² tri-national body of coastal and marine waters in the extreme western Caribbean Sea, including portions of the exclusive economic zones of Belize, Guatemala, and Honduras. For purposes of this paper, the Gulf extends from Gladden Spit (16.5° N, 88° W) on the Belize barrier reef to Punta Sal (15.9° N, 87.6° W) on the north coast of Honduras, including Bahía de Amatique and the entire Caribbean coast of Guatemala and the southern portion of the Belize barrier reef lagoon until Placentia (16.5° N, 88.4° W; Fig. 1.1). During the complex geologic history of the Gulf, volcanic and Paleozoic marine sedimentary rocks experienced successive periods of uplift and inundation resulting in metamorphism and karstification (Lara 1994). Volcanic activity is still evident from volcanic eruptions in Guatemala City, hot springs in Rio Dulce and southern Belize, and a recent earthquake (6.6 on the Richter scale). The tectonic plate boundary between North and South American lithospheric plates is marked by the Cayman Trench, which bisects the Gulf from northeast to southwest and stretches landward into the Motagua watershed at the Guatemala-Honduras border. As a result, water depths exceeding 2,000 m extend deeply into the Gulf of Honduras. The 1,000 m isobath parallels the Belize barrier reef for 50 km starting less than 3 km seaward of the reef crest at Gladden Spit and extending towards the southwest. At the Sapodilla Cays in the southern extreme of the barrier reef, the 200 m isobath continues 40 km south towards the Central American land mass. The northwestern edge of the trench in Belize exhibits an extremely steep wall adjacent to the reef and coastal lagoon, while the southeastern edge of the trench rises more gently but without interruption such that the Honduran coast is oceanic to the shore (Fig. 1.1).

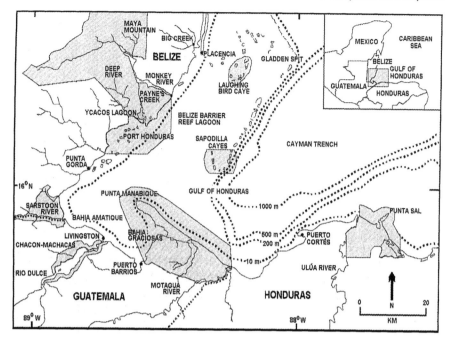

Fig. 1.1. The Gulf of Honduras with major cities, major rivers, and protected areas

1.2 Hydrometeorology and Oceanography

The Gulf is a receiving basin, where terrestrial drainage is distributed by marine processes and impacts ecosystem function. The surrounding watersheds (63,100 km^2) annually receive 3,000–4,000 mm of extremely seasonal rainfall (Portig 1976). A total of 16 rivers annually discharge 76 km^3 (average of 2,400 m^3 s^{-1}) of freshwater into the Gulf. The main rivers are Ulúa (mean of 690 m^3 s^{-1}), Chamelecón (370 m^3 s^{-1}), Motagua (530 m^3 s^{-1}), Rio Dulce (300 m^3 s^{-1}), and Sarstoon with 160 m^3 s^{-1} (Heyman and Kjerfve 1999). Peak transport of freshwater and fluvial sediments occurs in the wet season from July to October, and wet season discharge usually exceeds dry season discharge by a factor of 5–9. The flux of freshwater from the high rainfall areas in southern Belize, Guatemala, and Honduras gives rise to easterly flowing, density-driven surface currents that exit the inner Gulf of Honduras. In contrast, and in response to occasional southerly winds, deep, clear, nutrient-rich oceanic waters from the Cayman Trench enter the Gulf flowing westerly.

The Gulf of Honduras is a tropical water body with air temperatures varying seasonally from 23 to 28 °C (Portig 1976). Surface water temperatures at Gladden Spit vary from 27 to 31 °C and surface water salinities remain high (36–37) during the entire year, but at the end of the wet season or after a hurricane they may drop to 32. Inside the enclosed Amatique Bay, surface water salinities are 8–19 in December and even lower during the wet season (Reyes and Villagrán 1999). Northeasterly to easterly trade winds blow persistently at 3–8 m s^{-1} from December to May, giving rise to both seas and swells, 1–3 m high and with periods from 3 to 7 s. The wet season from July to October is characterized by spectacular thunderstorms with intense, gusty winds from varying directions, alternating with periods of calm. The tail end of North American frontal systems regularly sweeps across the Gulf of Honduras from November to April, giving rise to intense blows with winds from the north or northwest for 1–3 days.

Tropical storms and hurricanes regularly cross the Gulf of Honduras between August and October and have seriously impacted both Belize and Honduras, but generally have less impact on the Caribbean coast of Guatemala (Fig. 1.2). The frequency of tropical storms affecting the Gulf of Honduras is less than 20 storms per century in Bahía de Amatique but the incidence increases to 60 storms per century at the northeastern limit of the Gulf of Honduras (Gentry 1971). During the past 40 years, nine storms, in particular, have impacted the Gulf (Neumann et al. 1993). Two category 5 hurricanes, Hattie in October 1961 and Mitch in October 1998, were particularly destructive. Both originated in the western Caribbean north of Panamá, initially traveled north, turned sharply east, and finally veered south. Hurricane Abby (July 1960), Hurricane Fifi (September 1974), and Hurricane Greta (September 1978) traveled on a track from east towards west and slammed into Belize. Hurricane Francelia (September 1969), Tropical Storm Laura (November 1971), and Tropical Storm Kyle (October 1996) traveled on a very different track, towards the west or southwest, making landfall in Bahía de Amatique. While traversing the Gulf of Honduras from western Honduras on a track towards Belize City, Tropical Storm Gert impacted the Gulf in September 1993. Waves on the Belize barrier reef were hindcast to have had a significant wave height of 10.0 m and a period of 12.7 s preceding the arrival of Hurricane Greta (Kjerfve and Dinnel 1983).

The water circulation in the inner Gulf is dominated by cyclonic, counterclockwise rotating circulation gyres. These gyres are generated south of the Caribbean current as it flows from east to west and crosses the shallow banks between Honduras and Jamaica. These cyclonic gyres, characterized by a central water level depression of 20–30 cm, progress westward along the coast of Honduras towards the Belize barrier reef. A new cyclonic gyre is generated every few months and requires 2–3 months to reach the Belize barrier reef. Long-term current measurements along the shelf edge in the Gulf just north of Gladden Spit indicate a persistent reef-parallel

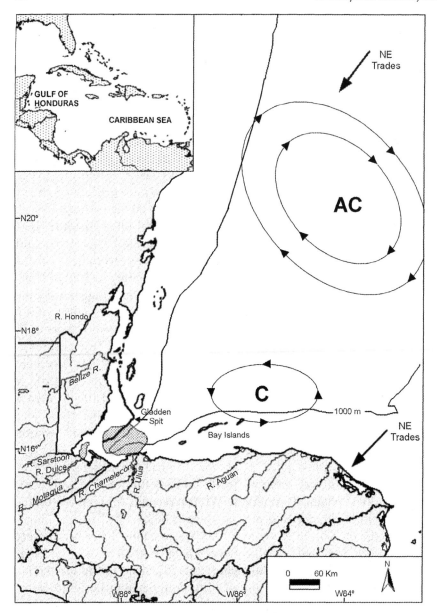

Fig. 1.2. Schematics of water circulation in the Gulf of Honduras, showing the predominance of westward progressing cyclonic meso-scale gyres (*C*; or eddies) just north of the coast of Honduras, and the northwestward progression of much larger anticyclonic meso-scale eddies (*AC*) further to the north. Whereas the cyclonic eddies collide with the Belize barrier reef, the anticyclonic eddies eventually progress into the Gulf of Mexico. As an example of the runoff from rain-filled rivers on the Guatemalan-Honduran border, the extent of a sediment-laden river plume is indicated as a shaded water mass, which interacts with the southern part of the Belize barrier reef

0.1–0.2 m s^{-1} flow towards the south with intermittent flow variability introduced by the cyclonic eddies. The cyclonic eddies are confined to an area south of latitude 18.5° N. In contrast, the Yucatan coast of Mexico is characterized by a northward flowing coastal current, resulting from westward traveling anticyclonic gyres (Fig. 1.2).

The Gulf of Honduras experiences a mixed, mainly semidiurnal microtide with a mean range of approximately 0.2 m. The semidiurnal constituent amplitudes measure 0.06 m for M_2, 0.03 m for S_2, and 0.02 m for N_2, and the main diurnal constituent amplitudes measure 0.03–0.07 m for K_1 and 0.03 m for O_1 (Kjerfve 1981). Although the astronomical tide is weak, tidal currents nevertheless are of substantial magnitude in tidal reef passages and near river mouths, and play a significant role in the spatial dispersion of sediment, nutrients, and larvae. The dominant semidiurnal (M_2) and diurnal (K_1) tidal constituents progress westward along the coast of Honduras and northward along the barrier reef in Belize (Kjerfve 1981). Meteorological tides at times completely dominate the astronomical tides, as when the storm surge associated with Hurricane Mitch (on 27 October 1998) raised the mean water level at Gladden Spit 2.8 m, as the center of the hurricane was progressing towards the south less than 100 km to the east.

1.3 Habitats, Diversity, and Productivity

The Gulf of Honduras includes multiple habitat convergence whereby freshwater habitats meet marine habitats, terrestrial areas meet aquatic habitats, and open ocean environments meet coastal areas. The proximity of estuarine river environments, mangrove wetlands, sea grass beds, the barrier reef, and deep oceanic waters provides for high biological diversity, recognizing that many species require several of these habitats during their ontogenetic development. The high productivity within the Gulf is maintained by river-transported terrestrial nutrients, leaf litter from mangrove wetlands, sea grass beds, coral reefs, and open-ocean nutrients. Large stands of mangroves (Sarstoon-Temash, Port Honduras-Payne's Creek, Punta Manabique, and near Punta Sal) and pervasive sea grass beds (Bahia Graciosas and Port Honduras) are the two most important primary producers and also provide habitat for most juvenile vertebrate and invertebrate marine species of the Gulf (Fig. 1.1). Also, the reefs are critical contributors to the productivity by providing important spawning areas for a variety of fishes. Finally, open ocean waters of the Caribbean occasionally contribute to the productivity of the system via deep water intrusions, sometimes transporting shoals of pelagic fishes towards inshore environments.

Important habitats within the Gulf are sandy beaches and mud-dominated areas at the mouth of the Sarstoon River, which contain the most valuable shrimp fishery in the Gulf. Sand beaches on the Belize coast, within the inner and barrier reef cays, in Punta Manabique, and along the coast of Honduras provide nesting areas for endangered green (*Chelonia mydas*), hawksbill (*Eretmochelys imbricata*), and some leatherback turtles (*Dermochelys coriacea*). Loggerhead turtles (*Caretta caretta*) are seen in coastal waters but generally nest further to the north. However, with the exception of these habitats, the 70-km-long Maya Mountain Marine Area Transect (MMMAT) includes the main coastal forms within the Gulf of Honduras (Heyman 1996; Heyman and Kjerfve 1999; Figs. 1.1, 1.3). The terrestrial area can be divided into the Maya Mountains, karstic limestone relief (Fig. 1.3A), hilly to undulating lowlands (Fig. 1.3B), and coastal flat lands which are highly varied in geologic and soil composition and contain patches of pine forests, open savannas, broadleaf forests, and swamp forests.

1.3.1 Coastal Lagoons

At the coast, rising sea level has resulted in inundation of flat coastal areas, leading to formation of expansive coastal lagoons (Fig. 1.3C). The Ycacos Lagoon is perhaps the best example of this habitat type, and includes a vast complex of dwarf and tall red mangroves *(Rhizophora mangle*; Fig. 1.3D) lining channels and creeks and expansive areas of shallow (1–2 m) open water with a mud bottom. The fringes of the creeks are colonized by taller red mangroves, with dwarf mangroves behind and in the shallow basins. The basins are separated from the coastal embayment by a thin siliceous sand berm (Fig. 1.3E). Buttonwood (*Conocarpus erectus*), white mangrove (*Laguncularia racemosa*), and black mangrove (*Avicennia germinans*) are also present on areas of slightly higher ground. The mangrove leaf litter contributes importantly to biotic production. Litter production reaches 1,187 and 1,570 g dry weight per m^2 per year at two river mouth sites and is thus more productive than many riverine mangrove sites in the Caribbean (Heyman 1996). The coastal lagoons serve as critical nursery habitats for a variety of marine and coastal species of fish and invertebrates. The mud bottom habitat provides for bottom dwelling crustaceans and mollusks, which serve as food for benthic feeding rays, fish, and birds. Cownose rays (*Rhinoptera* sp.) are rare or absent in the rest of Belize but aggregate during January and February in the Ycacos Lagoon. Permit fish (*Trachinotus falcatus*) also feed in the lagoon. The mangrove creeks provide refuge for juvenile fishes of many species, large black snappers (*Lutja-*

The Gulf of Honduras

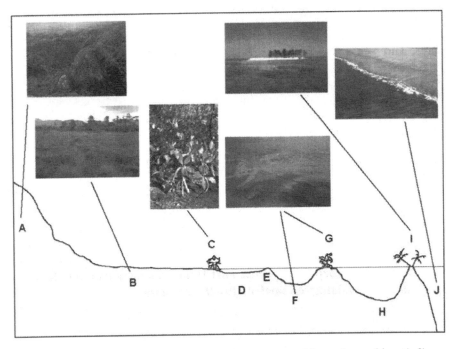

Fig. 1.3. A transect across the Gulf of Honduras: A Maya Mountains and karstic limestone hills, B hilly and undulating lowlands and coastal lowlands, C coastal mangroves, D coastal lagoons, E thin siliceous sand berm, F coastal embayment, G inner cays, H barrier reef lagoon, I barrier reef cays, J barrier reef

nus griseus), and the largest fish in the grouper family, the jewfish (*Epinephelus itajara*), which is listed by the IUCN as an endangered species. The area also supports American crocodiles (*Crocodylus acutus*) and Morelet's crocodiles (*Crocodylus moreleti*; Platt, pers. comm.).

1.3.2 Coastal Embayments

Port Honduras is a coastal embayment (Fig. 1.3F) which receives an annual freshwater input (2.5 km^3) from seven watersheds (Heyman and Kjerfve 1999). The rivers, draining directly to Port Honduras, carry mostly fine-grained clay and mud sediments which are easily re-suspended and thus limit water transparency. As a result, sedimentation and freshwater influence dictate the inshore to offshore gradient in benthic habitats. The embayment contains over 130 mangrove-dominated cays, arranged in three shore-parallel lines (Fig. 1.3G), which protect the coast from the full force of offshore winds and waves, and also filter seaward-moving erosional products from the rivers (Heyman and Kjerfve 1999). Shallow coral flats, abundant mud bottom habitats covered with fine-grained silts and clays, and sea grass beds represent benthic environments within the embayment (Fig. 1.3F). Lush turtle grass (*Thalassia testudinum*) beds are common but patchy in areas of high sedimentation near the mouths of the Ycacos Lagoon and Deep River and on many submerged banks. However, *Halodule* sp. tends to dominate close to the mouth of rivers (Heyman 1996). The sea grass beds are feeding grounds for numerous manatees, which represent a portion of the Caribbean's largest population of the Antillean subspecies of the West Indian manatee (*Trichecus manatus*). The manatee is listed as threatened by the IUCN and listed on Appendix 1 of the CITES convention (Auil 1998). The coastal embayment also supports the Atlantic bottlenose dolphin (*Tursiops truncatus*), and may contain the estuarine dolphin of the genus *Tucoxi*. Accurate identification of these dolphins by cetacean experts would represent a significant range-extension for this largely Amazonian species. Port Honduras remains a refuge for a few remaining endangered sawfish, *Pristis pectinata*.

1.3.3 Inner Cays

The inner cays are almost entirely covered by red mangroves (*R. mangle*). Most of these cays are perched on surrounding carbonate banks, which are broader in the north and northeast (Heyman 1996). The carbonate banks are prograding towards the windward, possibly as a result of increased turbulence and oxygenation. Several of the banks exist without associated cays. Nearly all of the banks deepen gradually to 2 m and then drop sharply into deeper waters (Fig. 1.3G). The shallow, hard-substrate banks serve a critical role in providing habitat for corals, algae, gorgonians, and associated fauna, which cannot survive in the deeper, soft-bottom, low-light environments which otherwise dominate the basin. The mixed community of sea grasses, algae, sponges, gorgonians, and sediment-tolerant corals provides diverse benthic habitats for a rich invertebrate fauna (115 species minimum) from ten phyla (i.e., Mollusca, Arthropoda, Echinodermata, Annelida, Porifera). As refuges for invertebrates, the banks serve as critical feeding habitats for a variety of fish, including the permit, *Trachionotus falcatus*, which is prized as a sport fish. The banks are also critical for recruitment of juvenile lobsters (*Panulirus argus*), queen conchs (*Strombus gigas*), and coral reef fishes.

1.3.4 Mid Lagoon Cays

The cays at the outer edge of Port Honduras have sand beaches and are surrounded by fringing coral reefs that have characteristics of both inshore coastal and offshore barrier reef environments (Stoddart et al. 1982). The presence of about 45 species and subspecies of hard corals within the Snake Cays (A. Harborn, pers. comm.) indicate that these reefs are dominated by common reef building species, such as *Montastrea annularis* and *Diploria strigosa*. These corals do not tolerate the low salinity and high sediment of inshore environments. However, in addition, three species of coral were identified that have not been documented along the barrier reef (*Oculina diffusa*, *O. varicosa*, and *Solenastrea bournoni*), but which are common on inshore reefs. The mid-lagoon cays harbor fish fauna of the offshore barrier reef but serve as a mid-point for many fish during their ontogenetic migration from inshore to offshore reefs, thus fish diversity is exceptional. Within the reefs of the Snake Cays, 118 species of fish (without cryptic blennies and gobies) were listed, notably the blue hamlet *Hypoplecturs gemma*, which previously was not known to exist outside of the Florida Cays (A. Harborn, pers. comm.). The Port Honduras basin harbors an abundance of commercially harvested species and is heavily fished for lobster (*Panularis argus*), conch (*Strombus gigas*), and finfish. The finfish include jewfish (*Epiniphelus itajara*), lane snapper (*Lutjanus syngaris*), mutton snapper (*L. analis*), snook (*Centropomus undecimalis*), barracuda (*Sphyraena barracuda*), tarpon (*Megalops atlanticus*), jacks (Carangidae), mackerel (Scombridae), grunts (Pomadasyidae), groupers (Serranidae), and other snappers (Lutjanidae). Although the mid-lagoon location is responsible for high biodiversity and productivity, it also renders these cays vulnerable to upland processes. During a 1997 coral bleaching event, as much as 80% of the live corals in less than 2 m depth were killed, when surface water temperature reached 31 °C and the salinity reached 18.

1.3.5 Barrier Reef

The Belize barrier reef is separated from Port Honduras by the 40-km-wide and up to 60-m-deep barrier reef lagoon, which is dominated by carbonate bottom sediment (Fig. 1.3H). The fauna is concentrated in shallow (5–20 m depth) bank areas that provide substrate for coral and gorgonian reefs, and seagrass beds. The Belize barrier reef is the largest reef tract in the Western Hemisphere and has been declared as a World Heritage Site.

The reef exhibits typical barrier reef structure with shallow back reef, reef crest, scattered cays (Fig. 1.3I), fore-reef slope, and spur and groove morphology on the deep fore-reef (Fig. 1.3J; Rützler and Macintyre 1982). Several promontories on the crest of the reef, including Gladden Spit, favor spawning aggregations of a variety of commercially important reef fishes (i.e., groupers, snappers, jacks). Similar to other reefs in the Caribbean, the reefs of southern Belize suffer a rapid decline in coral cover and fish abundance due to synergistic impacts of increased sea water temperature, freshwater intrusions, and overfishing.

1.3.6 Open Ocean Environment

Directly offshore from the barrier reef the water-depth Cayman Trench drops steeply to more than 2000 m (Figs. 1.1, 1.3J). The open ocean environment of the Gulf of Honduras provides habitat for species like Spanish mackerel (*Scomboromerous maculatus*), kingfish (*S. cavalla*), bonito (*Sarda sarda*), little tunny (*Euthynnus alletteratus*), blackfin tuna (*Thunnus atlanticus*), yellowfin tuna (*T. albacares*), bigeye tuna (*T. obesus*), wahoo (*Acanthocybium solandri*), sailfish (*Istiophorus playtpterus*), blue marlin (*Makaira nigricans*), and whale sharks (*Rhincodon typus*).

1.4 Land Use and Economic Activities

The Gulf coast is home to half a million inhabitants, with major population centers in Guatemala (Puerto Barrios 100,000 and Livingston 40,000), Honduras (Puerto Cortés 100,000), and Punta Gorda (4,000) as the largest coastal town in southern Belize. The majority of residents are of Mayan, Latin, Creole, and Garifuna descent, thus the region is a melting pot and rich in indigenous cultures and traditional ways of life. Major upland activities surrounding the Gulf of Honduras include agriculture, forestry extraction, cattle farming, aquaculture, fishing and industrial shipping.

The banana industry has been an active contributor to the development of the region, especially in northern Honduras and Guatemala. Bananas grow best in low-lying riparian and flood plain areas along rivers, and large-scale cultivation relies heavily on pesticides, fungicides, herbicides and extra nutrients. As a result, the banana industry has had severe deleterious effects on estuaries, coastal and marine habitats via sedimentation, changes in hydrology, and eutrophication (Readman et al. 1992; Castillo et al. 1997). Logging and cattle ranching are also major industries, especially in Guatemala. When lands are cleared for logging and cattle ranching, the result is often extreme climatic and hydrologic alterations. Flood retention, watershed discharge, and evapotranspiration are depressed, resulting in increased droughts and accelerated flooding. Small-scale, slash-and-burn agriculture is common and is an important contributor to both the overall economy and to local subsistence. This practice is common among the Mayans, and is accompanied by a minimum of environmental damage as long as riparian areas are respected and fallow periods sufficiently long (5–7 years). Increasing population and land pressure, however, are minimizing the sustainability of this method of farming.

Industrial shipping is one of the largest, and potentially most environmentally damaging industries in the Gulf of Honduras. The largest ports are Puerto Cortés in Honduras and Puerto Barrios (Cobigua Port and Santo Tomás de Castilla Port) in Guatemala. Big Creek is a smaller but also important shipping port in Belize. About 90 % of all imports to Guatemala enter through these two ports. Each month, one to three ships enter and exit with petroleum products, while four to seven ships move chemicals (FUNDAECO, unpubl. data). There is a high risk potential for a catastrophic spill, which quickly could cover all of Bahía de Amatique and cause severe ecological damage with impact on coastal tourism and fisheries.

Fishing represents an important contributor to the regional economy and a source of protein for subsistence-based coastal communities. Based on interviews of fishermen around the Gulf of Honduras, the artisanal and small-scale commercial fisheries harvest over 7,200 tons of fishery products, representing US$ 12.6 million of landings (TIDE 1998; FUNDAECO 1998; PROLANSATE 1998). Spiny lobster (*Panulirus argus*), conch (*Strombus gigas*), shrimp (*Paneaus* sp.), and sardines (*Anchoa* sp.) are the major earners. In addition, finfish, including snappers (Lutjanidae), groupers (Serranidae), mackerels (Scombridae), and jacks (Carangidae), have a high landing value. However, the uneven regional population distribution (Guatemala 95 inhabitants km^{-2}, Honduras 49 inhabitants km^{-2}, Belize 9 inhabitants km^{-2}; Windevoxhel et al. 1999) has led to the depletion of the fisheries resources of Guatemala and Honduras, so fishermen from these countries cross illegally into Belize, where fisheries resources are less exploited. The decline in commercially important species has been recognized by fishermen who are now in favor of increased management in the form of declaration of protected areas, bans on gill netting and shrimp trawling, increased enforcement, and closed fishing seasons in certain areas (TIDE 1998; FUNDAECO 1998; PROLANSATE 1998). Sustainable economic alternatives to fish harvesting may be sport fish guiding, which is a small but growing industry in the region.

The close proximity of all major Caribbean coastal and marine habitats in the Gulf of Honduras and their rich biological diversity warrants an enormous potential for a developing ecotourism industry. Tourism in the Gulf region has been increasing steadily, and forms an important component of the economy of Belize, Guatemala, and Honduras. In Belize, 22% of the gross domestic product (GDP) results from tourism (Heyman 1996). Many visitors to the region are ecotourists who come to experience natural environments and intact native cultures. On the Atlantic coast of Guatemala, tourist arrivals were estimated at 90,000 individuals during 1996 by the national tourism authority (INGUAT) who also estimated that tourism will double in the coming decade. In Belize, tourism represents the largest sector of the economy, and the Government has embraced the concept of ecotourism which has become a paradigm for national development.

1.5 Marine-Protected Areas

Each country bordering the Gulf of Honduras has acknowledged the importance of the coastal and marine ecosystems and has designated coastal protected areas. These nationally developed reserves can be viewed as a system of coastal and marine reserves which contribute to the health of the entire ecosystem. The reserves are generally based on a multiple-use concept, and are designed for biodiversity conservation, tourism, and limited sustainable use by local communities. In fact, marine reserves are becoming increasingly popular as people begin to realize the benefits to surrounding fisheries and ecotourism development.

The oldest (1989), functional marine reserve is the Parque Nacionál Janette Kawas in Punta Sal, Honduras (Fig. 1.1), which contains a system of lagoons and estuaries, high coastal ridges, sandy beaches, sea grasses, and nearshore coral reefs. Both the wetlands of Punta Sal and Punta Izopo to the east of Punta Sal are of international significance and have been declared as RAMSAR sites. A significant portion of the coastal and marine area in Guatemala at Punta Manabique was declared as a new preserve in May 1999 (Fig. 1.1). The reserve includes the Bahía Graciosas with contains the Gulf's largest sea grass beds and harbors dolphin, manatee, and juvenile fish of many species. Local fishermen have suggested that the bay be completely closed to fishing, to protect the area as an important spawning and nursery ground for commercially important species (FUNDAECO 1998). The Chacón-Machacas Reserve and the Rio Dulce National Park of the Rio Dulce watershed in Guatemala (Fig. 1.1) were designated as a reserve since

1985, although there has been little management and agricultural and tourism development is increasing.

In Belize, coastal and marine reserves include the Sapodilla Cays Marine Reserve (1996), and Port Honduras Marine Reserve (2000), and National Parks at Laughing Bird Cay and Payne's Creek (1994) (Fig. 1.1). The Sapodilla Cays Marine Reserve forms part of the Belize Barrier Reef World Heritage Site, and yet no management structure is in place. The reefs were exemplary in the region but are now suffering from freshwater inputs and high seawater temperatures. Nonetheless, the five inhabitable cays within the reserve attract tens of thousands of visitors each year to experience the exemplary fishing, diving, snorkeling, and camping resources available in the Sapodillas. The possibility of sustainable management is real, given the existing tourism traffic and regional desire for the conservation of this reserve. Payne's Creek National Park contains freshwater lakes, savanna grasslands, and the Ycacos Lagoon, which is world-famous for its permit fishing. In spite of a lack of management, limited use of the area has left it relatively pristine. The Port Honduras Marine Reserve encompasses 415 km^2 of the coastal waters in southern Belize and straddles a gradient from freshwater-dominated estuaries inshore to marine-dominated waters and coral reefs to the east. The embayment contains more than 130 mangrove-lined cays and protection and management should ensure protection for manatee and contribute to high regional fisheries productivity.

Marine reserves are administered by the Fisheries Department, and are designated as multiple use areas with limited fishing. National Parks, on the other hand, are administered by the Forestry Department and do not allow any extraction. In all cases, the national government is working in close collaboration with local communities and non-governmental organizations (NGOs) to develop co-management agreements to capitalize on local talents, and to overcome a lack of national resources for management.

1.6 Integrated Coastal Zone Management

Integrated coastal management efforts at the national levels in Central America have been limited by lack of information, restricted technical and financial resources, and sectoralism (Windevoxhel et al. 1999). The Coastal Zone Management Authority and Institute in Belize represents one of the more advanced national programs in Central America. Coastal planning in Guatemala and Honduras lags behind. Due to a lack of political will and organizational structure, there is so far only limited success in integrated coastal management on the multinational scale.

In addition to political will, lack of scientific information has limited the implementation of integrated coastal management. The relevance of landscape ecology and ecosystem structure to biodiversity is now better understood (e.g., Noss 1983). The understanding of the dynamics of marine systems lags behind the understanding of terrestrial ecosystems, partly because of the role of water circulation (Mann and Lazier 1996). Marine processes are not constrained by national boundaries, yet decisions on resource utilization can only be made at the national government level. International, national, and local collaboration is required to address the conservation and management of large marine ecosystems. There is a need for marine research to be conducted on large enough geographic scales to help national governments reach a common understanding of shared marine resources, and thus to make effective regional management decisions (Norse 1993).

Recognizing the need for multinational coastal management, the governments of Central America and the World Conservation Union (IUCN) identified the Gulf of Honduras as a critical area for conservation and management. The Gulf has therefore been a focal site for tri-national integrated coastal management under the PROARCA/Costas project. PROARCA operates under the auspices of the Central American Commission for Environment and Development (CCAD) and is largely funded by the US Agency for International Development. The 5-year project is implemented in collaboration between The Nature Conservancy, World Wildlife Fund, the University of Rhode Island's Coastal Resource Center, several Central American NGOs, and national governments. It was designed in direct response to the Alliance for Sustainable Development for Central America (ALIDES), an agreement signed by the seven Central American leaders.

Because of high interest by the non-government sector and lack of financial resources within national governments, many of the coastal and marine reserves are co-managed by national governments, community groups, and local non-government organizations (NGOs). In 1996, a group of nine local NGOs formed the Trinational Alliance for the Conservation of the Gulf of Honduras (TRIGOH). TRIGOH is coordinating the development and implementation of a tri-national system of coastal and marine protected areas, tri-national fisheries management, manatee protection, development of economic alternatives for local residents, port contingency planning, and regional research, monitoring, and planning.

Acknowledgements. This paper has benefited from the contributions of many colleagues and collaborators via thousands of conversations over the period of years. Rachel Graham provided extensive work on the graphics and editing. Nestor Windevoxhel, Marco Cerezo, Marcia Brown, and Janet Gibson provided critical reviews.

Funding was provided in part by the US Agency for International Development and The Nature Conservancy via the Mellon Ecosystem Research Project.

References

Auil N (1998) Belize Manatee Recovery Plan. UNDP/GEF Coastal Zone Management Project, Belize City, Belize

Castillo LE, Cruz E, Rupert C (1997) Ecotoxicology and pesticides in tropical aquatic ecosystems of Central America. Environ Toxicol Chem 16(1):41–51

FUNDAECO – Fundación para el Ecodesarollo y la Conservación de Guatemala (1998) La Voz de Los Pescadores de Guatemala. FUNDAECO, Puerto Barrios, Guatemala

Gentry RC (1971) Hurricanes, one of the major features of air-sea interaction in the Caribbean Sea. In: Symposium on investigations and resources of the Caribbean Sea and adjacent regions, preparatory to the CICAR. UNESCO Paris, pp 80–87

Heyman W (1996) Integrated coastal zone management and sustainable development for tropical estuarine ecosystems: a case study of Port Honduras, Belize. PhD Thesis, University of South Carolina, USA

Heyman WD, Kjerfve B (1999) Hydrological and oceanographic considerations for integrated coastal zone management in southern Belize. Environ Manage 24(2):229–245

Kjerfve B (1981) Tides of the Caribbean Sea. J Geophys Res 86:4243–4247

Kjerfve B, Dinnel S (1983) Hindcast hurricane characteristics on the Belize barrier reef. Coral Reefs 1:203–207

Kjerfve B, Ribeiro CHA, Dias GTM, Filippo AM, Quaresma VS (1997) Oceanic characteristics of an impacted coastal bay: Baía de Guanabara, Rio de Janeiro, Brazil. Continent Shelf Res 17(3):1609–1643

Lara ME (1994) Divergent wrench faulting in the Belize Southern Lagoon: implications for tertiary Caribbean plate movements and quaternary reef distribution. Am Assoc Petrol Geol Bull 77(6):1041–1063

Mann KH, Lazier JRN (1996) Dynamics of marine ecosystems: biological-physical interactions in the oceans. Blackwell, Boston

Neumann CJ, Jarvinen BR, McAdie CJ, Elms JD (1993) Tropical cyclones of the North Atlantic Ocean, 1871–1992. Historical climatology series 6-2. National Climatic Data Center and National Hurricane Center, Asheville, NC

Norse EA (1993) Global marine biological diversity – a strategy for building conservation into decision making. Island Press, Washington, DC

Noss RF (1983) A regional landscape approach to maintain diversity. Bioscience 33:700–706

Portig WH (1976) The climate of Central America. In: Schwerdtfeger W (ed) Climates of Central and South America. Elsevier, Amsterdam, pp 405–478

PROLANSATE – Fundación para la Protección de Lancetilla, Punta Sal y Texiguat (1998) La Voz del Los Pescadores de Honduras. PROLANSATE, Tela, Honduras

Readman JW, Wee Kwong LL, Mee LD, Bartocii J, Nilve G, Rodriguez-Solano JA, Gonzalez-Farias F (1992) Persistent organophosphorous pesticides in tropical marine environments. Mar Pollut Bull 24:398–402

Reyes AS, Villagrán JC (1999) Parámetros oceanográficos del Atlántico Guatemalteco durante el 3er crucero de investigación del recurso camarón, Deciembre, 1998. Centro de Estudios del Mar y Acuicultura CEMA-USAC, Dirección General de Investigación DIGI-USAC, Consejo Nacional de Ciencia y Tenología CONCYT

Rützler K, Macintyre IG (1982) The Atlantic barrier reef ecosystem at Carrie Bow Cay, Belize, I: structure and communities. Smithsonian Institution Press, Washington, DC

Stoddart DR, Fosberg FR, Spellman DL (1982) Cays of the Belize barrier reef and lagoon. Atoll Research Bulletin, vol 256. Smithsonian Institute, Washington, DC, pp 1–75

TIDE – Toledo Institute for Development and Environment (1998) The voice of the fishermen in Southern Belize. Toledo Institute for Development and Environment, Punta Gorda, Belize

Windevoxhel NJ, Rodríguez JJ, Lahmann EJ (1999) Situation of integrated coastal zone management in Central America; experiences of the IUCN wetlands and coastal zone conservation program. Ocean Coastal Manage 42:257–282

2 The Coastal Lagoon Ciénaga Grande de Santa Marta, Colombia

J. Polanía, A. Santos-Martínez, J.E. Mancera-Pineda, and L. Botero Arboleda

2.1 Introduction

Located on the central Caribbean coast of Colombia, the lagoon complex of Ciénaga Grande de Santa Marta (CGSM) is part of the eastern delta of the Magdalena River, Colombia's largest river. The delta and the CGSM rest on a coastal plain, which was formed by marine and fluvial sedimentary depositions during Holocene transgression and regression phases (10,000 B.P.), while the surrounding formations date back to the Cenozoic era. The lagoon complex (1,321 km^2) comprises the Ciénaga Grande (450 km^2), the Ciénaga de Pajarales (120 km^2), several smaller lagoons, creeks, and channels (150 km^2), and mangrove swamps. A narrow, continuous sandbar (Isla de Salamanca) borders the entire CGSM complex to the north. At the eastern end of Isla de Salamanca a 120-m-wide and 10-m-deep inlet (Boca de la Barra) connects the shallow (average depth 1.2 m) Ciénaga Grande lagoon to the Caribbean Sea (Fig. 2.1). The CGSM is a highly productive ecosystem but has suffered severe degradation in the last four decades. However, the system is likely to considerably recuperate if current restoration plans are successful.

2.2 Environmental Setting

The geomorphologic, geophysical and biological characteristics of the CGSM identify the region as a Type I setting (Thom 1984). The region is arid, with a high annual water deficit (mean 1,031 mm) and low annual precipitation (mean 531 mm; Wiedemann 1973), which decreases from the west (760 mm) to the east (450 mm) of the lagoon complex. Rainfall patterns are seasonal, with a dry season (249 mm) beginning in December, including a

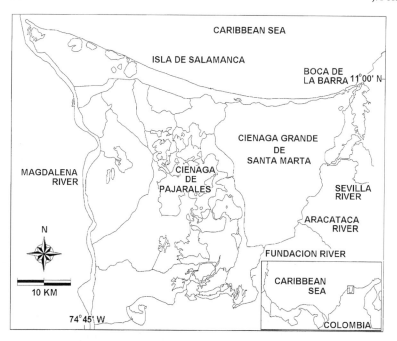

Fig. 2.1. Ciénaga Grande de Santa Marta in the Colombian Caribbean

short and very dry period in July/August, followed by a rainy season (1,268 mm) from September to November. Northeasterly trade winds (3.2–5.4 m s^{-1}) dominate during the dry season. Average air temperature is 27 °C, and relative humidity ranges from 77 to 83%. The sun shines between 169.7 and 2,703.4 h and the average annual irradiance is 2,140 cal cm^{-2}.

Surface currents (5–3 cm s^{-1}) from the east to the west, low-energy waves, and a tidal amplitude of 20–30 cm characterize the central Colombian coast. According to Kjerfve (1986), the CGSM can be defined as a restricted lagoon, owing to the geomorphology of the entrance channel, which acts as a filter, and the vertical mixing of water by winds. Hydrological patterns are a consequence of freshwater discharge by the tributaries. Prior to flood control and drainage projects in the 1970s, the Magdalena River (mean annual flow 7,000 m^3 s^{-1}) supplied most of the freshwater to the Ciénaga Grande complex. The runoff from the Magdalena River is highly seasonal, with high discharge in April, May, and October to December, and low discharge from January to March and in August. In contrast, runoff from tributary rivers (Sevilla, Aracataca, and Fundación) at the east of the CGSM is more constant. These rivers annually add a combined average freshwater flow of 19.25 m^3 s^{-1} from extensive agricultural areas at the foothills of the Sierra Nevada de Santa Marta in the east and southeast.

Circulation patterns in the lagoon are a consequence of the interactions between east-northeasterly trade winds and river discharge. Due to the opposing direction of winds and freshwater runoff, the mixing process of the different water masses is restrained, thus horizontal physico-chemical gradients are common. During the rainy season, the increased freshwater levels in the lagoon cause an outwelling current to the open sea; during the dry season, seawater enters through Boca de la Barra (Inderena-Sodeic 1987).

Sedimentary loads of the Magdalena River ($30-100 \times 10^3$ m^3 day^{-1}) and those from tributaries of Sierra Nevada (180×10^6 tons year^{-1}; Deeb Sossa 1993) have decreased the lagoon depth at several sites from 2.3 to an average of 1.5 m over the last 30 years (Bernal 1996). Muddy-clay sediments with a surficial component of detrital matter predominate in the lagoon, though in some areas gravel represents an important fraction, and sand and shell fragments of marine origin tend to increase toward the sandbar.

The CGSM can be considered a euhaline-mixohaline system, with mean annual water temperature of about 30 °C. Nevertheless, pronounced temporal and spatial salinity gradients are common, resulting from variable freshwater runoff, seawater intrusion, rainfall, and evaporation. During the dry season, salinity in the inlet is about 40 ‰ (parts per thousand), and still may exceed 30 ‰ 2 km up the tributaries. During the short wet season salinity values decrease to about 15–20 ‰ in the center of the lagoon, but during the long wet season salinity may be close to zero even near the inlet (Botero and Mancera-Pineda 1996). Furthermore, every 6 or 7 years the Magdalena River has high discharge and, for extended periods, the entire CGSM experiences salinities close to zero (Kaufman and Hevert 1973). Because of the lagoon shallow depth and constant winds, the water column of the CGSM is normally well mixed. Nevertheless salinity stratification may occur in some areas during the rainy seasons.

Dissolved oxygen concentrations in the lagoon vary between seasons and with location but they generally demonstrate an inverse relationship with salinity distributions. The maintenance of mean dissolved oxygen levels (6.2 mg l^{-1}) in the lagoon can be attributed to mixing of shallow waters by winds, while reduced dissolved oxygen levels (2 mg l^{-1}) are related to pulses of organic matter decomposition and to the introduction of urban or agricultural wastewater. In general, total suspended solids (TSS) in the water column show a clear seasonal pattern and vary as a consequence of interactions between sedimentary loads from the Magdalena River, intrusion of seawater, and the influence of winds. During the dry season, when the water column is well mixed, TSS may reach concentrations of 360 mg l^{-1} (mean Secchi depth 70 cm). During rainy seasons, due to stratification TSS concentration plummets to 15 mg l⁻¹. The organic matter fraction of TSS tends to increase from 28% in the inlet to 60% towards the Fundación River. However, rivers that drain Sierra Nevada de Santa Marta do not appear to be an important source of suspended solids for the CGSM. The concentration of nutrients (nitrogen, phosphate, silicate) fluctuates between months and years and follows steep horizontal gradients. The N:P ratio in the CGSM tends to increase with salinity; however, in the center of the lagoon the ratio increases during rainy seasons at low salinity. Although dissolved nitrogen concentrations have remained about constant in the CGSM during the last 10 years, chlorophyll concentrations have increased. As a result increasing phosphorus levels may have initiated a eutrophication process.

2.3 Biotic Components

2.3.1 Plankton

The phytoplankton community in the CGSM is comprised of pennate (91 species) and centric diatoms (48), dinoflagellates (34), Chlorophyta (31), Euglenophyta (21), and Cyanobacteria (17; Vidal 1995). Some of the typical brackish water species are continuously present at high numbers and may exceed densities of 2×10^6 cells m⁻³. The distribution of zooplankton varies between seasons and with depth. Microzooplankton species are responsible for high abundance during the rainy season, in which three species may account for 84% of abundance. Macrozooplankton and ichthyoplankton species have higher abundance during the dry season. Zooplankton density gradients in the water column may be pronounced during the wet seasons but vertical distribution is uniform during the dry season. Salinity changes and predation by ichthyoplankton influence successional changes between micro- and macrozooplankton components.

2.3.2 Invertebrates

Among the invertebrate fauna, mollusks are represented by approximately 98 species (66 genera and 48 families) of which 61 are of marine origin. Six species occur in mangrove sediments, *Melampus coffeus* being the most abundant among the three gastropods. The fiddler crabs *Uca rapax* and *U. vocator* are the most abundant crabs in areas of mangrove sediments. The abundance of the mangrove crab *Aratus pisonii* is controlled by salinity fluctuations, and reproductive effort and recruitment correlate with rainfall patterns. Biometric and life history traits of this crab appear to be closely related to mangrove productivity. The largest individuals of *A. pisonii* and the highest percentages of ovigerous females occur in arboreal mangroves, while small crabs with low reproductive effort occur in scrub mangrove habitats. Similarly, the large size and high fecundity of the land crab *Cardisoma guanhumi* can be ascribed to the high productivity of the CGSM. Some other species of invertebrates are of commercial importance, like the oyster *Crassotrea rhizophorae*, the conch *Melongena melongena*, and swimming crabs of the genus *Callinectes*.

2.3.3 Fishes

A total of 122 bony fish species (58% with a standard length of <350 mm) of 49 families, mainly Engraulididae (*Anchovia clupeoides, Anchoa parva, Anchoa* spp.), Mugilidae, Gerreidae, and Ariidae, as well as eight cartilaginous fish species, have been reported for the CGSM. Estuarine species represent the largest group (63% of the 122 recorded), 19 of these are coastal species with marine affinities and 9 freshwater species may appear seasonally. Many of the resident fishes (32), which include 24 estuarine species, are widely distributed in the lagoon and all occur near the inlet (Boca de la Barra; Santos-Martínez and Acero 1991), while eight coastal species are more or less restricted to the inlet. Species of the families Poeciliidae, Blenniidae, Eleotrididae, and Gobiidae are probably true residents, as might be species like *Arius proops, Lutjanus cyanopterus, Scomberomorus brasiliensis, Anchoa trinitatis,* and *A. hepsetus,* often considered to be frequent visitors. Most (62%) of the resident ichthyofauna, like Engraulididae (*A. clupeoides, Cetengraulis edentulus, A. parva*) either spawn during the entire year or have two spawning peaks. Most of these species seem to spawn inside the CGSM and Boca de la Barra (estuarine-resident), although some (i.e., *Mugil* spp.) spawn in adjacent marine waters (Santos-Martínez and Acero 1991). Several of the larger species (i.e., *Elops saurus, Tarpon atlanticus, Centropomus undecimalis, Caranx hippos, C. latus*) appear to use the lagoon only for recruitment since their larvae and juveniles occur during the entire year (Cataño and Garzón-Ferreira 1994). Although several species (i.e., *Mugil incilis, Eugerres plumieri, Cathorops spixii*) tolerate a broad range of salinities, changes in the salinity regime between different years influence their abundance and may favor the occurrence of stenohaline (*Triportheus magdalenae, Aequidens pulcher, Caquetaia krausii, Oreochromis niloticus*) and/or freshwater species (*Hoplias malabaricus, Ageneiosus caucanus, Sorubium lima, Pterygoplites undecimalis*).

2.3.4 Birds, Reptiles, and Mammals

The densities and diversity of birds (195 species) in and around the CGSM is noteworthy. The Ciénaga is an important stopover point in the migrations of Neartic Limicolae, some of which spend up to 70% of their life span on the Colombian coast (Botero and Marshall 1994). Mangroves are a major feeding and nesting ground for many of the 80 resident birds. Reproductive colonies of the brown pelican (*Pelecanus occidentalis*) and breeding populations of *Eudocimus ruber* are abundant in the CGSM. Twenty-one bird species depend on wetlands while four others, like the pied water-tyrant (*Fluvicola pica*), the prothonotary warbler (*Protonotaria citrea*), and blackhawks are terrestrial though often associated with water bodies. The most diverse family is the Ardeidae, with nine species of herons and egrets, followed by Tyrannidae, with seven species of flycatchers, and Charadriidae, with six species of shorebirds. Populations of the humming bird *Lepidopyga liliae* and the thrush *Molothrus armenti* are endemic to Isla de Salamanca. Large populations of the American crocodile (*Crocodylus acutus*) and the "babilla" (*Cayman crocodylus*) inhabit the CGSM. Snakes (e.g., *Corallus hortulanus*) and some tortoises are also common. Among the terrestrial mammals, the crab-eating raccoon (*Procyon cancrivorus*), prehensil-tailed porcupine (*Coendou prehensilis*), ocelot (*Felis pardalis*), ring-tail monkey (*Cebus olivaceus*), paca (*Agouti paca*), agouti (*Dasyprocta aguti*), brocket deer (*Mazama americana*), capybara (*Hydrochaeris hydrochaeris*), and bats (*Noctilio* and *Saccopteryx*) occur in mangrove forests and vegetation around the lagoon.

2.3.5 Mangroves

The vegetation of the sandy zones (i.e., Isla de Salamanca) is comprised of *Batis maritima, Sesuvium portulacastrum, Lemairocereus griseus, Prosopis juliflora, Acacia farnesiana, Pereskia colombiana,* and *Opuntia wentiana. Pistia stratiotes, Eichornia crassipes, Neptunia prostrata,* and *Typha dominguensis* are important aquatic species within the CGSM. However, the most conspicuous vegetation are extensive and patchy mangrove forests (50,000 ha) along the shores of the lagoons and channels. The dominant species is the black mangrove *Avicennia germinans* L., followed by the red mangrove *Rhizophora mangle* L., the white mangrove *Laguncularia race-*

mosa Gaertn., and the buttonwood mangrove *Conocarpus erectus* L. Three mangrove physiographies can be recognized. The lagoon is dotted with small islets of *R. mangle* and on the alluvial plain a 500–1,000-m-wide belt of *Avicennia-Laguncularia* fringes the shores of the lagoon. On the coastal sandbar, which separates the lagoon from the sea, all four mangrove species are present. The average height of these stands is 15 m, but 25-m-tall trees are common. The diameter at breast height ranges from 25 to 30 cm in *R. mangle*, from 10 to 40 cm in *A. germinans*, and from 12 to 13 cm in *Laguncularia racemosa* (Cardona and Botero 1995). About eight species of algae (i.e., *Cladophora, Vaucheria, Chaetomorpha, Oscillatoria lutea*) are associated with mangrove prop roots while hydrophilous plants (*Roystonea regia, Heliconia latisphata*) are common on poorly drained soils (slope <1%) near the mangroves. Mangrove forests are habitat for about 51 species of fish of which three (*Gymnothorax funerbris, Myrophis platyrhynus, Gambusia* sp.) are new additions to the list compiled by Santos-Martínez and Acero (1991). The barnacle *Balanus eburneus* is the most abundant (59% of all collected individuals) sessile organism though the crabs *Eurypanopeus dissimilis, Pachygrapsus gracilis*, and *Petrolisthes armatus*, the polychaete worm *Nereis virens*, and amphipods are also abundant. Especially prop roots of red mangrove stands are an important habitat for more than 44 invertebrates (e.g., bryozoans, mollusks, oysters) and three fishes (Santos-Martínez and Acero 1991; Cataño and Garzón-Ferreira 1994).

2.4 Trophic Structure

The CGSM is one of the most productive ecosystems among in- and offshore environments along the Caribbean coast (Hernández and Gocke 1990). The production of both phytoplankton (990 g C m^{-2} year^{-1}) and mangrove litter (15.7 tons ha^{-1} year^{-1}), 95% of which degrades within 16 weeks, contributes to the carbon budget in the lagoon. Sequential pulses of primary production by plankton communities and seasonal export of mangrove detritus sustain high secondary production in the CGSM; thus the proportion of plankton- or mangrove-derived carbon in tissues of shrimp and fish varies with seasons and distance from mangrove habitats. Among fish species, about 67% are carnivores, 15% omnivores, 10% plankton feeders, 6% ingest detritus, and 2% are herbivores. Some species are opportunistic (*Centropomus ensiferus, C. undecimalis*) and feed at all levels, while others (i.e., *Elops saurus, Tarpon atlanticus*) change their diet from insects to mainly juvenile of Mugilidae and Clupeidae as they grow older. Most of the birds (91 %) in the CGSM are secondary consumers, 19 species preferentially feed on insects, and 14 on fish and aquatic invertebrates. Only five species, which include the ruddy ground dove (*Columbina talpacoti*), orange-winged parrot (*Amazona amazonica*), brown-throated parakeet (*Aratinga pertinax*), and the hummingbird (Trochilidae), feed directly on seeds, fruits, and nectar (Botero and Marshall 1994).

2.5 Human Impact and Fisheries

During the last 40 years, human populations have profoundly modified the entire CGSM, leading to a general degradation of the ecosystem and to a substantial decrease in the abundance and diversity of fauna and flora. The discharge of ever increasing volumes of untreated sewage into the CGSM has elevated fecal coliform concentrations from 4,250 l^{-1} in 1972 to 46,000 l^{-1} in 1995, when *Salmonella* was detected in oysters (*Crassotrea rhizophorae*). Heavy metals (i.e., cadmium, zinc, lead, nickel) are added to the CGSM with suspended matter from the rivers of Sierra Nevada and during overflows of the Magdalena River. Although metal concentrations vary between seasons, copper concentrations in oysters have increased from 5.9–42.8 μg per g dry weight in 1982 to 50.2–179.1 μg per g dry weight in 1988.

The Colombian government constructed a coastal highway along the entire length of Isla de Salamanca between 1956 and 1960, thereby eliminating several important outlets through the Barrier Island and restricting water exchange with the Caribbean Sea to one bridged opening (100 m) at Boca de la Barra. To drain wetlands and to prevent floodwaters from reaching agriculture lands, earthen dikes were built along the eastern bank of the Magdalena River during the 1970s. As a result, the flow from several tributaries of the Magdalena River to the mangrove forests was reduced or eliminated. Furthermore, the diversion of water for agricultural irrigation has reduced freshwater runoff, while the erosion of deforested watersheds has led to ever-increasing sediment loads from the rivers. The modification of hydrologic patterns and sedimentation processes in the CGSM has induced a gradual increase in salinity and total suspended solids (Fig. 2.2) and has altered the composition, distribution, abundance, and diversity of organisms (Botero and Mancera-Pineda 1996). The discharge of wastewater, the addition of agricultural fertilizers, and phosphate-enriched runoff from the rivers of Sierra Nevada have also started a process of eutrophication in the lagoon complex. Changes in the structure of the phytoplankton community caused a reduction of species diversity by more than 10% while the density of some groups, like diatoms (2–20 × 10^6 cells l^{-1}) and

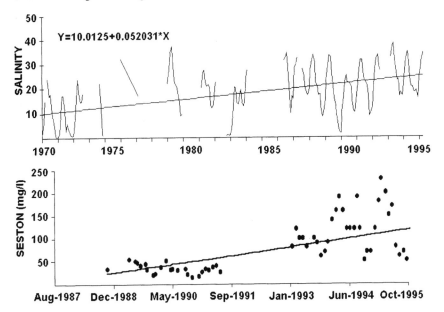

Fig. 2.2. Time series of salinity and seston in the Ciénaga Grande de Santa Marta

dinoflagellates (1.25–7.5 × 10⁶ cells l⁻¹), increased considerably. Frequent blooms of cyanobacteria (*Anacystis cyanea, Nostoc commune*; Hernández and Gocke 1990) have lowered oxygen concentrations, which together with high TSS appear to have led to high mortalities of the oyster (Squires and Riveros 1971; Hernández 1983; von Cosel 1985) and fish (Bula-Meyer 1985; Mancera and Vidal 1994; Botero and Mancera-Pineda 1996). The disappearance of the clam *Polymesoda solida* has been associated with an increase of salinity (Botero and Mancera-Pineda 1996).

During the last 40 years, more than 70% of the original mangrove forest area (350 km²) has disappeared and is still being lost at an increasing rate (1,531 ha year⁻¹; Cardona and Botero 1995) due to hyper-saline (>100‰) soils that now occur during most of the year (Botero and Mancera-Pineda 1996). Today most of the mangrove trees in western and northwestern areas of the CGSM are dead, while mangrove stands with intermediate damage occur in the central Pajarales Complex. The only large forests of healthy trees persist along the eastern and southern shores of the CGSM. The degradation and death of mangroves has profoundly altered the composition and diversity of associated fish and bird assemblages and the fauna on prop roots and sediments. Of more than 50 invertebrate species of the root-fouling community and 32 fish species in healthy forests, only 5 and 14 (i.e., *Poecilia* spp.) survive in damaged mangrove stands, respectively. Differences between bird assemblages of healthy (30) and dead (20) mangrove forests in terms of numbers of individuals and species are not significant, probably due to the mobility of birds and their use of dead trees as roost and nesting sites (Botero and Marshall 1994). However, while insectivorous bird species occur both in healthy and dead stands, primary consumers among the birds are absent from the dead mangrove forests.

More than 3,000 people are directly involved in fisheries, using cast nets (2,342), gill nets (1,692), and encircling gill nets (301). Today artisanal fisheries in the CGSM represent 47% of all catches in the Colombian Caribbean. Fisheries have traditionally exploited species like *Mugil incilis*, *Eugerres plumieri*, and *Cathorops spixii*, which account for more than 80% of the total annual catch in weight and in number of individuals. The degradation of nursery grounds, over-fishing, the catch of juveniles, the gradual reduction of mesh-sizes, and destructive fishing methods have reduced stocks of commercially important species. Large individuals of snook (*Centropomus undecimalis*), catfish (*Arius proops*), and mullet (*Mugil liza*) are less common. The landing statistics from CGSM indicate

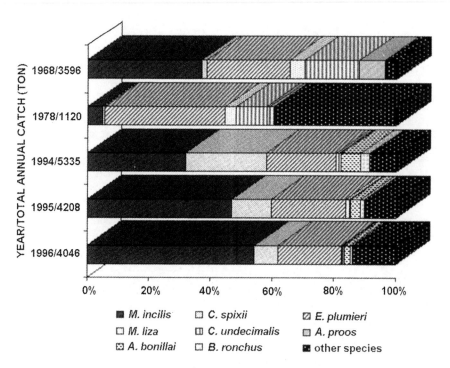

Fig. 2.3. Composition of annual catches in the Ciénaga Grande de Santa Marta

that the total catch and its composition vary between years (Fig. 2.3). Between 1993 and 1996, the total annual catch has decreased from 6,286 to 4,870 tons, of which about 82 % were fishes, 10 % oystera (*C. rhizophorae*), 4 % shrimp (*Penaeus* spp.), 3 % conch (*M. melongena*), and 0.5 % swimming crabs (*Callinectes bocourti, C. sapidus*). Catches of species like *Oligoplites saurus, O. palometa, Bairdiella ronchus,* and *Diapterus* spp., and juveniles of *E. saurus* and *T. atalanticus*, that previously were not exploited, have become more important. Fisheries of introduced exotic species like Tilapia (*Oreochromis* cf. *nilotica*; up to 37 tons), freshwater species (*Caquetaia kraussi*; 132 tons), and *Prochilodus magdalenae* (97 tons) seem to grow continuously (Santos-Martínez and Viloria 1998). Recent studies by Instituto de Investigaciones Marinas y Costeras de Punta Betín "José Benito Vives de Andréis" (INVEMAR) show that a healthy CGSM ecosystem could support a fishery production of about 10×10^3 tons year^{-1} if current levels of fishing technology and intensity were applied.

2.6 Management

The degradation of the Magdalena River delta and the CGSM ecosystem had severe economic implications, not only for the region but also for the entire country. Consequently, the Colombian government established policies to improve wastewater treatment and regulate fisheries among fishing and coastal villages. Furthermore, recent changes in the national constitution have given a local environmental agency (CORPAMAG) greater authority over CGSM issues. This agency, together with technical cooperation of the Colombian-German PRO-CIENAGA project, INVEMAR, and financial support from the Inter-American Development Bank, is developing a restoration program for the hydrologic regime of the CGSM, hoping to reach a point where the natural and assisted regeneration of mangrove forests will become viable. The diversion of the Magdalena River to its former course and the reopening of three original natural channels will eventually introduce a maximum total freshwater flow of approximately 150 m^3 s^{-1} into the CGSM. One of these channels adds freshwater (20 m^3 s^{-1}) to the northern part of the system since 1996, while the other two channels (about 100 m^3 s^{-1}) were opened recently. Since the salinity regime in the lagoon is a function of freshwater runoff and water exchange through the inlet, the restoration program also contemplates the construction of box-culverts at the Barrier Island (Isla de Salamanca) to increase water exchange between the lagoon and the sea. Ecological baseline assessments of the CGSM, to evaluate the effectiveness of the ongoing hydrological restoration program, are still lacking. However, the swift recuperation of mangrove forests after occasional strong overflow events of the Magdalena River gives reason for hope. Once the salinity of the soil and water column has decreased and the transparency and dissolved oxygen concentration have increased, mangrove stands are likely to begin to recover and populations of invertebrate and fish should respond rapidly to the new conditions.

Acknowledgements. The first author greatly thanks M. Sc. John Berry for helping to improve the manuscript.

References

Bernal G (1996) Caracterización geomorfológica de la llanura deltáica del río Magdalena con énfasis en el sistema lagunar de la Ciénaga Grande de Santa Marta, Colombia. Bol Invest Mar y Costeras 25:19–48

Botero l, Mancera-Pineda JE (1996) Síntesis de los cambios antrópicos ocurridos en los últimos 40 años en la Ciénaga Grande de Santa Marta (Colombia). Rev Acad Col Cienc Fis, Ex y Nat 20(78):465–474

Botero L, Marshall M (1994) Biodiversity within the living, dying, and dead mangrove forests of the Cienaga Grande de Santa Marta, Colombia. Final Report, Mote Marine Laboratory, Sarasota, Florida

Bula-Meyer G (1985) Florecimientos nocivos de algas verdes-azules en dos lagunas del Departamento del Magdalena. Rev Ing Pesquera 5(1/2):89–99

Cardona PP, Botero L (1995) Litterfall, flowering and fruiting phenology of *Avicennia germinans* (L.) in a stressed mangrove forest in the Colombian Caribbean. Estudio ecológico de la CGSM-Delta Exterior del Río Magdalena, vol 1. INVEMAR, Santa Marta

Cataño M, Garzón-Ferreira SJ (1994) Ecología trófica del sábalo *Megalops atlanticus* (Pisces: Megalopidae) en el área de Ciénaga Grande de Santa Marta, Caribe Colombiano. Rev Biol Trop 42(3):673–684

Deeb-Sossa SC (1993) Análisis de sedimentación en algunos caños de la Ciénaga Grande de Santa Marta. Estudios y Asesorías Ing. Consultores, Santafé de Bogotá

Hernández CA (1983) Estado actual de los bancos naturales de *Crassostrea rhizophorae* (Guilding, 1828) en el norte de la Ciénaga Grande de Santa Marta. Thesis, Universidad Nacional de Colombia, Bogotá

Hernández CA, Gocke K (1990) Productividad primaria en la Ciénaga Grande de Santa Marta, Colombia. An Inst Invest Mar, Punta Betín 19/20:101–119

INDERENA-SODEIC (1987) Estudios y diseños complementarios para la construcción de las obras de recuperación de la región deltáico-estuarina del Río Magdalena, en especial del área del Parque Nacional Isla de Salamanca. Inderena, Bogotá

Kaufman R, Hevert F (1973) El régimen fluviométrico del Río Magdalena y su importancia para la Ciénaga Grande de Santa Marta. Mitt Inst Colombo-Alemán Invest Cient Santa Marta Invemar 7:21–137

Kjerfve B (1986) Comparative oceanography of coastal lagoons. In: Wolfe DA (ed) Estuarine variability. Academic Press, New York, pp 63–81

Mancera JE, Vidal LA (1994) Florecimiento de microalgas relacionado con mortandad masiva de peces en el complejo lagunar Ciénaga Grande de Santa Marta, Caribe colombiano. An Inst Invest Mar, Punta Betín 23:103-117

Santos-Martínez A, Acero A (1991) Fish community of the Ciénaga Grande de Santa Marta (Colombia): composition and zoogeography. Ichthyol Explor Freshwaters 2(3):247-263

Santos-Martínez A, Viloria EM (1998) Evaluación de los recursos pesqueros de la Ciénaga Grande de Santa Marta y el Complejo Pajarales, Caribe Colombiano: Estadística pesquera. Final Project Report, Evaluación de los Principales Recursos Pesqueros de la Ciénaga Grande de Santa Marta, Caribe Colombiano, INVEMAR-Colciencias-GTZ-Procienaga

Squires HJ, Riveros GC (1971) Algunos aspectos de la biología del ostión (*Crassostrea rhizophorae*) y su producción potencial en la Ciénaga Grande de Santa Marta. Estudios e Investigaciones Inderena 6:5-21

Thom BG (1984) Coastal landforms and geomorphic processes. In: Snedaker S, Snedaker J (eds) The mangrove ecosystem: research methods. UNESCO, Paris, pp 3-17

Vidal LA (1995) Estudio del fitoplancton en el sistema lagunar estuarino tropical Ciénaga Grande de Santa Marta, Colombia, durante el año 1987. MSc Thesis, Universidad Nacional de Colombia, Bogotá

von Cosel R (1985) Moluscos de la región de la Ciénaga Grande de Santa Marta (costa del Caribe de Colombia. An Inst Invest Mar, Punta Betín 15/16:79-370

Wiedemann HU (1973) Reconnaissance of the Cienaga Grande de Santa Marta, Colombia: physical parameters and geological history. Mitt Inst Colombo-Alemán Invest Cient Santa Marta Invemar 7:85-119

3 The Maracaibo System, Venezuela

G. Rodríguez

3.1 Introduction

The Maracaibo Basin was formed during the Mio-Pliocene when the Venezuelan Andes and the Sierra de Perijá attained their maximum development (Gonzalez de Juana et al. 1980). Within this basin, the Maracaibo System acts as an assemblage of interactive brackish water bodies (>220 km^2), comprised of the Gulf of Venezuela, Tablazo Bay and the Maracaibo Strait, which connect Lake Maracaibo in the interior of the basin to the Caribbean Sea (Fig. 3.1). Eustatic sea level changes during the Holocene have induced either fresh or brackish water conditions in the lake. After the last transgression (about 9,000 years B.P.), when sea levels were about 20 m above present levels, numerous islands and bars were formed at the mouth of the estuary.

3.2 Environmental Factors

The topographic features of the Maracaibo Basin strongly determine the airflow of trade winds and convergence processes over the lake and consequently influence the regional climate. The mean annual air temperature in the basin is 28.3 °C with a difference of approximately 40 °C between the hottest and coldest month. The arid and semiarid zones around the Gulf of Venezuela have distinct dry and rainy seasons, with 90% of the rainfall (about 200 mm) occurring from May to October when the trade winds decrease. Further inside the basin, differences between dry and rainy seasons decrease and rainfall increases (annual mean 1,400 mm), owing to almost permanent low pressure over the Catatumbo River mouth. The highest precipitation (annual mean 3,500 mm) is recorded in the upper Catatumbo River watershed.

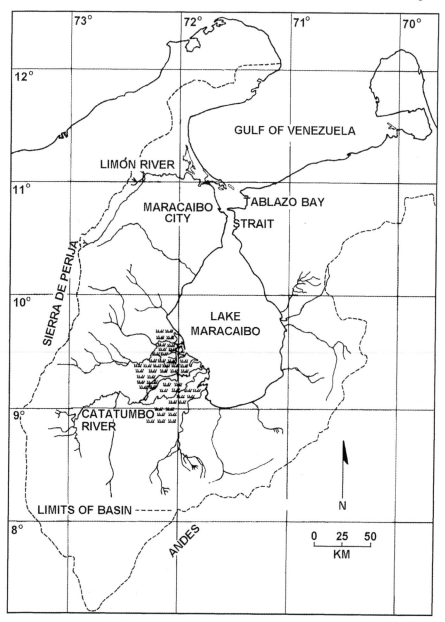

Fig. 3.1. The Maracaibo System with Lake Maracaibo, Strait of Maracaibo, Tablazo Bay, and the Gulf of Venezuela

3.2.1 Tides and Circulation

The tides in the Maracaibo System change from mixed-diurnal to mixed-semidiurnal types due to the size and morphology of the system. The tidal amplitude varies between the outer Gulf of Venezuela (20 cm), the entrance of Tablazo Bay (110 cm), the southern Maracaibo Strait (20 cm), and the lake (2.5–6 cm; Pelegrí and Avila 1986). The circulation pattern in the Gulf of Venezuela is influenced by winds, bottom topography, and outflowing water from Tablazo Bay, whereas Coriolis forces seem to have little effect. The low-salinity water of the southern portion of the gulf flows towards the center where it merges with Caribbean water, causing a cyclonic circulation at the west and anticyclonic circulation at the southeast of the gulf. Upwelling of deep water occurs at the northwestern section, where high-salinity and low-temperature waters flow to the north and reach the surface (Ginés 1982). The dynamics of water in Tablazo Bay and the strait are dominated by tides, which generate alternate currents (12 h) along the north-south axis, with a maximum of 1.6 m s^{-1} at the mouth of the estuary and 0.45 m s^{-1} at the southern end of the strait. Tidal currents and the diurnal component are almost absent from Lake Maracaibo. The cyclonic circulation pattern in the lake, facilitated by the quasi-circular shape, is initiated by the action of winds and by large freshwater runoff from rivers at the southeastern corner of the lake (57 % from Catatumbo River; Escam 1991). As a consequence, the water in the lake circulates like a gigantic vortex in a counter-clockwise direction (Emery and Csanady 1973), forming a cone-shaped hypolimnion (Parra Pardi 1983). Near the center of the lake, in an area of approximately 40 km in diameter (Battelle 1974; Parra Pardi 1983), surface current velocities can reach up to 0.5 m s^{-1} and, though velocities rapidly decrease with depth, the direction does not change. The residence time of water in the lake is about 6 years (Escam 1991). The surplus volume of freshwater (49.1×10^9 m^3 year^{-1}) discharges from the epilimnion of the lake along the eastern coast of the strait at about 500 m^3 s^{-1} during the dry, and up to 2,600 m^3 s^{-1} in the rainy season. The final discharge of freshwater from the Maracaibo System creates a considerable estuarine zone in the open sea (Rodríguez 2000). In contrast, a deficit in water balance, usually in February and March, is compensated for by inflowing seawater from the Gulf of Venezuela.

3.2.2 Physico-chemical Characteristics

Water temperatures follow a seasonal pattern, with an extreme minimum mean temperature of 29.0 °C at 1 meter depth in the lake in February and a maximum temperature of 32.5 °C in September. Highest water temperatures occur in the strait, occasionally exceeding air temperature by 20 °C. The water column in the Gulf of Venezuela is only slightly stratified, with a maximum temperature difference of 1.4 °C at 20 m in March.

The lowest salinity (0.02) in the Maracaibo System occurs at the mouth of the Catatumbo River in the lake. About 7 km off the coast, surface water (<1 m) salinity is 1.3 but values increase to 2–3 at deeper layers (Hyne et al. 1979). The brackish water, which enters the lake through the strait, moves close to the bottom towards the deeper portions and contributes to the saline cone-shaped hypolimnion in the center of the lake. The cone is a permanent feature of the lake (Parra Pardi 1983; Escam 1991) though its volume, which may represent up to 25 % of the total lake volume, changes seasonally with input of brackish water from the strait or freshwater discharge from the rivers. The counterclockwise circulation of the cone moves brackish water towards the epilimnion where it becomes completely mixed. Salinity distributions in the strait, Tablazo Bay, and the southern part of the gulf are typical of partially mixed estuaries. The tidal movements produce cyclic oscillations and periodic mixing, which result in vertical and longitudinal gradients. Seasonal maximum and minimum salinities occur from February to May and from June to August, respectively. Salinities in the Limón River estuary at the northwest corner of Tablazo Bay may vary between 0.1 and 8 in the wet and dry season, respectively (Rodríguez 1973). Salinity stratification in the navigation channel of Tablazo Bay and the strait is related to the intrusion of a saline wedge from the gulf. The discharge of water from the estuarine region (i.e., strait and Tablazo Bay) principally effects the southern part of the gulf, where an isohaline (30) usually forms in a SW-NE direction at 6–7 m depth. Pronounced stratification, with a salinity of 2 at 12 m depth, occurs at the SW corner of the gulf. In the outer gulf (approximately 120° N) salinity differences between surface and 60 m depth are less than 0.1.

Anoxic conditions may occur in a large area (up to 50 km across) in the hypolimnion of the lake when the water column becomes completely stratified (Gessner 1953; Parra Pardi 1979). The oxygen content in the epilimnion varies little (Escam 1991). Lowest concentrations under limited inflow of saline water are common in December, when the base of the cone is reduced and the apex approaches the surface. Under these conditions, the volume of the hypolimnion is reduced and the water is mixed (Redfield and Doe 1964). In the strait, variations in oxygen content are rarely more than 1.5 ml l^{-1} (Parra Pardi 1977) and surface water is frequently supersaturated due to algae blooms. Although the oxygen content in the water column drops to 40% saturation at 10 m and 30% at 14 m, anoxic conditions have not been reported.

3.2.3 Nutrients

The resuspension of bottom sediments is the principal source of nutrients in the gulf (0.2–0.4 mg atom per l P at surface; 1.0 mg atom per l P at bottom), Tablazo Bay, and the strait. Elevated phosphorus levels in the lake are largely due to terrestrial runoff. The saline wedge in the strait is an important source of nitrogen compounds, which are added through sewage discharge from Maracaibo City. Owing to sedimentation of organisms and slow outward circulation of the cone-shaped hypolimnion, phosphorus concentrations tend to increase from surface (1.0–10.32 mg atom l^{-1}) to depth (15.0–158 mg atom l^{-1}; Battelle 1974; Parra Pardi 1979; Escam 1991). Phosphorus levels in the water column are likely to the control by bottom sediments and adsorption processes to them (López-Hernández et al. 1980). The concentrations of nitrogen compounds follow similar distribution patterns, with ammonia concentrations reaching 6 mg atom l^{-1} at 10 m depth and exceeding 180 mg atom per l at the base of the cone (Parra Pardi 1979). Based on the nitrogen balance in the lake and on N/P ratios (Parra Pardi 1979), nitrogen is a limiting factor under most conditions in the lake.

3.3 Biological Components

3.3.1 Benthic Communities

The species composition of littoral communities in the Maracaibo System largely depends on salinity and on substrate characteristics (Rodríguez 1973). Dunes with a sparse cover of *Hypomoea pes-caprae* fringe the extensive sandy shores of the Gulf of Venezuela. The drift lines are characterized by the presence of talictrid amphipods. Intertidal beaches are dominated by the clam *Donax striatus* which in the mid-shore zone occurs together with some nereid, capitellid, and spionid polychaetes, and on the lower shore together with *Donax denticulatus, Oliva caribacensis*, and the anomuran crustacean *Lepidopa*.

Inside Tablazo Bay, euryhaline forms progressively displace marine elements of the gulf. The west coast of the bay, the strait, and the northern shores of the lake are occupied by halophytic vegetation typical of littoral dunes and by scarce mangrove growth. In general, these shores have a poor invertebrate fauna. The upper intertidal is inhabited by the fiddler crab *Uca leptodactyla*, while an association of the abundant brackish water clams *Polymesoda solida, Ruppia maritima, Uca maracoani*, and *Callinectes sapidus* occupies the lower intertidal. Encrusting gastropods (*Littorina nebulosa, Thais aemastoma floridana*) are typical on the trunks of occasional mangrove trees. Muddy shores with dense mangroves (*Rhizophora mangle*) along the eastern shore of the bay and in the lower Limón River tend to be void of halophytic vegetation. Intertidal organisms, like *Balanus*, the gastropod *Neritina reclivata*, and the red alga *Caloglossa leprieurii* are abundant on the trunks of mangroves, while the mud crab *Rhithropanopeus harrisii* and mangrove crabs (*Metasesarma rubripes, Aratus pisonii*) occur among the branches. The invertebrate fauna of muddy intertidal bottoms is comprised of the pulmonate gastropod *Melampus coffeus* and high densities of fiddler crabs (*Uca cumulanta* with 330 burrows m^{-2} and *Uca rapax* with 113 burrows m^{-2}) and the swamp ghost crab *Ucides cordatus* (10 burrows m^{-2}). In coarser sediments, *Uca rapax* is substituted by *Uca leptodactyla* (85 burrows m^{-2}; Rodríguez 1973). The ubiquitous clam *Polymesoda solida* is also present in the sublittoral of densely forested shores. The seagrass *Ruppia maritima* covers lower intertidal bottoms and *Polymesoda solida* and the blue crab *Callinectes sapidus* are abundant components in the submersed meadows. Well-developed mangrove stands are only found on the northeastern shore of Maracaibo Lake and north and south of the Catatumbo River (E. Medina, pers. comm.). Intertidal invertebrates of oligohaline waters, like the freshwater sponge *Spongilla aspinosa*, the gastropod *Neritina reclivata*, and the mangrove crab *Metasesarma rubripes*, settle on artificial peers and platforms and on rare natural hard substrates.

The sublittoral community of the Maracaibo System is not well known; however, the fact that 165 species of mollusks have been recorded suggests a rich and diverse subtidal fauna. Dominant components of the lower subtidal community are the mytilid pelecypod *Mytella maracaiboensis* (up to 158 g dry weight m^{-2}), the tubicolous amphipod *Corophium rioplatense*, and polychaetes, like *Heteromastus filiformis* (980 ind. m^{-2}), *Sigambra* sp. (1,380 ind. m^{-2}), *Streblospio* sp. (4,000 ind. m^{-2}), *Capitella capitata* (160 ind. m^{-2}), and *Nereis succinea* (360 ind. m^{-2}). The density of polychaetes has decreased during the last decade (Morales 1987), partly due to substrate changes as a result of natural turbulence and continuous dredging operations. The subtidal benthic community of Tablazo Bay extends into the lake, though *Mytella* disappears and densities of polychaetes are much lower.

3.3.2 Plankton Communities

Similar to the distribution of benthic species, the composition of the phytoplankton community in the Maracaibo System, with 125 species of diatoms (Hustedt 1956), clearly follows a salinity gradient. Phytoplankton in the southern part of the gulf is composed of dinoflagellates (i.e., *Noctiluca scintillans, Porocentrum micans*) and 19 diatoms, such as *Chaetoceros afinis, C. decipiens, C. seriacanthus, Rhizosolenia calcar-avis, R. imbricata*, and *R. robusta*. About 50 diatom species (i.e., *Sceletonema costatum, Chaetoceros subtilis*) and cyanobacteria (*Anabaena spiroides, Microcystis aeruginosa*) are found in the strait. The dinoflagellates *Ceratium furca, C. massiliense, C. tripos*, and *Gymnodinium simplex* have been recorded in Tablazo Bay where Chlorophyta (i.e., *Scenedesmus quadricauda, S. dimorphus, Ankistrodermus falcatus*) are poorly represented. The distribution of the dinoflagellates *Ceratium fusus, C. tripos*, and *Dinophysis caudata* extends from the bay into the southernmost part of the lake. Twelve species of diatoms and the green algae *Ankistrodermus falcatus* are only found in the lower Limón River. The phytoplankton community in the lake is represented by abundant (>500 cells ml^{-1}) Chlorophyta (*Chodatella longispina, Stichococcocus bacillaris, Ankistrodemus gracilis, Selenastrum gracile, Scenedemus quadricauda, S. dimorphus, Tetraedron caudatum, Pediastrum tetras, Agmenellum quadriduplicatum*), cyanobacteria (*Anacystis cyanea, Oscillatoria submembranacea, Spirulina subsalsa, Anabaena spiroides*), and about 37 diatoms (i.e., *Chaetoceros similoides, Melosira ambigua, Plagiogramma tenuissimum, Nitzchia palea*; Battelle 1974).

The zooplankton in the Gulf of Venezuela is typically composed of neritic species (i.e., *Lucifer faxoni, Centropages furcatus, Acartia lilljeborgi*, and stomatopod larvae) and, together with pelagic elements (i.e., the pteropod *Cresseis acicula*, the tunicate *Salpa democratica*, and siphonophores), zooplankton biomass may reach 40 mg m^{-3}. Tablazo Bay acts as an interface between euhaline species of the gulf and oligohaline species of the lake. The polihaline waters at the mouths of the bay constitute the distributional limit for several neritic species as well as for meroplanktonic species of the bay. The abundance of allochthonous plankters, like *Sagitta hispida, S. enflata, Bowmaniella brasiliensis, Aegathoa linguifrons, Acetes americanus, Lucifer faxoni, Cresseis acicula*, larvae of *Penaeus*, carideans, brachyurans, and fishes in surface waters at the mouth of the bay roughly follows tidal cycles; thus, neritic species form a front that sways back and forth with tides. Only the post-larvae of *Penaeus* move continuously towards the bay (except in July and September), which suggests the presence of an active movement, possibly crawling on sediments, to prevent massive return to the gulf (Rodríguez 1973).

The vertical displacement of zooplankton in the water column can be attributed to turbulence (Rodríguez 1984). The zooplankton of the bay is typically estuarine with copepods like *Labidocera fluviatilis* and *Acartia tonsa*, which reaches high densities (up to 4,241 ind. m^{-3}). Total zooplankton biomass varies between 17 and 31 mg m^{-3}. During the dry season, neritic species of coastal waters in the gulf (*Lucifer faxoni*, *Centropages furcatus*, *Sagitta hispida*, and medusae) are common. Although their tidal dependence is less pronounced, both neritic and estuarine species follow salinity cycles, with maximum density during inflowing seawater. In contrast, density changes of cirriped larvae tend to reflect their reproductive period (Rodríguez 1973). The estuarine character of the zooplankton community decreases towards the strait and neritic organisms tend to disappear altogether (Rodriguez 1973). In the lake, zooplankton is dominated by calanoid and cyclopoid copepods (i.e., *Acartia tonsa*, *Metacyclops distans*, *Mesocyclops ellipticus*, *Notodiaptomus maracaibensis*; Escam 1991) with low biomass (7–10 mg m^{-3}). Several species of the Cladocera are common (*Diaphanosoma*, *Moinodaphnia*, *Ceriodaphnia*, *Bosmina*) and others are abundant (i.e., *Alona davidi* and *Moina micrura*, 191 ind. l^{-1}). Rotifers have high specific diversity (*Brachionus*, *Keratella*, *Epiphanes*, *Monostyla*, *Asplanchna*, *Hexarthra*, *Conochilus*) but generally do not represent more than 15% of the zooplankton community.

3.3.3 Vertebrates

The estuarine ichthyofauna of the Maracaibo System consists of three different species assemblages. Marine euryhaline species migrate between the estuary and the sea or are permanently found in Tablazo Bay. A large group of estuarine-resident species lives and reproduces only in mixohaline waters (i.e., the endemic species *Cynoscion maracaiboensis*). Alloxenic species are usually found close to the margins of the lake and the Engraulidae *Anchoa argenteus* (gemminate of *A. spinifer*) and *Anchovia nigra* (gemminate of *A. clupeoides*) are endemic to oligohaline waters of the lake. The sardine *Astyanax fasciatus*, catfishes (Pimelodidae), *Prochilodus reticulatus*, *Potamorhina laticeps*, and *Mylossoma acanthogaster* migrate to the middle and upper reaches of rivers between January and March, followed by *Doradops zuloagai* in June. Seaward migrations of these species initiate in April and last until September. Among the freshwater ichthyofauna (113 species), Siluriformes predominate and about 40 species are endemic to rivers in the Maracaibo Basin (Pérez Lozano and Taphorn 1993). Several species of the Catatumbo River display an annual migratory cycle (Galvis et al. 1997) that parallels maximum river discharge in April-May and September-November. Fish populations feed and reproduce in the productive lowland swamps and initiate their upriver migration when freshwater runoff decreases.

Five species of Cetacea have been recorded from the Gulf of Venezuela and other vertebrates like *Cayman sclerops fuscus* and *Crocodylus acutus* occur at the mouth of rivers at the southern lake coast. The freshwater dolphin *Sotalia fluviatilis* is a common visitor of the lake and the strait, but the manatee *Trichechus manatus* has only been cited once (Agudo et al. 1994). The west coast of Tablazo Bay is the only nesting site of the flamingo *Phoenicopterus ruber* in Venezuela.

3.4 Production Processes

Primary production in the outer gulf shows considerable fluctuations, though pigment concentrations are consistently high (0.5 mg m^{-3}) in upwelling areas at the southeastern coast (Ginés 1982; Müller-Karger et al. 1989). In Tablazo Bay and the strait, primary production follows a bimodal distribution with maximum productivity (5.56 g C per m^2 and day) in March (Rodríguez and Conde 1989). Total chlorophyll *a* concentrations in lake surface waters vary between 13 and 45 µg l^{-1} (1–3 g C per m^2 and day) and tend to decrease at depths between 10 and 25 m, except in areas around the Catatumbo River mouth (14 µg l^{-1}; Parra Pardi 1979). At the northeastern side of the lake, where oil production concentrates, chlorophyll *a* (223 µg l^{-1}) and primary productivity (10.5 g C per m^2 and day) may reach elevated levels (Battelle 1974; Rodríguez and Conde 1989). The visible effect of high productivity in the lake is the formation of algal surface blooms, which suggests a process of eutrophication despite the occurrence of blooms since 1937 (Escam 1991).

Similar to phytoplankton, zooplankton follows large temporal and spatial fluctuations in the Gulf of Venezuela. Dense populations of tunicates, especially *Salpa democratica*, occur in March and August and larval forms of *Lucifer faxoni* may represent 80% of the community (Rodríguez 1973; Ginés 1982). Biomass in the gulf varies between 3 and 63 mg m^{-3}, with a maximum in August (120 mg m^{-3}). Lower zooplankton biomass in the center of the gulf may be due to the deflection of nutrient-rich waters toward

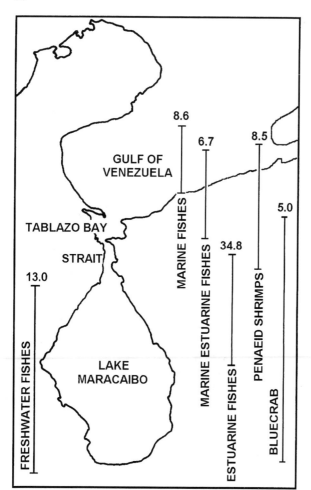

Fig. 3.2. Fishery production (× 1000 tons) in the Maracaibo System

the southwestern corner (Rodríguez 1973). At the center of the lake zooplankton concentration tends to be low (159 ind. l^{-1}) while higher concentrations are usually associated with large river discharge at the southwest corner (345 ind. l^{-1}) and with phytoplankton-rich waters near the northeastern coast (753 ind. l^{-1}; Battelle 1974; Ginés 1982; Escam 1991).

3.5 Environmental Issues

Annual fisheries in the Maracaibo System represent about 18 % of the total Venezuelan catch. Artisanal fisheries in the lake account for 71,700 tons, whereas artisanal and industrial trawling fisheries in the gulf produce 4,200 and 17,000 tons, respectively. Gulf fisheries are largely composed of demersal marine fishes, the two most important species being pigfish (*Orthopristis ruber*) and cutlassfish (*Trachiurus lepturus*). Seasonal migrants into Tablazo Bay, like sábalo (*Megalops atlanticus*), sharks (*Mustelus* spp.), snappers (*Lutjanus* spp.), and the croaker *Micropogon furnieri* account for 10% of the catch. Artisanal fisheries in the lake are composed of *Micropogon furnieri* (2,700 tons) and of estuarine species (7,800 tons) like catfishes which comprise the largest component (20%) of the yield, followed by the "corvina del lago" (*Cynoscion maracaiboensis*), mullets (*Mugil* spp.), and snooks (*Centropomus* spp.). Freshwater fisheries are important and haloxenic species like *Potamorhina laticeps* (manamana) and *Prochilodus reticulatus* (bocachico) yield 5,150 and 3,000 tons, respectively. Invertebrates, especially several species of shrimp (*Penaeus*), represent one of the most important fishery resources of the system, with a yield of 3,000 tons in the gulf and 5,000 tons in the lake. Estuary fisheries of the blue crab (*Callinectes sapidus*) total 2,000 tons and the clam *Polymesoda solida* is an important resource (Fig. 3.2).

A population in excess of 5.5 million and expanding agriculture and industry in the Maracaibo Basin have led to increasing freshwater demands and decreasing freshwater quality during the last decades. Furthermore, continuous dredging of almost 20% of the 360-km-long Maracaibo Channel, which represents a strategic waterway between the gulf and the lake for 70% of Venezuela's oil production, has induced major physiographic changes. The effect of a deeper navigation channel, freshwater diversion from the river basins, and dammed-off areas in the northern part of the lake appear to have induced a gradual process of salinization in the lake after 1970 (Fig. 3.3). Despite environmental regulations to establish a "minimum eco-

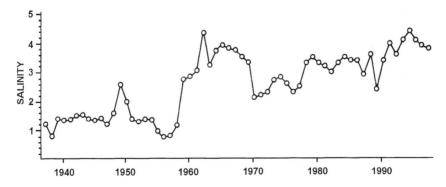

Fig. 3.3. Mean annual surface salinities on the eastern shore of Maracaibo Lake

logical flow", the reduction of river flow has created barriers for the migration of fishes. Urban sewage (34 m^3 day^{-1}) is transported with tides from the strait into the lake and nutrients are added through runoff from agricultural areas. The nutrients are trapped in denser saline waters at the base of the cone-shaped hypolimnion and cause anoxic conditions near the bottom, initiating a process of eutrophication. The addition of sewage and agricultural runoff also give rise to elevated bacterial and organochloride pesticide contamination (Escam 1991) with particular concern for the shellfish industry. Heavy metals, like mercury leaching from sedimentation lagoons of a petrochemical complex between 1975 and 1982, are of major concern. Additional environmental hazards arise from the production of oil (10,000 wells), distribution through pipelines (about 40,000 km) at the bottom of the lake, transport by tankers, dump sites of crude oil and solid wastes, gas processing, and petrochemical industries.

3.6 Management Considerations

Environmental management consideration in the lake has focused on the availability of freshwater resources for human and agricultural use as well as on water quality standards appropriate for urban and industrial use, fisheries, navigation, and recreation (Conde and Rodriguez 2000). Although stringent laws regulate potential environmental hazards and present petroleum operations are in line with new regulations, vital economic activities like petroleum exploitation, estuarine fisheries, and navigation will continue to pose environmental problems. Since 1981, the Institute for the Control and Conservation of the Lake Maracaibo Basin (ICLAM) is addressing management issues of the lake. Furthermore, the Colombian and Venezuelan Governments have established a "Commission for Shared Basins" because 15% of the Lake Maracaibo Basin is part of Colombian territory. Environmental problems, like oil spills following guerrilla attacks on pipelines in the Catatumbo River Basin, can now be controlled by oil companies of both countries in the upper reaches of the river to prevent oil pollution of important nursery grounds for fish in lowland swamps (Rodríguez-Miranda et al. 1996).

References

Agudo AI, Viloria AL, Coty JR, Acosta RJ (1994) Cetofauna (Mammalia:Cetacea) del Estado Zulia, Venezuela noroccidental. Anartia 5:1–23

Battelle (1974) Study of effects of oil discharges and domestic and industrial wastewaters on the fisheries of Lake Maracaibo, Venezuela. 1. Ecological characterization and domestic and industrial wastes. Richland, Washington

Conde JE, Rodríguez G (2000) Integrated coastal zone management in Venezuela: the Maracaibo system. In: Salomons W, Turner K, Spoudiolus A, Lacerda LD, Ramachandran R (eds) Integrated coastal zone management: principles and practice. Springer, Berlin Heidelberg New York (in press)

Emery KO, Csanady GT (1973) Surface circulation of lakes and nearly land-locked seas. Proc Natl Acad Sci USA 70:93–97

Escam (1991) Plan maestro para el control y manejo de la cuenca del Lago de Maracaibo, vol 1. Mimeo Report, Caracas

Galvis G, Mojica IJ y, Camargo M (1997) Peces del Catatumbo Asociación Cravo Norte. Ecopetrol, Occidental de Colombia, Shell, Bogotá

Gessner F (1953) Investigaciones hidrográficas en el Lago de Maracaibo. Acta Científica Venezolana 4:173–177

Ginés H (1982) Carta Pesquera de Venezuela. 2. Areas Central y Occidental. Fundación La Salle de Ciencias Naturales, Monografía 17:1–226

Gonzalez de Juana C, Iturralde JM, Picard C (1980) Geología de Venezuela y de sus cuencas petrolíferas. Ediciones Foninves, Caracas

Hustedt F (1956) Diatomeen aus dem lago de Maracaibo in Venezuela. In: Gessner F, Vareschi V (eds) Ergebnisse der Deutschen Limnologischen Venezuela-Expedition 1952, vol 1. VEB Deutscher Verlag der Wissenschaften, Berlin

Hyne NJ, Laidig LW, Cooper WA (1979) Prodelta sedimentation on a lacustrine delta by clay mineral flocculation. J Sed Petrol 49:1209–1216

López-Hernández D, Herrera T, Rotondo F (1980) Phosphate adsorption and desorption in a tropical estuary (Maracaibo System). Mar Environ Res 4:153–163

Morales F (1987) Revisión de los macroinvertebrados benticos del Sistema de Maracaibo. Trabajo Especial de Grado. Universidad del Zulia, Maracaibo, 142 pp

Müller-Karger FE, McClain CR, Fisher TR, Saias WE y, Varela R (1989) Pigment distribution in the Caribbean Sea: observations from space. Prog Oceanogr 23:23–64

Parra Pardi G (1977) Estudio integral sobre la contaminación del Lago de Maracaibo y sus afluentes, parte I. Estrecho de Maracaibo y Bahía El Tablazo. Ministerio del Ambiente y de los Recursos Naturales Renovables, Caracas

Parra Pardi G (1979) Estudio integral sobre la contaminación del Lago de Maracaibo y sus afluentes, parte II. Evaluación del proceso de eutroficación. Ministerio del Ambiente y de los Recursos Naturales Renovables, Caracas

Parra Pardi G (1983) Cone shaped hypolimnion and local reactor as outstanding features in eutrophication of Lake Maracaibo. J Great Lakes Res 9:439–451

Pelegrí JL, Avila RG (1986) Las mareas como sistemas cooscilantes en los golfos de Venezuela y Paria. Revista Técnica de Intevep 6(1):3–15

Pérez Lozano A, Taphorn D (1993) Relaciones zoogeográficas de las cuencas del Río Magdalena y el Lago de Maracaibo. Biollania 9:95–105

Redfield AC, LAE Doe (1964) Lake Maracaibo. Verh Int Verein Theor Angew Limnol 15:100–111

Rodríguez G (1973) El Sistema de Maracaibo: biología y ambiente. Instituto Venezolano de Investigaciones Científicas, Caracas

Rodríguez G (1984) Ecological control of engineering works in the Maracaibo estuary. Water Sci Tech 16:417–424

Rodríguez G (2000) The Maracaibo system: a physical profile. In: Perillo G (ed) Estuaries of Latin America. Springer, Berlin Heidelberg New York (in press)

Rodríguez G, Conde JE (1989) Producción primaria en dos estuarios tropicales de la costa caribeña de Venezuela. Rev Biol Trop 37:213–216

Rodríguez-Miranda D, Rodíguez-Grau J, Restrepo M, Rodríguez G (1996) Monitoreo biológico y químico de la cuenca del Río Catatumbo. Ecopetrol-Intevep-PDVSA

4 The Orinoco River Delta, Venezuela

J.E. CONDE

4.1 Introduction

The Orinoco River watershed (1.1 × 10^6 km^2) is the third largest in the world and the prevailing hydrographic unit in Venezuela. Between its origin at the Sierra Párima and Sierra Unturan near Brazil and the Atlantic Ocean, the Orinoco drains more than 2,000 rivers (Depetris and Paolini 1991; Bonilla et al. 1993; Cressa et al. 1993). The distinctive triangular-shaped Orinoco Delta (20,000 km^2) includes a cluster of ecosystems (Pannier 1979; Lewis et al. 1990; Colonnello 1996) and encompasses more than 300 tributary channels (caños) and several independent streams. Most channels (e.g. the important Caños Macareo and Mánamo) originate as branches of the Orinoco River at Barrancas. Together with the rivers of the northeastern plain (San Juan and Guanipa) and the Rio Grande channel, the delta discharges waters into the Gulf of Paria, the straits Boca de Serpientes and Boca de Dragones and the open sea (Pannier 1979; Fig. 4.1). The delta itself and the plume are distinct estuarine units; their limits probably depending on interactions brought about by seasonal variations in runoff and intrusion of tidal fronts.

4.2 Environmental Setting

The annual precipitation in the Orinoco River watershed can be lower than 1,000 mm in the northern lowlands and as high as 8,000 mm in southern high relief areas (Lewis and Saunders 1989). In the delta, the climate is perhumid tropical (Thornthwaite's classification), with annual rainfall varying between 1,200 and 2,800 mm in the Upper Delta and the coastal perimeter, respectively (Lewis and Saunders 1989). The pluvial regime is markedly seasonal with two dry and two wet periods. The principal dry

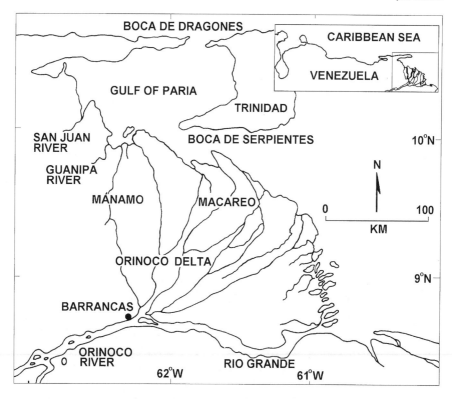

Fig. 4.1. Orinoco River Delta and main tributaries

season between January and April is followed by a period of heavy rains from May to August. A second drier period occurs in September and October and a second wet period in November and December. Although air temperatures may vary considerably during the day, moderate but steady northeasterly trade winds (Heinen et al. 1995) maintain mean annual temperatures between 25.5 and 26.6 °C, with a relative humidity of 60–80 % (Wilbert 1994–1996). Mean annual evaporation ranges from 1,380 to 1,928 mm (Pannier 1979).

The Orinoco River watershed is geologically diverse (Schubert et al. 1986; Colveé et al. 1990), ranging from Precambrian formations in the Guayana Shield (Díaz de Gamero 1996) to the delta, which dates from the Holocene. The modern delta is composed of fluvial-marine argillaceous sediments (Pannier 1979; Colonnello 1996) which are supplied by the Orinoco River (86–150 × 10^6 t year^{-1}; Bonilla et al. 1993; Cressa et al. 1993) and probably by long-shore transport from the Amazon River (Van Andel 1967). Prominent landmasses in the delta are usually comprised of levees with sandy soils and central depressions, where peat, grasses and sedges cover clay and silt deposits (Colonnello 1996). The delta has a slope of less than 1%. The Lower Delta (-1 to 1 m above sea level) is under the influence of semidiurnal tides (T=12.4 h; Herrera et al. 1981; Herrera and Masciangoli 1984), while the Middle (1-2.5 m) and Upper (2.5-7 m) Delta are seasonally flooded during longer and shorter periods, respectively. Owing to well-defined periods of precipitation in the basin, the freshwater discharge through the Orinoco River, with a mean annual flow of 36,000 m^3 s^{-1} (1.1 × 106 m^3 year^{-1}), is highly seasonal. During the rainy periods, low salinity waters (<5) extend throughout the Gulf of Paria and surround Trinidad (Rodríguez 1974). Brackish surface waters may disperse all the way to Puerto Rico some 1,200 km away (300,000 km^2; Müller-Karger et al. 1989), owing to NW flow of the Guyana Current and trade wind-induced NW and W drift currents (mean annual surface velocity 20 cm s^{-1}; Herrera et al. 1981; Herrera and Masciangoli 1984). In contrast, the dry season with reduced runoff limits brackish waters to a narrow strip along the delta coast and western coast of the Gulf of Paria. The effect of tides on water levels (Δ60 cm) is felt as far as 250 km upstream (Colonnello 1995) and marine diatoms occur in several tributaries (Varela and Varela 1983).

Although temperature, pH and dissolved organic carbon concentrations vary little in the course of the year, conductivity decreases with discharge and average particulate organic carbon concentrations increase from the dry (0.84 mg l^{-1}) to the wet season (4.02 mg l^{-1}; Paolini et al. 1983). Physicochemical parameters and nutrient concentrations also differ significantly between channels in the delta. Conductivity values tend to be higher (32.5-78.1 µS) in the Mánamo channel (Macareo 19.5-20.9 µS), but mixohaline conditions (8) only occur at the mouth. As a consequence of freshwater flow controlled by a cofferdam, suspension loads in the Mánamo channel (11.5-45.8 mg l^{-1}) are lower than those in the Macareo channel (66.6-149.5 mg l^{-1}; Paolini et al. 1983). The N/P values (Mánamo 7.1, Macareo 6.8) suggest, apart from plankton, contribution of re-suspended sediments and terrestrial material. Carbon and nitrogen levels in the sediments are significantly higher in the Mánamo (11% C, 1.58% N) than the Macareo channel (2.4% C, 0.36% N) or even in the Orinoco River (Paolini 1995).

4.3 Biological Components

4.3.1 Phytoplankton Community

The phytoplankton community in the lower Orinoco River and Delta is composed of Bacillariophyceae (94), Dinophyceae (6), Cyanophyceae (29), Euglenophyceae (5), Chrysophyceae (1), Xanthophyceae (1), Euchlorophyceae (24) and Zygophyceae (38) (Varela et al. 1983). While marine diatoms, *Melosira granulata* being the most abundant and widely distributed species, and dinoflagellates are more common at the mouth of channels and along the coast, five species of marine diatoms have also been reported for the lower Orinoco River. The species of most other groups prevail under limnic conditions in the delta. In general, the diversity of the plankton community is higher in tributary channels of the delta than in the Orinoco River (Varela and Varela 1983). Seasonal biomass differences are a function of variations in phytoplankton biomass (Lewis 1988); however, since primary production rates are low, autotrophy does not contribute significantly to nutrient concentrations and carbon balance.

4.3.2 Vegetation

About 100 aquatic vascular plant species (e. g., *Cabomba aquatica*, *Utricularia* spp., *Eichhornia crassipes*, *Pistia stratiotes*, *Limnobium laevigatum*, *Nymphaea rudgeana*, *Ludwigia helminthorhiza*) and many grasses and sedges (e. g., *Leersia hexandra*, *Oxicaryum cubense*) have been recorded from the delta, though species that form rooted meadows (*Ludwigia octovalvis*, *Polygonum acuminatum*, *Oxicaryum cubense*) prevail. *Hymenachne amplexicaulis* and *Mimosa pigra* are restricted to the shores (Colonnello 1996). Although eight sea grass species have been reported for the coast of Venezuela (Vera 1992), none has been reported for the delta. The absence of sea grasses and benthic algae might be a consequence of high suspension loads and lack of suitable substrates.

The dominant terrestrial vegetation of the delta is mangroves, marsh forests and sand dune vegetation (Pannier 1979). More than 73% (183,500 ha) of Venezuela's mangrove forests are located between the Gulf of Paria and the Orinoco Delta (Conde and Alarcón 1993) and about 100,000 ha (MARNR 1991) fringe the coast, tributaries and rivers of the delta, occasionally more than 80 km upriver (Heinen et al. 1995). The dominant species are the red mangrove *Rhizophora mangle*, the black mangroves *Avicennia germinans* and *Laguncularia racemosa*, though *R.*

harrisonii and *R. racemosa* have also been reported for the mid-delta (Pannier 1979; MARNR 1991; Conde and Alarcón 1993). *Rhizophora* forests tend to form a 100-m-wide belt along the margins of channels, with some individuals growing up to 20 m high (dbh 25 cm). Mangroves are replaced landward by tall stands of the herbaceous *Montrichardia arborescens*, followed by halophobic species, like the abundant *Pterocarpus officinalis*, *Crinum erubescens* and *Tabebuia* aff. *aquatilis*, which is endemic to swamp forests of the Orinoco and Amazonian rivers (Colonnello and Medina 1998).

In the areas where tidal flooding is reduced, mangroves give way to mixed marsh forests, marsh palm groves (Morichales) and herbaceous marshes (Rabanales). Extensive stands of tall mixed marsh forests, or mixed swamp forest (Bacon 1990), often develop directly behind fringing mangroves in the Lower Delta. The dominant trees include *Symphonia globulifera*, *Pterocarpus officinalis*, *Bombax aquaticum* and the palms *Euterpe oleracea* and *Manicaira saccifera*. On levees, *Erythrina glauca* forms an open mixed forest with some palms (*Euterpe, Mauritia*) and heavy undergrowth (Bacon 1990). Marsh palm groves in the lower and mid-delta are represented by dense stands of the palm *Mauritia flexuosa*. Associated palms are *Euterpe* spp. and *Manicaria* spp. as well as dense herbaceous vegetation in waterlogged areas. Herbaceous marshes are dominated by the rábano (*Montrichardia arborescens*), which occurs together with the giant fern *Acrostichum aureum*, *Paspalum fasciculatum* and several aquatic species. Accreting banks of tidal creeks are often invaded by sand dune vegetation (*Paspalum repens*, *Montrichardia arborescens*, *Avicennia germinans*, *Rhizophora mangle*), though on older banks which resist erosion, mangrove species are replaced by mixed marsh forests (Pannier 1979).

4.3.3 Vertebrate Fauna

Although more than 450 species of fishes have been reported for the Orinoco River (Novoa 1990), in the delta extensive surveys are still lacking. Nevertheless, 97 species have been cited for main channels and tributaries and for other water bodies in marsh forests and palm groves of the lower delta. Juveniles (e. g. *Piaractus brachypomum*, *Hoplosternum littorale*, *Serrasalmus* spp., *Hoplias malabaricus*) as well as a large number of small species are generally associated with the smaller water bodies and mats of floating vegetation (Ponte 1995). The structure of fish assemblages is likely to be a direct influence of transparency, depth and size of water bodies (Rodríguez and Lewis 1997). Some of the commercially important fish species in the delta are *Colossoma macropomum*, *Piaractus brachypomum*,

Mylossoma duriventris, Pseudoplatystoma fasciatum, P. tigrinum, Brachyplatystoma rousseauxi, Hoplosternum littorale and *Plagioscion squamosissimus* (Novoa et al. 1984). Several aquatic mammals are widely distributed in the delta. The most common mammal is the freshwater dolphin *Inia geoffrensis* (tonina) while other species, like the river dolphin *Sotalia fluviatilis*, the manatee *Trichechus manatus*, the giant otter (*Pteronura brasiliensis*) and the long-tailed otter (*Lontra longicaudis*) are rare (Meade and Koehnken 1991; Linares 1998).

The Orinoco Delta has a rich bird fauna (178), including 39 species of aquatic birds and 13 Limicolae (Lentino and Bruni 1994). The red (*Eudocimus ruber*) and white ibis (*Eudocimus albus*), as well as *Ara ararauna, Todirostrum maculatum* and *Buteogallus aequinoctialis*, are among the most common species. The last two species and *Coccizus melacoriphus* are typical mangrove inhabitants. Reptiles in the delta are represented by the spectacled caiman (*Caiman crocodilus*), the tegus (*Tupinambis nigropunctatus*), turtles (*Podocnemis unifilis, Phrynus gibbus, Chelonia mydas*), and the iguanids *Plica umbra* and *Spilotes pullatus* (MARNR 1991). Only one species (the agouti *Dasyprocta guamara*) of the 96 mammals recorded is endemic to the delta and San Juan River floodplain (Linares 1998). Typical terrestrial species are the tapir (*Tapirus terrestris*), giant anteater (*Myrmecophaga tridactyla*), jaguar (*Panthera onca*), ocelot (*Leopardus pardalis*), paca (*Agouti paca*), kinkajou (*Potos flavus*), racoon (*Procyon cancrivorous*), capibara (*Hydrocharis hydrocharis*) and the red howler (*Alouatta seniculus*) and ring-tail monkeys (*Cebus olivaceous*).

4.4 Human Impact and Exploitation

For more than 7,000 years Warao-like people have lived in the vicinity of the Orinoco Delta. Today, approximately 21,000 Warao Indians live in groups of 50–100 individuals and represent the principal ethnic group in the delta (Wilbert 1994–1996). The Warao (canoe or sandbank people) are semi-nomadic harvesters or sustain a semi-agricultural way of life. They cut mangrove timber to meet their basic needs and exploit seven species of palms for food, medicine and raw materials. About 40% of their time is spent fishing the more than 38 species which are part of their diet (Heinen 1988). As long as they follow traditional ethno-ecological codes, Warao people interact with the delta in an ecologically sound way because they do not overexploit resources and hunt fish, rodent and bird species with high reproductive capacity and short gestation time (Ponte 1995; Wilbert 1994–1996).

During many decades the absence of roadways has protected the Orinoco Delta from destructive human activities, apart from illegal wildlife extraction and unauthorized logging. Environmental impacts, however, have increasingly been felt over the last decades and are likely to augment in coming years. For example, a gold rush has attracted some 40,000 Garimpeiro miners to the Middle Orinoco River basin; the sediments and mercury contamination from mining activities will find their way into the waterways and the biota of the delta. Furthermore, the Orinoco River and the Río Grande channel in the delta are among the most important fluvial waterways in the world and are the main shipping route (21×10^6 tons year^{-1}) for minerals (i.e., bauxite, iron, aluminum) to the Atlantic Ocean. Continuous dredging activities in the channel are likely to modify physico-chemical and biological characteristics in the delta and nearshore waters.

In 1965, the construction of a cofferdam, aiming at land reclamation at the Manamo channel, turned vast areas into barren land (cat clays) unsuitable for agricultural use. Additionally, the obstruction of waterways modified the hydrological regime (Colonnello 1996), and the advance of saltwater into the oligohaline Manamo channel led to the expansion of mangroves at rates six times faster (6–7 ha year^{-1}) than before (Colonnello and Medina 1998). In the near future, however, oil prospecting and drilling might be the major threat to the delta. Although concessions for oil extraction in the delta were granted as early as 1928, owing to the inherent difficulties of working in swamps and channels and high operating costs, activities soon ceased. New exploration rights have been awarded over the last decade and at present production is approximately 100,000 barrels per day. During the following years, oil extraction in an area of 213,000 ha could produce up to 160,000 barrels per day.

The exploitation (silviculture) of mangrove timber in the northern Orinoco Delta during the 1970s has resulted in ecological consequences that still need to be assessed and, although exploitation was abandoned, the community is far from having recovered (Ernesto Medina, pers. comm.). The semi-industrial extraction of palm hearts (*Euterpe* spp.) has been steadily expanding (239,841 kg year^{-1}) in past years. The alternative extraction of palm heart from plantations of *Bactris gasipaes* might eventually attenuate the large-scale exploitation of *Euterpe oleracea*. Apart from palm hearts, the pulp of *Euterpe oleracea* fruits is used for juice, ice cream and liquor.

More than 60 species of fishes are commercially exploited in the Orinoco River and the delta. Fisheries are largely artisanal, using small wooden canoes (curiaras) carved from tree trunks. Since catches are landed at different ports in the delta and the lower Orinoco River, fisheries statistics for the delta are not precise. About 45% of the total catch is comprised of catfish (*Arius herzbergii, Cathorops spixii*), followed by *Mugil brasiliensis, M. curema* and *Centropomus ensiferus* which represent 36%. Other species are *Piaractus brachypomum, Prochilodus mariae* and freshwater catfish (*Brachyplatystoma*, Paulicea; Novoa and Ramos 1990). Between 1966 and 1976, mean annual landings from the Orinoco River oscillated between 1,204 and 3,584 tons (Novoa and Ramos 1978) but increased to an annual average of 12,463 tons (1985-1989; Novoa and Ramos 1990), which is still below the estimated maximum sustainable yield of 45,000 tons (Novoa 1982).

4.5 Management Considerations

Despite the legal protection of almost half of the Orinoco Delta, owing to the Mariusa National Park (2,650 km^2) and the Orinoco Delta Biosphere Reserve (8,765 km^2), the delta also has highest conservation priority among coastal areas of Venezuela (Díaz Martín et al. 1995). Several non-governmental and indigenous organizations have called attention to environmental hazards of oil-related activities (Red Alerta Petrolera) in the delta. In 1989, based on hydrological, biological, ecological, economic and social factors and fishing gear employed, the Venezuelan Ministry of Agriculture and Livestock (Ministerio de Agricultura y Cría) regulated fisheries in the Orinoco system. As a consequence, fisheries management in the delta established larger fishing grounds, new periods for fishing activities, and a mesh-size of nets which permitted catch of smaller fish (Novoa and Ramos 1990). Specific legal instruments control the catch of species, like the sapoara (*Semaprochilodus laticeps*), which is an important food source for Orinoco dwellers. In the face of growing environmental problems, an integrated coastal zone management program, which regulates and monitors unsustainable activities, is urgently called for.

Acknowledgements. I am grateful to Giuseppe Colonnello, Nelda Dezzeo, Saúl Flores, Dieter Heinen, Elizabeth Olivares, Jorge Paolini, Milagro Rinaldi, Delfina Rodríguez, Gilberto Rodríguez and Erika Wagner for making bibliographic material available, reading the manuscript, or providing critical comments. The secretarial contribution of Berta Sánchez was most valuable. Special thanks are given to Carlos Carmona.

References

Bacon PR (1990) Ecology and management of swamp forests in the Guianas and Caribbean region. In: Lugo AE, Brinson M, Brown S (eds) Forested wetlands. Ecosystems of the world, no 15. Elsevier, Amsterdam, pp 213-250

Bonilla J, Senior W, Bugden J, Zafiriou O, Jones R (1993) Seasonal distribution of nutrients and primary productivity on the Eastern Continental Shelf of Venezuela as influenced by the Orinoco River. J Geophys Res 798:2245-2257

Colonnello G (1995) La vegetación acuática del delta del río Orinoco (Venezuela). Composición florística y aspectos ecológicos (I). Mem Soc Cienc Nat La Salle 55:3-34

Colonnello G (1996) Aquatic vegetation of the Orinoco River Delta (Venezuela). An overview. Hydrobiologia 340:109-113

Colonnello G, Medina E (1998) Vegetation changes induced by dam construction in a tropical estuary: the case of the Mánamo River, Orinoco Delta (Venezuela). Pl Ecol 139:145-154

Colveé GPE, Szczerban D, Talukdar SC (1990) Estudios y consideraciones geológicas sobre la cuenca del río Caura. Una porción del Escudo de Guayana venezolano. In: Weibezahn FH, Alvarez H, Lewis WM Jr (eds) El río Orinoco como ecosistema. Editorial Galac, Caracas, pp 11-53

Conde JE, Alarcón C (1993) Mangroves of Venezuela. In: Lacerda LD (ed) Conservation and sustainable utilization of mangrove forests in the Latin America and Africa regions, part I: Latin America. The International Society for Mangrove Ecosystems (ISME) and The International Tropical Timber Organization, Okinawa, Japan, pp 211-243

Cressa C, Vásquez E, Zoppi E, Rincón JE, López C (1993) Aspectos generales de la limnología en Venezuela. Interciencia 18:237-248

Depetris PJ, Paolini JE (1991) Biogeochemical aspects of South American rivers: the Paraná and Orinoco. In: Degens ET, Kempe S, Richey JE (eds) Biogeochemistry of major world rivers. Wiley, New York, pp 105-125

Díaz de Gamero ML (1996) The changing course of the Orinoco River during the Neogene: a review. Palaeogeogr Palaeoclimatol Palaeoecol 123:385-402

Díaz Martín D, González E, Hernández D (1995) Prioridades de conservación en las áreas marino-costeras de Venezuela. Fundación para la Defensa de la Naturaleza, Caracas

Heinen HD (1988) Los Warao. In: Coppens W (ed) Los aborígenes de Venezuela. CONICIT-Fundación La Salle de Ciencias Naturales. Monte Avila Editores, Caracas, pp 585-689

Heinen HD, San José JJ, Caballero Arias H, Montes R (1995) Subsistance activities of the Warao indians and anthropogenic changes in the Orinoco Delta vegetation. Scientia Guaianæ 5:312-334

Herrera LE, Masciangioli P (1984) Características de las corrientes frente al delta del Orinoco, sector occidental del océano Atlántico. Rev Téc INTEVEP 4:133-144

Herrera LE, Febres GA, Avila RG (1981) Las mareas en aguas venezolanas y su amplificación en la región del delta del Orinoco. Acta Cient Venez 32:299-306

Lentino M, Bruni AR (1994) Humedales costeros de Venezuela: Situación Ambiental. Sociedad Conservacionista Audubon de Venezuela, Caracas

Lewis WM Jr (1988) Primary production in the Orinoco River. Ecology 69:679-692

Lewis WM Jr, Saunders JF III (1989) Concentration and transport of dissolved and suspended substances in the Orinoco River. Biogeochemistry 7:203-240

Lewis WM Jr, Weibezahn FH, Saunders JF III, Hamilton SK (1990) The Orinoco River as an ecological system. Interciencia 15:346-357

Linares OJ (1998) Mamíferos de Venezuela. Sociedad Conservacionista Audubon de Venezuela, Caracas

MARNR (1991) Ministerio del Ambiente y de los Recursos Naturales Renovables. Conservación y manejo de los manglares de Venezuela y Trinidad Tobago. FP-11-05-81-01 (2038), Sector Delta del Orinoco, Territorio Delta Amacuro. (PT) Serie Informe Técnico DGSIIA/IT/256, Caracas

Meade RH, Koehnken L (1991) Distribution of the river dolphin, tonina *Inia geoffrensis*, in the Orinoco River basin of Venezuela and Colombia. Interciencia 16:300-312

Müller-Karger FE, McClain CR, Fisher TR, Esaias WE, Varela R (1989) Pigment distribution in the Caribbean Sea: observations from space. Prog Oceanogr 23:23-64

Novoa D (1982) Los recursos pesqueros del río Orinoco y su explotación. Corporación Venezolana de Guayana, Caracas

Novoa D (1990) El río Orinoco y sus pesquerías: estado actual, perspectivas futuras y las investigaciones necesarias. In: Weibezahn FH, Alvarez H, Lewis WM Jr (eds) El río Orinoco como ecosistema. EDELCA-Fondo Editorial Acta Científica Venezolana-CAVN-Universidad Simón Bolívar, Caracas, pp 387-406

Novoa D, Ramos F (1978) Las pesquerías comerciales del río Orinoco. Corporación Venezolana de Guayana, Caracas

Novoa D, Ramos F (1990) Las pesquerías comerciales del río Orinoco: su ordenamiento vigente. Interciencia 15:486-490

Novoa D, Ramos F, Cartaya E (1984) Las pesquerías artesanales del río Orinoco: su ordenamiento vigente. Mem Soc Cienc Nat La Salle 44:163-215

Pannier F (1979) Mangroves impacted by human-induced disturbances: a case study of the Orinoco delta mangrove ecosystem. Environ Mgmt 3:205-216

Paolini J (1995) Particulate organic carbon and nitrogen in the Orinoco River (Venezuela). Biogeochemistry 29:59-70

Paolini J, Herrera R, Nemeth A (1983) Hydrochemistry of the Orinoco and Caroní Rivers. Mitt Geol Paläont Inst Univ Hamb 55:223-236

Ponte JV (1995) Contributions of the Warao Indians to the ichthyology of the Orinoco Delta, Venezuela. Scientia Guaianæ 5:371-392

Rodríguez G (1974) Some aspects of the ecology of tropical estuaries. In: Golley FB, Medina E (eds) Tropical ecological systems. Springer, Berlin Heidelberg New York, pp 313-333

Rodríguez MA, Lewis WM Jr (1997) Structure of fish assemblages along environmental gradients in floodplain lakes of the Orinoco River. Ecol Monogr 67:109-128

Schubert C, Briceño HO, Fritz P (1986) Paleoenvironmental aspects of the Caroní-Paragua river basin (Southeastern Venezuela). Interciencia 11:278-289

Van Andel TJH (1967) The Orinoco Delta. J Sed Petrol 37:297-310

Varela M, Varela R (1983) Microalgas del Bajo Orinoco y Delta Amacuro, Venezuela. II. Bacillariophyceae. Dinophyceae. Mem Soc Cienc Nat La Salle 43:89-111

Varela R, Varela M, Fariña AC (1983) Microalgas del Bajo Orinoco y Delta Amacuro, Venezuela. I. Cyanophyceae, Euglenophyceae, Chrysophyceae, Xanthophyceae, Euchlorophyceae, Zygophyceae. Mem Soc Cienc Nat La Salle 43:59-88

Vera B (1992) Sea grasses of the Venezuelan coast: distribution and community components. In: Seeliger U (ed) Coastal plant communities of Latin America. Academic Press, San Diego, pp 135-140

Wilbert W (1994-1996) *Manicaria saccifera* and the Warao in the Orinoco Delta: a biogeography. Antropologica 81:51-66

5 The Itamaracá Estuarine Ecosystem, Brazil

C. Medeiros, B. Kjerfve, M. Araujo, and S. Neumann-Leitão

5.1 Introduction

In northeastern Brazil (7°34'–7°55' S and 34°48'–34°52' W), the tropical Itamaracá estuary system (824 km²) was formed during the early Holocene. A fault which gave origin to the Santa Cruz Channel was flooded by seawater and connected with the South Atlantic Ocean by the Catuama and Orange entrance (Fig. 5.1). Offshore regions typically have a series of two or three parallel sandstone reefs (20–60 m wide) which consist of cemented sand (20–80 % of quartz and biogenic detritus) and become exposed during low waters, as well as eight to ten permanently submerged reef lines (Mabesone 1964). Most of the exposed and submersed reefs were formed during Holocene sea level still-stands (Morais 1970) and some are still being formed. Remnant reefs on land were presumably formed during temporary still-stands of the Quaternary regression phase (Ottmann 1960). The discontinuous sandstone ridges, as well as reefs and sand banks north of the Catuama and south of the Orange entrance, form a semi-open elongated lagoon-like environment ("inner sea") which restricts the exchange of water between the Santa Cruz Channel and offshore areas (Medeiros and Kjerfve 1993; Fig. 5.1).

5.2 Environmental Setting

The wind regime in the Itamaracá area is determined by the position of the semi-stationary anticyclone of South America (Ratisbona 1976). Prevailing winds are southeast trade winds with a mean velocity of 3.2 m s^{-1} that tends to increase (4.0 m s^{-1}) during July. Owing to continental warming, barometric pressure decreases introduce a northerly component, and prevailing winds (mean 2.6 m s^{-1}) during the dry season (September to January) blows from the northeast mainly in December–January. The climate is tropical,

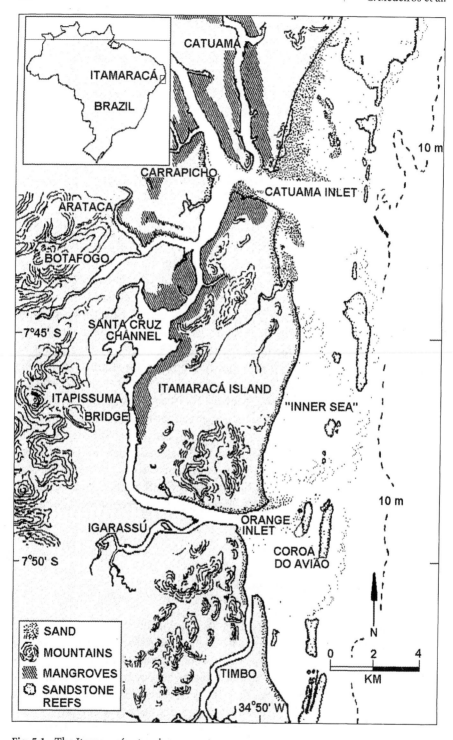

Fig. 5.1. The Itamaracá estuarine ecosystem

hot and humid (Köppen classification type Aws). The mean annual air temperature is 26 °C (mean range 2.8 °C) with a maximum of 34 °C and a minimum of 18 °C (Ratisbona 1976). The advance of maritime air masses over the continent causes rainfall, which gradually decreases towards the inland. Annual precipitation in the Itamaracá drainage area is 1,500–1,700 mm (Medeiros 1991). During the rainy season (February to August), with mean monthly rainfall of 180–212 mm, about 50 % of the annual precipitation occurs between April and June (400 mm month^{-1}; Passavante 1981) and the hydrological balance is strongly positive. Low mean monthly rainfall (40–51 mm), evaporation exceeding precipitation, and a negative hydrologic balance characterize the dry season (September to January).

5.3 Hydrodynamics

Most of the freshwater is added to the northern branch of the Santa Cruz Channel (94 % during the dry season and 70 % during the rainy season) through the Catuama, Carrapicho, do Congo, Arataca, Botafogo, and Igarassu rivers, the last three being the major freshwater sources (Fig. 5.1). Seasonal differences in precipitation (4.5×) and freshwater runoff (50×) are significant, with total average river discharge varying between 55.9 m^3 s^{-1} at the peak of the rainy season and 0.8 m^3 s^{-1} at the peak of the dry season. The overall freshwater balance (river discharge and direct rainfall) in the Itamaracá system yields a freshwater input of 0.28 m^3 s^{-1} during peak summer and 57.7 m^3 s^{-1} during winter conditions, representing a 206-fold freshwater input to the system (Medeiros and Kjerfve 1993).

Tides in the Santa Cruz Channel are strongly semidiurnal ($F=0.08$–0.12). The mean tidal range along the channel varies between 1.0 and 1.8 m (spring 1.4–2.2 m, neap 0.5–1.1 m) with lower ranges near the nodal point about 1.7 km north of Itapissuma Bridge (Medeiros 1991). Water elevation responds largely to tidal forcing with main diurnal and semidiurnal tidal constituents and shallow water overtides explaining 78 to 95 % of the total variability. The remaining variability of water elevation is due to meteorological forcing and runoff (Medeiros and Kjerfve 1993). The amplitude and phase of semidiurnal tides are identical at both entrances, with an amplitude of 88 % and a phase lag of 15 min at Itapissuma Bridge (Fig. 5.1). Tides propagating through the northern Catuama channel (1.5 km wide and 15 m deep) are subject to funneling and shoaling, and tidal amplitude attenuation is more pronounced. Tides propagating through the southern Orange channel are only affected by shoaling, and phase shifts are more prominent. The southern channel branch is more uniform in width and tidal damping occurs mainly in response to friction.

Tropical systems, like the Santa Cruz Channel, have weak vertical salinity stratification and normal levels of turbulence distribute temperatures uniformly from surface to bottom (Medeiros and Kjerfve 1993; Araujo et al. 2000a,b). Bottom stirring by tidal currents is the most important component of vertical water column mixing, while water surface evaporation and energy input from surface wind stress contribute much less. Although precipitation has little effect on stratification, diurnal surface heating may cause vertical stratification during the dry season and stratification can be expected to develop during higher freshwater input. Compared to tropical systems with a single inlet, the differential advection of horizontal densities is seasonally attenuated and horizontal density gradients along the estuary axis are small (Araujo et al. 2000b).

The residual circulation in the Santa Cruz Channel is primarily due to river discharge, intensity of tides, and channel shape and depth. In general, residual circulation (0.01–0.19 m s^{-1}) flows from the Itapissuma Bridge northward to the Catuama and southward to the Orange entrance. However, during high freshwater discharge and spring tides the entire channel has northward residual circulation. At both entrances, residual currents tend to be highest during high runoff. In contrast, during low freshwater discharge nearshore water enters the Catuama entrance in the north, but net inflow through the southern Orange entrance is absent.

5.4 Hydrology

Overall salinity in the Itamaracá estuary is lower in the rainy (27) than in the dry season (34). During the dry season, hypersaline conditions (37) at both entrances are due to evaporation, evapotranspiration by mangroves, and reduced exchange between channel and reef shelf waters with lower salinity. The estuary is well mixed and surface to bottom salinity differences are less than 1. The lowest salinity (30) occurs at the innermost part of the channel during low water neap tide (Medeiros and Kjerfve 1993). During the rainy season, the northern channel branch tends to be stratified ($\Delta S/S_0 > 0.1$) at neap tide. Especially in deeper areas (15 m), at the confluence of the Botafogo and Arataca rivers, large freshwater discharge and reduced tidal mixing energy favor stratification with a surface to bottom salinity gradient (12). Stratification is less pronounced during high tidal energy of spring tides (Medeiros and Kjerfve 1993).

Total suspended solids (TSS) tend to be lower during the rainy (20 ppm) than in the dry season (40 ppm). The exchange of both TSS and particulate organic carbon (POC) between channel and reef shelf water is principally controlled by tidal dynamics. During the dry season, a large part of the TSS remains in the channel and tidal trapping promotes its homogeneous distribution. Reduced dynamics in the channel and "inner sea" favor a net import of TTS (35.12 tons per tidal cycle, POC 2.34 tons per tidal cycle) and the accumulation of particulate matter within the system. During the rainy season, despite the large riverine input of suspended matter, the retention by mangrove swamps and the water exchange between nearshore and shelf areas appears to remove TSS from the system. The high river discharge flushes the system and net fluxes cause a net export of TSS (260 tons per tidal cycle) and POC (6.9 tons per tidal cycle) through the Orange entrance (Broce 1994).

Independent of the season, net longitudinal salt flux and transport of TSS in the Santa Cruz Channel are largely dominated by advective transport, as is the salt flux towards the ocean. The tidal wave transport is a secondary mechanism of salt and TSS transport and counteracts advective transport. Both advection and tidal wave transport contribute to the removal of estuarine TSS. Since cross-sectional shear is much smaller than vertical shear, cross-sectional variability of salinity, TSS, and current distribution contributes little to net longitudinal salt and TSS fluxes in the Itamaracá system (Medeiros 1991).

Water temperatures in the Santa Cruz channel lack vertical and horizontal gradients but vary between 30.1 °C (28.8–30.9 °C) in the dry season and 26.8 °C (25.7–27.7 °C) in the rainy season. The Secchi disk depth in the channel (mean 1.9 m) and at the confluence of tributaries (0.3–0.7 m) decreases during the rainy season. The pH varies little in both channel entrances (Orange 7.3, Catuama 8.2; Flores Montes 1996). Dissolved oxygen is near or above saturation levels (95–130%), though lower values (2.5–4.5 ml l^{-1}) occur in the central area of the channel during the rainy season. River discharge is the principal source of nutrients in the Santa Cruz Channel, followed by sediment resuspension, mangrove litter, waste input, terrestrial runoff, and atmospheric input. Nitrate (0.02–6.73 µmol l^{-1}) and nitrite (0.001–0.730 µmol l^{-1}) concentration tends to be higher at the mouth of tributaries at low tide during the rainy season (Flores Montes 1996). Phosphorus levels in the channel are directly related to river discharge and surface runoff, with average concentrations between 0.71 and 2.4 µmol l^{-1} during the dry and 0.4 to 0.6 µmol l^{-1} during the rainy season (Macêdo et al. 1973).

5.5 Biota

Mangrove forests (approx. 28 km^2) occupy the lowlands along the inner portion of the Santa Cruz Channel and the lower part of tributaries. The dominant species is the red mangrove (*Rhizophora mangle*), followed by white mangrove (*Laguncularia racemosa*) and the less common black mangrove (*Avicennia schaueriana*) and button mangrove (*Conocarpus erectus*; Silva 1995). Enhanced mangrove growth and size of trees in areas around the nodal point of tides may be due to deposition of fine-grained, nutrient-rich sediment from freshwater runoff. A large part of the organic matter production originates from *Rhizophora mangle* litter (7 tons dry weight per ha and year). Additionally, high net productivity of associated primary producers (i.e., the red alga *Bostrichia* 52.4 mg C h^{-1} m^{-2} and periphyton 208.3 mg C h^{-1} m^{-2}) contributes significantly to total annual carbon production.

The phytoplankton community in the Santa Cruz Channel is composed of 78 species, 85% of which are diatoms. Phytoplankton density is highest in the Orange inlet (4.46 × 10^6 cells l^{-1}) during the rainy season. Phytoplankton biomass decreases rapidly from the northern (796 mg m^{-3}) and southern inlet (1,052 mg m^{-3}) towards oceanic regions (<60 mg m^{-3}). Owing to high solar radiation and nutrient concentrations (N limiting factor, N:P ratio < 10:1), phytoplankton primary productivity in the Santa Cruz channel is high (2.25–39.0 mg C h^{-1} m^{-3}). Lower mean productivity values occur in the northern Catuama (10.2 mg C h^{-1} m^{-3}) and southern Orange entrance (18.8 mg C h^{-1} m^{-3}), decreasing even further (<1.0 mg C h^{-1} m^{-3}) beyond the reef line over the adjacent continental shelf. Similarly, elevated chlorophyll *a* concentrations in both channel inlets (mean 12.0 mg m^{-3}) decrease towards nearshore (0.34 mg m^{-3}) and offshore (0.01 mg m^{-3}) regions.

A large number of benthic organisms (19% Foraminifera) compose the zooplankton community. Meroplankton (i.e., larvae of bivalves, gastropods, polychaetes) comprise about 20% of the microzooplankton but copepods (27 species) and Tintinnina (15 species) are the dominant euryhaline marine species (Eskinazi-Leça 1974). The abundance of microzooplankton in the Orange inlet (22,449–244,216 org. m^{-3}) tends to decrease towards oligotrophic shelf waters. Macrozooplankton (49 taxa) is characterized by the dominance of copepods, Brachyura zoeae (Decapoda), followed by *Lucifer* sp., Chaetognatha, Larvacea, fish eggs, gastropods, and Upogebidae zoeae. Meroplankton represents approximately 35% (Paranaguá and Nascimento 1973). Ctenophores prevail during dry periods and polychaetes associated with rotifers are common in upstream areas and at the mouths of tributaries. The productivity of the plankton community decreases from estuarine areas towards the ocean. However, despite the hydrographic barrier between estuarine and shelf waters, the export of diatoms, larvae of bivalves, gastropods, and cirripedes is significant. Especially Brachyura zoeae are exported at high rates (1.5×10^8 zoeae day^{-1}; Wehrenberg 1996) to coastal waters (10-20 km offshore) and are occasionally more abundant than copepods (Schwamborn 1997). Since both copepods and zoeae are important food sources for various species of fish (Vasconcelos et al. 1984; Morgan 1990; Sautour et al. 1996), their export is likely to fuel the marine food web.

Around the islands and in the "inner sea" extensive and productive (approx. 3 tons dry weight per ha and year) sea grass meadows of *Halodule wrightii* (Kempf 1970) constitute a substrate for epifaunal organisms (mollusks and crustaceans). The associated macrofauna is dominated by the mollusk *Tricolia affins*. The most abundant taxa among decapod crustaceans are pangurid hermit crabs (Panguridae) and the shrimp *Hippolyte curacaoensis* (Caridae). The approximately 70 species of fish in the Santa Cruz Channel (365 kg ha^{-1} year^{-1}) are represented by the families Mugilidae (*Mugil curema, M. liza, Mugil* sp.), Centropomidae (*Centropomus parallelus, C. undecimalis*), Gerreidae (*Eugerres brasilianus, Diapterus olisthostomus*), Clupeidae (*Opisthonema oglinum*), Hemirhamphidae (*Hypohamphus unifasciatus, Hemirhampus brasiliensis*), Carangidae (*Caranx fatus*), and Ariidae (*Arius parkeri, A. proops*; Eskinazi 1972; Azevêdo and Guedes 1973).

The bird fauna (71 species of 30 families) of the Itamaracá estuarine system is represented by resident and migratory species. Common residents (i.e., *Butorides striatus, Conirostrum bicolor, Chloroceryle americana*, and the kingfisher *Ceryle torquata*) are related to mangrove habitats. Other species, like wading birds (*Jacana jacana, Porphirula martinica*), river swallow (*Tachycineta albiventer*), seagull (*Larus maculipenis*), sea tern (*Sterna eurygnatha*), tyrant flycatcher (*Pitangris sulphuratus*), and tanager (*Thraupis sayaca*) are found on riverine and estuarine flood plains, reefs, and beaches. Migratory birds (i.e., *Pluvialis squatarola, Charadrius semipalmatus, Arenaria interpres, Calidria pusilla, C. alba*) stay from August to April in the Santa Cruz Channel area to feed and change their plumage (Azevêdo 1992). The manatee (*Trichechus manathus*), a native species of north and northeastern Brazil, with an estimated population of 400 individuals, occurs in sea grass beds around Itamaracá Island but is in danger of extinction.

5.6 Trophic Relations

The Itamaracá estuarine ecosystem displays strong trophic interactions between the pelagic and benthic compartment. While meroplanktonic larvae of benthic organisms (i.e., decapods) appear to feed on phytoplankton, mangrove detritus is a significant food source for the copepod-rich holoplankton (13–40%; Schwamborn 1997) as well as for the crabs *Aratus pisone*, *Ucides cordatus*, and *Goniopsis cruentata*. High primary production frequently causes hypoxia in bottom substrates, resulting in a meiobenthic community with low diversity and strong dominance (143 ind. cm^{-2}) of opportunistic species (polychaetes, nematodes, oligochaetes) which feed on decomposing organic matter and associated bacteria and fungi. Plant detritus and associated microflora and microfauna also comprise the diet of ostracods and copepod Harpactidoida and of macroconsumers, such as mollusks (*Littorina australis*, *Tagellus plebeus*, *Neritina virginea*), crustaceans (*Uca leptodactyla*, *U. rapax*, *U. maracoani*, *U. thayeri*), and shrimps (*Penaeus schimiti*, *P. subtilis*). Suspended detritus particles in the water column are a food source for the oyster *Crassostrea rhizophorae* as well as for shrimps. The fish species *Arius parkeri*, *Arius proops*, *Diapterus olisthostomus*, and *Eugerres brasilianus* feed largely on invertebrates, such as polychaetes, mollusks, and crustaceans. Important carnivores are *Centropomus undecimalis*, *C. parallelus*, and *Caranx fatus*. The planktivorous fish *Hemirhamphus brasiliensis* and *Hypohamphus unifasciatus* feed upon *Acartia lilljeborgi*, *Paracalanus crassirostris*, *Oithona hebes*, *Oithona oswaldocruzi*, *Euterpina acutifron*, *Oikopleura dioica*, and *Sagitta tenuis* among others.

5.7 Environmental Problems

As most estuaries, the Itamaracá estuarine ecosystem is exposed to multiple pressures from industrial pollution, domestic sewage discharge, urban expansion and land reclamation, and fisheries. Raw domestic sewage is discharged from Itapissuma (population 19,000) into the Santa Cruz Channel, thus coliform contamination in mangrove oysters tends to be high. However, one of the major environmental problems in the estuary has been the release of mercury (22–35 tons) into the Botafogo River by a chlor-alkali factory between 1964 and 1987 (Meyer 1996). The impact of Hg pollution was aggravated by the concomitant discharge of chlorine and acid wastes from sugarcane mills, which increased Hg dissolution and therefore impeded Hg removal from the water column and facilitated its distribution throughout the Itamaracá estuary. Mercury emission decreased considerably after the construction of a precipitation basin in 1986; nevertheless, leakage continued until 1991 due to a technical failure. Today, the distribution of Hg in oysters (0.27–2.21 ppm based on dry mass), suspended matter (0.43–5.56 ppm), and sediments (0.04–6.2 ppm, <62 µm fraction 0.3–20.5 ppm) in the Santa Cruz Channel is largely a function of proximity to the original source and channel hydrodynamics (Meyer 1996; Meyer et al. 1998). Since only 10% of the total released Hg has remained in the system, it appears that reduced trapping by mangroves, owing to acidic water conditions, resuspension and current-controlled export via suspended matter, and transformation processes in intertidal sediments contributed to removal processes.

Both sea grass and mangrove habitats, which are reproduction and feeding grounds for important fishery resources, have been reduced during the last decades. Mangrove areas in Santa Cruz Channel (36.2 km^2 in 1974, 28.1 km^2 in 1988) suffered the impact of real estate expansion (5.2 km^2 in 1974, 33.0 km^2 in 1988) compounded by increasing demands for freshwater and waste disposal sites. Non-commercial fisheries constitute the major local economic activity for about 5,000 fishermen in Itamaracá. Although overall catch has increased from 200 tons (Macêdo 1974) to 400 tons over the last 20 years (Börner 1994), overfishing and destructive fishing methods have led to declining fisheries stocks in the Santa Cruz Channel. Owing to high demand for seafood by the growing tourist industry, shrimp stock are most affected, but oysters, crabs, and blue crabs are also in high demand and are fished throughout the year. Under these circumstances, extensive and semi-intensive commercial aquaculture (i.e., mullet, shrimp, mangrove oyster) would offer an economic alternative for local communities and should reduce the fishing pressure on natural stocks. The destruction of mangrove areas and the construction of seawalls and jetties to protect real estate developments (hotels and marinas) have altered the geomorphology around Itamaracá Island. Whereas 10 years ago Coroa do Avião near the Orange inlet (Fig. 5.1) was a small sandbank and only exposed during low tide, it is now a permanently exposed, vegetated island. As a consequence, the inlet channel has been displaced northwards and the southern shore of Itamaracá Island is being eroded.

Despite the protection offered by state and federal laws, mangrove areas are still rapidly diminishing. However, part of the Itamaracá drainage basin now includes a Reserve for the Biosphere of the Atlantic Forest and a large area of the Itamaracá estuary is under state protection. Finally, the Brazilian Institute for the Environment (IBAMA) maintains a center for the reproduction and management of endangered manatees on Itamaracá Island.

Acknowledgements. The authors thank Drs. Ignacio Salcedo and Luis Salcedo for their valuable comments and proofreading.

References

Araujo M, Medeiros C, Ribeiro C, Freitas I (2000a) Testing surface boundary conditions for turbulence modeling in tidal systems. Turbulent diffusion on the Environment 25–38
Araujo M, Medeiros C, Ribeiro C, Freitas I, Bezerra MO (1998b) Turbulence modeling in a stirring tropical estuary. Turbulent diffusion on the Environment 59–66
Azevêdo SH Jr (1992) Anilhamento de aves migratórias na Coroa do Avião, Igarassú, Pernambuco, Brasil. Caderno Omega, Universidade Federal Rural de Pernambuco, Série Ciências Aquáticas no 3
Azevêdo SB, Guedes DS (1973) Novas ocorrências de peixes para o Canal de Santa Cruz (Itamaracá, Pernambuco). Ciência Cult, S Paulo 25(6):353
Börner R (1994) Fischereibiologische Untersuchungen von den Fischbeständen des "Canal de Santa Cruz", Pernambuco, Brasilien. MSc Dissertation, Universität Bremen, Germany
Broce DAS (1994) Importação e exportação de carbono orgânico particulado através da barra sul do canal de Santa Cruz, Itamaracá. MSc Thesis, Universidade Federal Rural de Pernambuco, Brazil
Eskinazi AM (1972) Peixes do Canal de Santa Cruz – Pernambuco – Brasil. Trab Oceanogr Univ Fed Pernambuco 13:283–302
Eskinazi-Leça E (1974) Composição e distribuição do microfitoplâncton na região do Canal de Santa Cruz (Pernambuco – Brasil). Thesis, Instituto de Biociências, Universidade Federal Rural de Pernambuco, Brazil
Flores Montes MJ (1996) Variação nictimeral do fitoplancton e parâmetros hidrológicos no canal de Santa Cruz, Itamaracá, PE. MSc Thesis, Universidade Federal Rural de Pernambuco, Brazil
Kempf M (1970) Nota preliminar sobre os fundos costeiros da região de Itamaracá (norte do Estado de Pernambuco, Brasil). Trab Oceanogr Univ Fed Pernambuco 9/11:95–110
Mabesone JM (1964) Origin and age of the sandstone reefs of Pernambuco (Northeastern Brazil). J Sed Petrol 34(4):715–726
Macêdo SJ (1974) Fisioecologia de alguns estuários do Canal de Santa Cruz Itamaracá-PE. MSc Thesis, Universidade de São Paulo, Brazil
Macêdo SJ, Lira MEF, SILVA JE (1973) Condições hidrológicas do canal de Santa Cruz, Itamaracá, PE. Boletim de Recursos Naturais, SUDENE 11(1,2):55–92
Medeiros C (1991) Circulation and mixing in the Itamaracá estuarine system, Brazil. PhD Thesis, University of South Carolina, USA
Medeiros C, Kjerfve B (1993) Hydrology of a tropical system: Itamaracá, Brazil. Estuar Coast Shelf Sci 36:495–515
Meyer U (1996) On the fate of mercury in the northeastern Brazilian mangrove system Canal de Santa Cruz, Pernambuco. ZMT-Contributions 3, University of Bremen, Germany
Meyer U, Hagen W, Medeiros C (1998) Mercury in a northeastern Brazilian mangrove area, a case study. The potential of the mangrove oyster *Crassostrea rhizophorae* as bioindicator for mercury. Mar Biol 131:113–121

Morais JO (1970) Contribuição ao estudo dos "beach-rocks" do Nordeste do Brasil. Trab Oceanogr Univ Fed Pernambuco 9/11:79-94

Morgan SG (1990) Impact of planktivorous fishes on dispersal. Hatching, and morphology of estuarine crab larvae. Ecology 71:1639-1652

Ottmann F (1960) Une hypothèse sur l'origine des "arrecifes" du Nordeste brésilien. Compte Rendu Sommaire des Séances de la Société Géologique de France 7:175-176

Paranaguá MN, Nascimento DA (1973) Estudo do zooplâncton da região estuarina de Itamaracá. Ciência Cult, S Paulo 25(6):198

Passavante JZO (1981) Estudos ecológicos da região de Itamaracá, Pernambuco – Brasil, XIX Biomassa do nano e microfitoplâncton do Canal de Santa Cruz. Trab Oceanogr Univ Fed Pernambuco 16:105-156

Ratisbona LR (1976) The climate of Brazil, vol 12. In: Schwerdtfeger W (ed) World survey of climatology – climates of Central and South America. American Elsevier, New York, pp 219-293

Sautour B, Artigas F, Herbland A, Laborde P (1996) Zooplankton grazing impact in the plume of dilution of the Gironde estuary (France) prior to the spring bloom. J Plankt Res 18(6):835-853

Schwamborn R (1997) Influence of mangroves on community structure and nutrition of macrozooplankton in northeast Brazil. ZMT-Contributions 4, University of Bremen, Germany

Silva JDV (1995) Parâmetros oceanográficos e distribuição das espécies e bosques de mangue do estuário do Rio Paripe-PE. MSc Thesis, Universidade Federal Rural de Pernambuco, Brazil

Vasconcelos AL, Guedes DS, Galiza EMB, Azevedo-Araújo DS (1984) Estudo ecológico da região de Itamaracá-Pernambuco-Brasil. XXVII. Hábitos alimentares de alguns peixes estuarinos. Trab Oceanogr Univ Fed Pernambuco 18:231-260

Wehrenberg T (1996) Zum Einfluss von Tageszeit und Gezeiten auf Zusammensetzung und Transport des Makrozooplanktons in den Mündungsbereichen des Mangrovenästuars "Canal de Santa Cruz", Pernambuco, Brasilien. MSc Thesis, University of Bremen, Germany

6 The Abrolhos Reefs of Brazil

Z.M.A.N. Leão and R.K.P. Kikuchi

6.1 Introduction

The Abrolhos reefs represent the most extensive and richest area of coral reefs in Brazil and in the Southwestern Atlantic. Among the first scientific reports on Abrolhos coral reefs are those from the Darwin visit to the Abrolhos Archipelago; however, it was the Thayer Expedition led by Louis Agassiz that produced the seminal work with a first detailed description of coral zonation on the reefs (Hartt 1870). The Abrolhos reef ecosystem, which occupies approximately 6,000 km² of the northern Abrolhos Bank, is comprised of two reef arcs (Coastal and Outer Arcs) almost parallel to the coast, as well as of volcanic islands, sand shoals and channels. The Abrolhos Bank is an enlargement (up to 200 km off Caravelas) of the southern part of the generally narrow (average width 50 km) Eastern Brazilian continental shelf (Fig. 6.1A). The shelf around the Abrolhos reefs is shallow (<30 m) with a slope of 0°08' and the shelf edge at about 70 m. The topography of the inner shelf is smooth and flat while narrow channels and sandbanks are common on the middle and outer shelves. A shallow (Caravelas Channel 15 m) and a deeper (Abrolhos Channel 20–30 m) channel separate the coast from the coastal reef arc and the outer reef arc, respectively (Fig. 6.1A). Topographic features of the shelf evolved during the last Pleistocene regression when the shelf was subaerially exposed and the surface deeply eroded by the fluvial drainage system discharging into the Abrolhos Depression. Depositions of terrigenous sediments were later replaced by marine biogenic carbonate depositions (Vicalvi et al. 1978) which mainly occur on the middle and outer shelves but extend onto the inner shelf around the reefs, while terrigenous sediments are confined to the inner shelf (Leão 1982; Leão et al. 1988). Volcanic accretions to the shelf formed five islands (Abrolhos Archipelago), which consist of sedimentary Cretaceous rocks, and basalt sills and dykes, originated from Tertiary intrusive and volcanic activity. Until today the Abrolhos reef ecosystem has an unchallenged scientific importance because it differs in many aspects from other well-established coral reef models.

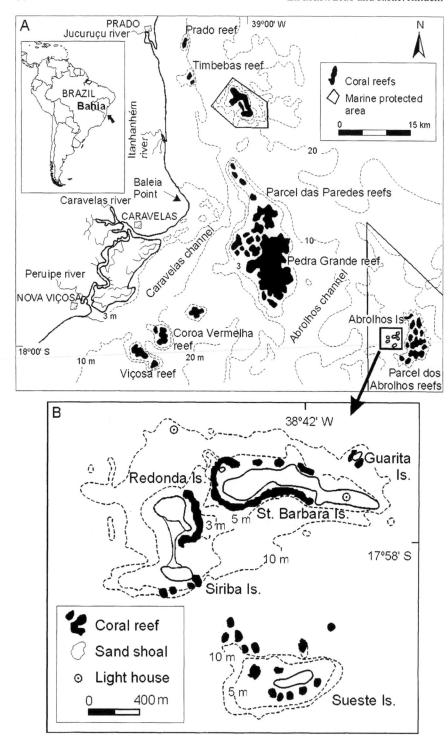

6.2 Environmental Setting

Abrolhos is located in the southern part of the trade wind area. The northward migration of the South Atlantic anticyclone cell in summer causes NE and E winds from October to March. The southward migration of the anticyclone cell in winter favors winds from the SE dominant between April and September and the northward advance of polar cold fronts enhances these winds and adds an SSE component (Nimer 1989). The climate of the Eastern Brazilian coastal region is tropical-humid with average air temperature between 24 °C in winter and 27 °C in summer. July is the coldest month and March the warmest. The mean annual rainfall in coastal regions near Abrolhos is 1,750 mm, with March, April and May concentrating about 35 % of the precipitation (Nimer 1989). The average monthly sea surface temperature ranges from 24.5 °C in August to 27.5 °C in March, though temperature anomalies of up to 1.5 °C may occur during strong El Niño events (Servain et al. 1987).

The waters of the Brazil Current flow over the Abrolhos Bank in a north to south direction and tides are semidiurnal with a range of 1.6–1.7 m. Owing to interactions between tides and current flow, which favor N-S circulation in the Caravelas Channel, the exchange of dissolved and suspended materials between the Coastal and the Outer Arc is more significant than that between coast and reefs (Meyerhöfer and Marone 1996). The principal wave trains coincide with the wind regime. Spring and summer are dominated by NE-E waves (significant height of 1 m and period of 5 s) which cause southward longshore sediment transport north of Baleia Point. Dominant waves from the SE-SSE (significant height of 1.5 m and period of 6.5 s) in fall and winter lead to northward longshore sediment transport south of Baleia Point (Fig. 6.1A). Despite the humid tropical climate, elevated runoff of fresh water and terrigenous sediments only occurs during winter storms. During these storms, large amounts of resuspended bottom sediments reach the reefs and cause turbid waters. Therefore, and contrary to most of the world's coral reefs, Abrolhos is an example of reefs surrounded by sediments with considerable fractions of siliceous sand and mud.

Fig. 6.1. Location of the Abrolhos Reefs (**A**) and Abrolhos Archipelago with fringing reefs and isolated *chapeirões* (**B**)

6.3 Reef Organisms

6.3.1 Corals

The Brazilian coral fauna displays low diversity, lacks branching acroporids common to most Atlantic reefs, and major reef builders are endemics and evolved from an isolated coral fauna of the Tertiary (Leão 1982). This situation may have been caused either by the west and northward flow of the Equatorial Current, which impedes the dispersion of the planula larval stage of many modern Caribbean coral species, or by the environmental characteristics of Brazilian reef areas. Despite optimal temperature, salinity and water depth for a diverse coral fauna in Abrolhos reefs, a stressful light regime, owing to high water turbidity in winter, and deposition of siliciclastic sediments may have deterred the establishment of most Caribbean coral species but not the growth of endemic reef-building species. While all ahermatypic Abrolhos reef corals are related to Caribbean species, most hermatypic frame-building corals tend to be endemic species (Verrill 1868; Laborel 1969a,b; Castro 1994). Of the 17 stony corals (scleractinians) in Abrolhos (18 Brazilian species), the common *Mussismilia hispida*, *Mussismilia braziliensis*, *Mussismilia harttii*, *Favia leptophylla*, *Favia gravida* and *Siderastrea stellata* are endemics. *Mussismilia braziliensis* is restricted to the coast of Bahia State and *Mussismilia hispida* occurs between 3°S and 30°S. Several species have large corallites, like *Mussismilia braziliensis*, *M. harttii*, *M. hispida*, *Favia leptophylla* and *Montastrea cavernosa*, the first two being important Abrolhos reef framebuilders. Owing to self-cleaning mechanisms, corals with large polyps generally tend to be more resistant to sedimentation and stressful light (Fishelson 1973; Dodge et al. 1974; Rogers 1990), and thus may favor development of hermatypic corals and the growth and calcification of the reef under these conditions. Cosmopolitan species (*Porites astreoides*, *Porites branneri*, *Agaricia agaricites*, *Agaricia fragilis*, *Montastrea cavernosa*, *Madracis decactis*) play a secondary role in the construction of Abrolhos reefs and several small and rare corals (*Scolymia welsii*, *Phyllangia americana*, *Astrangia braziliensis*, *Stephanocoenia michelini*) are insignificant reef builders. Most frame-building corals from Brazilian reefs have massive and encrusting forms. Branching forms, which are dominant on reef crests and fore reef slopes of most North Atlantic reefs, are absent from Abrolhos reefs, though *Mussismilia harttii* exhibits dichotomous groups of corallites. The shallow water forms of *Montastrea cavernosa* tend to be hemispheric, becoming fringed forms at depths greater than 5 m, and eventually flattened and encrusting in even deeper waters. However, morphological variability within this species cannot be entirely explained by environmental parameters (Amaral 1994).

Among hydrocorals, two of the three species of millepores (*Millepora braziliensis, M. nitida*) are considered endemics (Verrill 1868; Laborel 1969a; Amaral 1997) while *Millepora alcicornis* occurs along the entire tropical coast of Brazil and in the Caribbean. Millepores exhibit branching and encrusting growth forms. Encrusting forms are found on reef tops or in the axes of gorgonians. Delicate, finger-like branched forms (i.e. *Millepora nitida*) are characteristics of low energy environments and thick, massive branched forms (i.e. *Millepora braziliensis*) of high energy zones. A small hydrocoral, *Stylaster roseus*, forms colonies a few centimeters high with a thick base covered by small pointed branches among branches of *Millepora* (Castro 1994). A rare hydrozoan, *Solanderia gracilis*, thrives in the deeper and shaded zones of the Abrolhos reefs (Belém et al. 1982).

The 11 octocorals (*Phyllogorgia dilatata, Plexaurella grandiflora, Plauxaurella regia, Muriceopsis sulphurea, Muricea flamma, Neospongodes atlantica, Lophogorgia punicea, Carijoa risei, Heterogorgia uatumani, Ellisella barbadensis, Ellisella elongata*) are among the best-known class of organisms in the Abrolhos reefs (Castro 1989, 1990a,b). Several species are considered endemic to Brazil (*Phyllogorgia dilatata, Plexaurella grandiflora, Plauxaurella regia, Muricea flamma, Neospongodes atlantica, Lophogorgia punicea*), while *Carijoa risei, Heterogorgia uatumani, Ellisella barbadensis* and *Ellisella elongata* also occur on reefs in the North Atlantic. The abundant Brazilian blade sea fan (*Phyllogorgia dilatata*) and *Plexaurella grandiflora* are common in shallow reef areas, characteristically yellow colonies of *Muriceopsis sulphurea* are frequent among concentrations of soft algae, *Neospongodes atlantica* occurs on soft bottoms at the reef base, and *Lophogorgia punicea* grows on the walls of Abrolhos reefs. The four antipatharians (*Antipathes* three species, *Cirripathes* one species; Castro 1994) form flat, fan-shaped colonies, with branches arranged like brush bristles (*Antipathes*), or constitute branched colonies up to several meters long (the wire coral *Cirripathes*).

6.3.2 Algae

Unicellular symbiont zooxanthellae are found in shallow well-illuminated waters and an abundant algal flora covers reef bottoms and grows on different reef structures. The abundant calcareous green alga *Halimeda* is an important producer of inter-reef sediments, with up to 20% of the coarse sand fraction around coastal reefs and about 70% of the sediment around fringing reefs of the Abrolhos Archipelago (Leão 1982). *Udotea* and *Peni-*

cillus are also important constituents of the Abrolhos reefs flora and contribute to the production of fine (mud-size) inter-reef sediments. Abundant (32–79 %) calcareous red crustose algae (*Lithothaminion, Lithophyllum, Sporolithon, Porolithon*; Figueiredo 1997) are among the major reef-framework builders in Abrolhos and may represent 20 % of the internal structure of coastal reefs (Leão 1982). Foliose macroalgae, like *Sargassum* sp., followed by *Padina sanctae-crucis, Dictyota cervicornis, Lobophora variegata* and *Dictyopteris plagiograma*, cover more than 90 % of the surface of Coastal Arc reefs (Amado-Filho et al. 1997). Algae cover decreases on offshore reefs, possibly due to higher herbivore activity (Coutinho et al. 1993). Accelerated foliose algal growth may restrain coral development and put the reef structures at risk, especially if important algae consumers (herbivore fishes) are overfished and/or if nutrients increase due to organic sewage discharge. Filamentous turf algae (i.e. *Sphacelaria tribuloides* and *Ceramium* spp.; Figueiredo 1997) may overgrow coralline substrates and cover up to 80 % of the Archipelago fringing reef surface, indicating high herbivore pressure (Coutinho et al. 1993).

6.3.3 Other Biota

Two species of corallimorpharian sea anemones (*Discosoma carlgrenic, D. sanctithomae*) grow attached to reef structures and true sea anemones (*Condylactis gigantea, Bellactis ilkalyseae, Alicia mirabilis, Lebrunia danae, L. coraligens*) are abundant. The endemic soft-bodied zoanthid, *Palythoa caribbaeorum*, often covers large areas of reef substrate. Sponges, which bioerode coral skeletons, occur on shallow reefs as do bioeroding polychaete worms, both acting as sediment producers. Other polychaete worms, like the common *Spirobranchus*, construct calcareous tubes protruding from the surface of living corals or feed on coral polyp tissue (i.e., *Eurithae complanata*), thus leaving the coral skeleton exposed to new larval settlement. Also, bioeroding mollusks are reef sediment producers and the mollusk *Cyphoma macumba* inhabits colonies of the Brazilian blade sea fan *Phylogorgia dillatata* where it feeds on cnidarians (Castro 1994). A large number of crustaceans hide in reef crevices and lobster fishing is widespread on reefs outside the limits of the national park. Among the echinoderms, the abundant herbivorous sea star *Oreaster reticulatus* and sea urchins play an important role on the reefs because they open space for new coral growth.

The Abrolhos Bank is the southernmost area in the Atlantic Ocean inhabited by large permanent populations of coral reef fishes. Most of the 95 species in the coastal region and reefs are related to the Caribbean fish

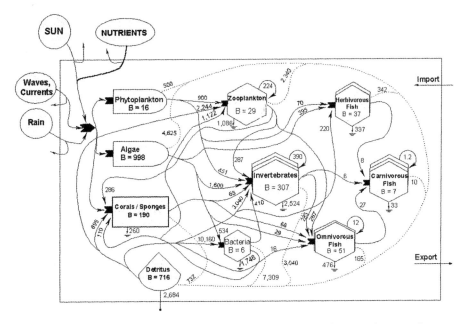

Fig. 6.2. Diagrammatic illustration of a simplified food chain of major elements from Abrolhos Islands fringing reefs (B biomass in g wet weight m^{-2}). *Arrows* indicate flow of matter (g m^{-2} year^{-1}), *dotted lines* represent detritus flow and *small vertical arrows* at the bottom of boxes indicate loss by respiration. Inflow and outflow do not match in some boxes because minor flows are not represented (modified after Telles 1998)

fauna and belong to two different communities, with only three species in common (Nunam 1979). About 39% of the species are herbivorous (Scaridae, Acanthuridae, Kyphosidae), 54% omnivorous (Haemulidae, Balistidae, Pomacanthidae, Lutjanidae, Pomacentridae), and 7% carnivorous (Serranidae, Carangidae, Sphyraenidae). Ecological modeling of fringing reefs around the Abrolhos Archipelago demonstrates that the fish to total biomass (Bf/Bt) ratios are high compared to reefs elsewhere, possibly owing to lack of fishing and that the principal energy transfer (74%) to highest trophic levels occurs via the detritivore path (Fig. 6.2; Telles 1998).

Marine turtles visit the Abrolhos reefs for feeding and reproduction. During summer, loggerhead (*Caretta caretta*) and green turtles (*Chelonia midas*) lay eggs on the sandy beaches of the Abrolhos islands and hawksbill turtles (*Eretmochelys imbricata*) feed on reef invertebrates. The humpback whale (*Megaptera novaeangliae*) migrates between June and November from subantartic waters to warm and shallow waters around the Abrolhos Archipelago, which represent the largest reproduction and breeding area for this species in the Southwest Atlantic. Seabirds, like the blue-faced booby (*Sula dactylatra*) and brown booby (*S. leucogaster*), the frigate bird (*Fregata magnificens*), the sooty tern (*Sterna fuscata*), the brown noddy (*Anous stolidus*), and the red-billed tropic bird (*Phaeton aethereus*) nest on the islands of Abrolhos and migrating birds are attracted to this area for resting and feeding.

6.4 Reef Types

Brazilian coral reefs display distinct forms of growth and morphology. Although vaguely comparable to algal cup reefs of Bermuda and algal ridges in the southwestern Caribbean, the mushroom-like growth forms of the Abrolhos coral reefs represent an unique ecological system. The Coastal Arc (about 10–20 km offshore) of Abrolhos reefs is formed by isolated coral pinnacles and different-sized bank reefs, while the Outer Arc (about 70 km offshore) to the east of the Abrolhos Islands is formed by isolated giant coral pinnacles (Figs. 6.1A, 6.3).

6.4.1 Coastal Arc

Abrolhos reefs are coral pinnacles or "chapeirões" (Hartt 1870) of irregular shape and laterally expanding tops. They are mushroom-shaped, roughly circular or, less often, elongate, and with overhangs often more pronounced at the windward side. They may vary between about 1–25 m in height and 1–50 m across the top. The mushroom-like growth results from the death of the upper part of parent corals, therefore directing new growth towards the edge of the colony. Lateral growth is enhanced when reef columns reach the sea level. Where closely spaced, pinnacles can fuse at the top, forming large bank reefs (Fig. 6.3A,B). For example, the 17-km-long Pedra Grande Reef of the Parcel das Paredes is a result of coalescence between numerous pinnacles and filling of open spaces at the upper reef with biogenic sediment, though narrow channels are still open at deeper parts. In consequence, coastal reefs of Abrolhos do not form classical barrier reefs, instead they are isolated shallow bank reefs of different shapes and sizes. Owing to patchiness, reduced dimensions, and position on the shelf, back reef zones and fore reef slopes are absent.

The reef top, located between about 10-m depths and low water levels, is the upper part of a pinnacle or a bank reef. Photophile forms of massive corals (*Mussismilia braziliensis* and *Favia leptophylla*), small colonies of *Porites*, hydrocorals (i.e. *Millepora*), and gorgonians (i.e. *Plexaurella regia*)

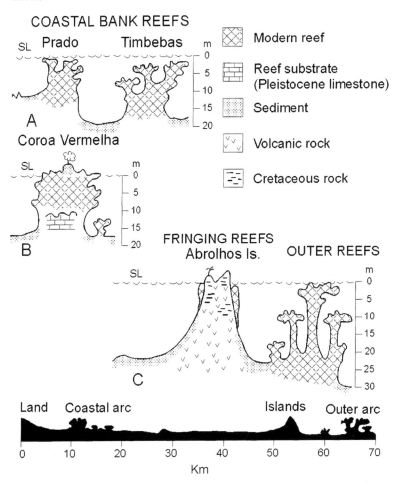

Fig. 6.3A–C. Idealized section across the Abrolhos area. A Northernmost bank reefs of the Coastal Arc; B the Coroa Vermelha Reef of the Coastal Arc; C Abrolhos Archipelago with fringing reefs and outer isolated *chapeirões* of the Parcel dos Abrolhos (*SL* sea level)

inhabit the expanded top. At low water, intertidal flats of bank reefs are completely exposed. The surface is irregular with numerous small, shallow and sandy or deep and rocky pools, probably representing areas between pinnacles that were not entirely filled. Carbonate sand of skeletal debris accumulates in mounds of all sizes. Zoanthids cover the margins of pools and bind coral rubble, small corals and hydrocoral colonies are present in deeper pools. *Siderastrea stellata* and *Favia gravida* are common species in shallow intertidal pools and resistant to variations in temperature, salinity and water turbidity. A carpet of calcareous, fleshy red and green algae and zoanthids covers most of the exposed reef surface together with patches of the prolific green alga *Penicillus* sp. and isolated laminated cyano-algal structures (soft oncolites).

The reef edge is a 2- to 3-m-thick irregular overhang which projects from the pinnacle. At the border of reef structures, small isolated pinnacles will eventually become part of the main reef body through lateral growth and trapping of sediment in open spaces. An irregular rim which rises about 30 cm above the flat top of bank reefs is constructed by encrusting organisms at the windward side where it tends to dissipate the energy of waves (Leão 1982). Coralline algae and encrusting mollusks (i.e. *Spiroglyphus* = *Dendropoma*) are major components of northern windward reef rims, while millepores (*Millepora alcicornis*) are also an important constituent of windward edges of southern reefs. *Millepora braziliensis* is located immediately below the zone of *Millepora alcicornis* (Laborel 1969b). Large colonies of *Millepora alcicornis* (<10 m depth) are sometimes followed by gorgonians (*Plexaurella grandiflora, Phyllogorgia dilatata, Muriceopsis sulphurea*). Several species of hard corals, mostly *Mussismilia braziliensis, M. hispida* and *Siderastrea stellata*, are also present (Castro 1994).

On upper reef walls *Mussismilia braziliensis, Millepora* sp. and gorgonians dominate, at about 2 m depth *Mussismilia harttii*, gorgonians and millepores are the major organisms, and at deeper reef walls millepores dominate. At greater depths, only small sciaphile forms (e.g. black corals and the octocoral *Carijoa riisei*) thrive in darker environments which are protected by the overhangs. On the leeward side of reefs, large colonies of *Montastrea cavernosa, Scolymia wellsi* and *Carijoa riisei* are common and large clumps of *Mussismilia harttii* occur close to the bottom sediments on the windward side (Castro 1994). On muddy bottoms (10–20 m depth) between pinnacles, prolific sea grass growth alternates with calcareous green algae (*Penicillus, Halimeda, Udotea*) and the free form of the coral *Meandrina braziliensis*.

6.4.2 Fringing Reefs of the Abrolhos Archipelago

Rather than a reef structure, fringing reefs of the five Abrolhos Islands are a veneer of reef organisms on stable volcanic or sedimentary rocks (Figs. 6.1B, 6.3C). They are formed by the growth of corals, cementation by coralline algae and other encrusting organisms, and the filling of cavities with cemented sediment. The reef flat extends about 50–60 m from the shore and does not exceed 5 m from the bottom to low water level (Fig. 6.3C). In protected areas (i.e., around Redonda Island), the shallow part of fringing reefs is composed of coral colonies (*Mussismilia braziliensis, M. hispida, Favia leptophylla*), bushes of *Mussismilia harttii*, small colonies of *Favia gravida* and *Siderastrea stellata*, and sporadic growth of *Agaricia agaricites*. *Montastrea cavernosa* occupies the deepest part of these reefs. *Millepora alcicornis* and *M. nitida* grow among the hard corals. Crustose coralline algae often cover the dead parts of reefs and branching types are common. An encrusting community that is comprised of reef-building organisms, however, covers most island slopes, though limestone formations do not exceed a few centimeters and a reef flat is absent.

6.4.3 Outer Arc

A reef area (Parcel dos Abrolhos, about 5 × 15 km), composed of abundant coral pinnacles (Figs. 6.1A, 6.3C), extends to the east of the Abrolhos Archipelago. Contrary to the Coastal Arc of reefs, the isolated pinnacles of the Outer Arc do not coalesce to form bank reefs and are not exposed during low tides. Depth differences between coastal reef arc (15 m) and outer reef arc areas (25 m), common reef growth initiation at around 7,000 years B.P., and similar growth rates may explain why reefs of the Coastal Arc reached the sea surface earlier than those of the Outer Arc. However, the morphology and the distribution of coral species on the reef walls follow a similar pattern in coastal and outer reefs. At the top of "chapeirões" large columns of *Mussismilia braziliensis* are common and some may reach heights of up to 2 m and may be topped by a living colony of approximately 1 m in diameter. Smaller colonies of *Favia leptophylla* and *Siderastrea stellata* occur between the columns. On the top of these reefs *Millepora alcicornis* surrounds central areas of corals. The vertical distribution of corals on pinnacle walls follows zonation from photophile forms in the upper portions to shade-loving species (sciaphile) which attain maximum development under overhangs in lower zones.

6.5 Environmental Impacts and Management

Natural disturbances of the Abrolhos reef environment are limited to millennial sea level oscillations, water temperature fluctuations, and episodic input of fine sediments. Holocene sea level oscillations had a profound effect on the evolution of Abrolhos coral reefs. Lower sea levels during regression periods exposed reef tops to marine erosion, dissolution, and to extensive bioerosion (Leão 1982), as well as approximated reefs to the coast, thus subjecting them to increased influence by siliciclastic sediments (Leão and Ginsburg 1997). Furthermore, anomalous high sea surface temperatures affected 51–88 % of *Mussismilia* colonies in the Abrolhos area during the summer of 1994. During the 1997 El Niño-Southern Oscillation (ENSO) event, maximum mid-March surface temperatures in 1998 were about 1 °C higher than normal (28.5 °C) in areas around the Abrolhos reefs (13–18° S) and up to 80 % of *Porites branneri, Mussismilia hispida, M. harttii, Porites asteroides* and *Agaricia agaricites* colonies were affected (bleached or partially bleached).

The rapid deforestation of the Atlantic coastal rain forest has significantly increased runoff, and untreated sewage discharge from expanding urban centers (i.e., Prado, Caravelas, Nova Viçosa) has led to abnormally high nutrients and algal growth at the expense of corals (Leão 1994, 1996). Accompanying the expanding marine tourism industry in Brazil, the number of visitors of Abrolhos National Marine Park has increased over 400 % between 1988 and 1992 (Leão et al. 1994). Since activities of tourists lack orientation and control, boat anchors, non-biodegradable garbage disposal, removal of reef organisms, diving and fishing may seriously damage the reef ecosystem. Reef walking on the exposed surface of fringing reefs in protected areas of the Abrolhos Archipelago has been considered a major cause of reef damage. The exploitation of coral by both artisanal and commercial fisheries as building material has existed since the 17th century and the extraction of *Millepora alcicornis* for aquarium decoration may be a major cause for the disappearance of this species (Pitombo et al. 1988). Industrial projects are a constant threat to Abrolhos coral reefs if industrial wastes are not properly disposed of and signals of heavy metal contamination in the southern part of Abrolhos might be due to chemical effluents of a paper plant (Amado-Filho et al. 1997). Finally, offshore drilling in southern Bahia State represents a threat to the reef ecosystem of Abrolhos National Marine Park, owing to boat collision with the reefs, higher water turbidity, or unpredictable oil spills.

Although our knowledge on the present conditions of Abrolhos reefs is still scarce and some areas are virtually unknown, only reefs in coastal areas with little urban development or relatively inaccessible offshore reefs can be considered well preserved or pristine. Environmental institutions concerned with the preservation of the coral reefs in Brazil have only recently been founded, the Abrolhos National Marine Park, comprising the Abrolhos Archipelago and the Outer Arc of *chapeirões* and the Timbebas reef of the Coastal Arc, being one of the oldest (Fig. 6.1A). Management plans and conservation programs for the Abrolhos National Marine Park have already been put into action (IBAMA/FUNATURA 1991). However, the park only represents one-fourth of the total Abrolhos reef area and no restriction for recreational or even commercial use of reefs exists for the remaining area, despite the declaration of the Bahia State Constitution (Article 215, Chap. VIII) that coral reefs are areas of permanent protection.

Acknowledgements. The authors thank Marcelo D. Telles and Jose E. Moutinho for their help with illustration.

References

Amado-Filho GM, Andrade LR, Reis RP, Bastos W, Pfeiffer WC (1997) Heavy metal concentrations in seaweed species from the Abrolhos reef region, Brazil. In: Lessios HA, Macintyre IG (eds) Proceedings of the 8th international coral reef symposium, Panamá, 2:1843–1846

Amaral FMD (1994) Morphological variation in the reef coral *Montastrea cavernosa* in Brazil. Coral Reefs 13:113–117

Amaral FMD (1997) Milleporidae (Cnidaria, Hydrozoa) do litoral brasileiro. PhD Thesis, Universidade de São Paulo, Brazil

Belém MJC, Castro CB, Rohlfs C (1982) Notas sobre *Solanderia gracilis* Duchassaing & Michelin, 1846 do Parcel de Abrolhos, Bahia. Primeira ocorrência de Solanderiidae (Cnidaria, Hidrozoa) no litoral brasileiro. An Acad Bras Ciênc 54(3): 585–588

Castro CB (1989) A new species of *Plexaurella* Valenciennes, 1855 (Coelenterata, Octocorallia), from the Abrolhos reefs, Bahia, Brazil. Rev Bras Biol, Rio de Janeiro 49:597–603

Castro CB (1990a) Revisão taxonômica dos Octocorallia (Cnidaria, Anthozoa) do litoral sul-americano: da foz do rio Amazonas à foz do rio da Prata. PhD Thesis, Universidade de São Paulo, Brazil

Castro CB (1990b) A new species of *Heterogorgia* Verrill, 1868 (Coelenterata, Octocorallia) from Brazil, with comments on the type species of the genus. Bull Mar Sci 4:411–420

Castro CB (1994) Corals of Southern Bahia. In: Hetzel B, Castro CB (eds) Corals of Southern Bahia. Editora Nova Fronteira, Rio de Janeiro, pp 161–176

Coutinho R, Villaça RC, Magalhães CA, Guimarães MA, Apolinario M, Muricy G (1993) Influência antrópica nos ecossistemas coralinos da região de Abrolhos, Bahia, Brasil. Acta Biol Leopold 15:133–144

Dodge RE, Aller RC, Thompson J (1974) Coral growth related to resuspension of bottom sediments. Nature 247:574–576

Figueiredo MO (1997) Colonization and growth of crustose coralline algae in Abrolhos, Brazil. In: Lessios HA, Macintyre IG (eds) Proceedings of the 8th international coral reef symposium 1:689–694

Fishelson L (1973) Ecological and biological phenomena influencing coral species composition on the reef tables at Eilat (Gulf of Aqaba, Red Sea). Mar Biol 19:83–196

Hartt CF (1870) Geology and physical geography of Brazil. Fields Osgood, Boston

IBAMA/FUNATURA (1991) Plano de Manejo do Parque Nacional Marinho dos Abrolhos. Instituto Brasileiro do Meio Ambiente e dos Recursos Naturais Renováveis/ Fundação Pró-Natureza Brasília. Aracruz Celulose SA

Laborel JL (1969a) Madreporaires et hydrocoralliaires recifaux des côtes brésiliennes. Systematique, ecologie, repartition verticale et geographie. Ann Inst Oceanogr Paris 47:171-229

Laborel JL (1969b) Les peuplements de madreporaires des côtes tropicales du Brésil. Ann Univ Abidjan Ser E II Fasc 3

Leão ZMAN (1982) Morphology, geology and developmental history of the southernmost coral reefs of Western Atlantic, Abrolhos Bank, Brazil. PhD Thesis, University of Miami

Leão ZMAN (1994) Threats to coral reef environments. In: Hetzel B, Castro CB (eds) Corals of Southern Bahia. Editora Nova Fronteira, Rio de Janeiro, pp 177-181

Leão ZMAN (1996) The coral reefs of Bahia: morphology, distribution and the major environmental impacts. An Acad Bras Ciênc 68:439-452

Leão ZMAN, Ginsburg RN (1997) Living reefs surrounded by siliciclastic sediments: the Abrolhos coastal reefs, Bahia, Brazil. In: Lessios HA, Macintyre IG (eds) Proceedings of the 8th international coral reef symposium 2:1767-1772

Leão ZMAN, Araujo TMF, Nolasco MC (1988) The coral reefs off the coast of eastern Brazil. Proceedings of the 6th international coral reef symposium Australia 3:339-347

Leão ZMAN, Telles MD, Sforza R, Bulhões HA, Kikuchi RKP (1994) Impact of tourism development on the coral reefs of the Abrolhos area, Brazil. In: Ginsburg RN (Compiler) Global aspects of coral reefs: health, hazards and history. RSMAS, University of Miami, Florida, pp 254-260

Meyerhöfer M, Marone E (1996) Transport mechanisms of biogenous material, heavy metals and organic pollutants in east Brazilian waters, small scale investigations. In: Ekau W, Knoppers B (Compilers) Sedimentation processes and productivity in the continental shelf waters off east and northeast Brazil. JOPS-II, Center for Tropical Marine Ecology, Bremen, pp 33-43

Nimer E (1989) Climatologia do Brasil. Instituto Brasileiro de Geografia e Estatística, Rio de Janeiro

Nunam GW (1979) The zoogeographic significance of the Abrolhos area as evidenced by fishes. MSc Thesis, University of Miami, USA

Pitombo FB, Ratto CC, Belém MJC (1988) Species diversity and zonation pattern of hermatypic corals at two fringing reefs of Abrolhos archipelago, Brazil. In: Choat JH, Barnes D, Borowitzka MA, Coll JC, Davies PJ, Flood P, Hatcher BG, Hopley D, Hutchings PA, Kinsey D, Orme GR, Pichon M, Sale PF, Sammarco P, Wallace CC, Wilkinson C, Wolanski E, Bellwood O (eds) Proceedings of the 6th international coral reef symposium Australia 2:817-820

Rogers CS (1990) Responses of coral reefs and reef organisms to sedimentation. Mar Ecol Prog Ser 62:185-202

Servain J, Seva M, Lukas S, Rougier G (1987) Climatic atlas of the tropical Atlantic wind stress and sea surface temperature: 1980-1984. Ocean-air interactions 1. Gordon and Breach Science, Philadelphia, pp 109-182

Telles MD (1998) Modelo trofodinâmico dos recifes em franja do Parque Nacional Marinho dos Abrolhos, Ba. MSc Thesis, Universidade do Rio Grande, Brazil

Verrill AE (1868) Notes of the radiate in the Museum of Yale College, with descriptions of new genera and species. 4. Notes of the corals and echinoderms collected by Prof. C.F. Hartt at the Abrolhos reefs, Province of Bahia, Brazil. Conn Acad Arts Sci Transact 1:351-371

Vicalvi MA, Costa MPA, Kowsmann RO (1978) Depressão de Abrolhos: uma paleolaguna Holocênica na plataforma continental leste brasileira. Bol Tec Petrobrás 21:279-286

7 The Cabo Frio Upwelling System, Brazil

J.L. VALENTIN

7.1 Introduction

Upwelling processes are most common and most intense on the eastern side of the Atlantic and Pacific oceans. The Cabo Frio upwelling system (23°S, 42°W) is an anomaly in that it is on the west side of the Atlantic Ocean on the coast of the state of Rio de Janeiro, Brazil (Fig. 7.1). The name "capo frigidio" (cold cape) first appeared on marine charts of Portuguese navigators at the beginning of the 15th century. A change in coastal direction from north-south to east-west at Cabo Frio and the proximity of the 100-m isobath lead to a topography which promotes upwelling of deep South Atlantic central water (SACW). However, the principal mechanism of upwelling is north-northeast wind. The process of SACW upflow follows two stages (Moreira da Silva 1973). The first stage begins in mid-August/September (late winter) and depends on the position of the main flow of the Brazil Current in relation to the continental slope and on variations in frequency and intensity of NE winds. During strong upwelling, SACW advances beyond the continental slope and invades the shelf to depths of about 50–80 m where is remains until April (mid-fall). The second stage of upwelling, mainly from September to March, is influenced by changes in intensity and direction of local winds, which cause alternating warming and cooling of near-surface waters. Two or 3 days of northeasterly winds (10 m s^{-1}) bring SACW to the surface and cold front passages (8- to 10-day cycles), which cause the wind to shift to the west, reverse the process (downwelling). Upwelled surface water forms a narrow belt (<1 km) around Cabo Frio with the plume flowing west-southwest (50 cm s^{-1}), owing to NE winds and the Coriolis effect. However, upwelling is not limited to the cold surface water zone, but south of Cabo São Tomé the penetration of the deep SACW over the shelf causes stratification, with thermocline differences of up to 10 °C within a few meters of depth during much of the year.

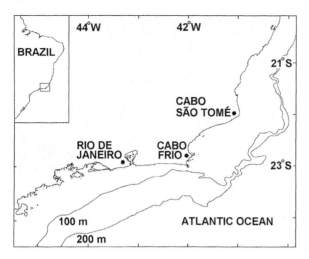

Fig. 7.1. The Cabo Frio upwelling region

7.2 Climate and Hydrology

The climate of the Brazilian coast in general, and the Cabo Frio region in particular, depends on the displacement of the tropical anticyclone in the South Atlantic and the polar anticyclone with a cold air mass originating in southern Argentina. The low-pressure zone between the two anticyclones corresponds to a cold front. The position and speed of displacement of these fronts determines the meteorology of the Cabo Frio region. During frontal passages, southwesterly winds and precipitation prevail while northeasterly winds (>6 m s^{-1}) and low precipitation (<800 mm year^{-1}) dominate in periods between fronts. Since the duration of inter-frontal phases varies from a few days in winter to several weeks in summer, they confer a seasonal and cold, dry microclimate with strong winds and frequent fogs to the Cabo Frio region.

Coastal water, tropical water of the Brazil Current, and upwelling SACW constitute the water masses of the Cabo Frio upwelling system, thus nutrient concentrations display large temporal and spatial variations. At the boundary between waters of the Brazil Current and SACW (300 m depth), represented by the 18 °C isotherm and the 36 isohaline, nitrate and orthophosphate concentrations are about 4 and 0.4 µM, respectively (Rodrigues 1977). Nutrients increase in deeper waters (about 400 m, 12.0 °C, salinity 35.3) to about 18 µM NO$_3$-N and 1.3 µM PO$_4$-P (Gonzalez-Rodriguez et al. 1992). Nitrite and ammonia are virtually absent but increase slightly at the base of the thermocline where accumulation of organic matter favors remineralization. Silicate tends to increase with depth (>15 µM SiO_4-Si), though surface values (5–10 µM) also indicate occasional addition through runoff from the Paraiba River north of Cabo Frio, which transports large quantities of silicate during the rainy season (>140 µM; Valentin et al. 1978). The dissolved oxygen levels (annual mean 4-5 ml l^{-1}) are stable and range from a minimum of 3 ml l^{-1} at the bottom (50 m depth) to a maximum of 8 ml l^{-1} during algal blooms.

7.3 Biological Community

7.3.1 Plankton

The upwelling phenomenon has a direct impact on the composition of species and trophic structure (Fig. 7.2). The vertical distribution and intensity of phytoplankton blooms is controlled by the depth and intensity of the thermocline in the euphotic layer proximate to the coast controls. Despite elevated nitrate levels and rarely limiting irradiance (1,500 µE m^{-2} s^{-1}), chlorophyll (0.5–6.0 µg l^{-1}) and primary production (2–14 mgC m^{-3} h^{-1}) tend to be lower than in upwelling systems elsewhere, probably owing to low initial biomass and reduced surface heating of upwelled water. Furthermore, biomass tends to be effectively dispersed by advection currents and strong winds and reduced by high zooplankton grazing pressure (50%; Druehl and Yoneshigue-Braga 1976). Although biomass and primary production levels are similar during post-upwelling phases in summer and non-upwelling in winter, they are still high when compared to oligotrophic tropical water of the Brazil Current and the abundant, probably resuspended organic detrital matter is an important energy source for higher trophic levels, especially zooplankton.

The presence of different water masses induces spatial and temporal variation in the composition and abundance of zooplankton, and community changes correlate with the upwelling cycles (Valentin and Moreira 1978; Valentin et al. 1987). The copepod species *Calanoides carinatus*, *Rhincalanus cornutus*, *Aetideus giesbrechti*, and *Heterorhabdus papilliger* are typical of deep SACW (<18 °C, salinity <36). Warm tropical water (>24 °C, salinity >36) of the Brazil Current with high diversity is dominated by copepods like *Clausocalanus furcatus*, *Mecynocera clausi*, *Corycella gracilis*, *Oithona setigera*, *Corycaeus typicus*, *Oncaea conifera*, *Undinula vulgaris*, *Calocalanus pavo*, and *Lucicutia flavicornis*, and the cladoceran *Evadne spinifera*. Zooplankton densities tend to be >10 org. l^{-1} (annual mean 66 mg ash-free dry weight m^{-3}) and may reach 100 org. l^{-1} during

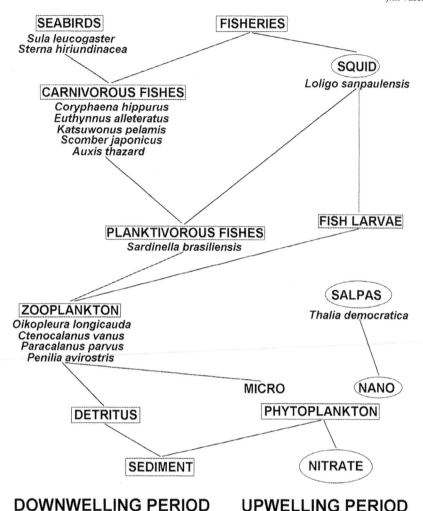

Fig. 7.2. Simplified pelagic food web of the Cabo Frio region. *Encircled variables* prevail during upwelling periods

post-upwelling at the end of summer (220 mg dry weight m^{-3}). Dense populations of filter-feeding salps (*Thalia democratica*) accompany high primary production in the narrow near-surface mixing zone. Salps contribute to a rapid decrease of primary biomass and energy loss for other trophic levels (Rocha 1998). They efficiently compete with zooplankton herbivores, which leads to reduced zooplankton biomass in detriment of zooplankton-feeding populations of larval sardines (Katsuragawa et al. 1993).

Simultaneous peaks of phyto- and zooplankton abundance in the Cabo Frio upwelling system indicate equilibrium between these trophic levels and an advanced state of system maturity, generally attributed to oligotrophic waters. The process, which synchronizes phyto- and zooplankton variations, depends on phytoplankton bloom-induced spawning of herbivores during upwelling. Drift currents carry the larvae to the open sea where they undergo ontogenic vertical migration and return as adults to coastal surface waters if the interval between consecutive upwellings approximately corresponds to the duration of their life cycle. The synchronization increases the rate of grazing and reduces peaks of primary biomass. The quali-quantitative variations in phytoplankton translate into changes of zooplankton species composition as a result of different feeding regimes. A model of zooplankton succession in Cabo Frio waters (Valentin 1988) suggests that during periods of phytoplankton maxima, opportunistic herbivores (*Paracalanus parvus, Oikopleura longicauda, Thalia democratica, Calanoides carinatus*) and opportunistic carnivores (*Sagitta enflata*) dominate. During post-maxima periods, facultative herbivores, predators, and detritivores are *Ctenocalanus vanus, Creseis acicula, Penilia avirostris, Doliolum* spp., and the genera *Eucalanus* and *Clausocalanus*. Detritivores and omnivores are represented by *Temora stylifera, Centropages furcatus, Euterpina acutifrons, Nannocalanus minor, Oithona plumifera*, and the genus *Conchaecia* while carnivores are *Sagitta enflata*, siphonophores, and copepods of the families *Euchaetidae, Candaciidae*, and *Pontellidae*.

7.3.2 Benthos

Large temperature variations (10 °C) within hours, high nutrient input, and phytoplankton bloom impact the structure of the benthic community. Owing to a wide variety of physical and biological conditions (Boudouresque and Yoneshigue 1983), the macroflora of rocky shore communities of the Cabo Frio region displays high diversity with a large number of Rhodophyta (64.5 %), Phaeophyta (18.6 %), and Chlorophyta (16.8 %) (Yoneshigue 1985). The species richness among Phaeophyta at the tropical latitude of Cabo Frio is unusual and characterizes the local flora as intermediate with warm-temperate affinities, confirming the biogeographic importance of the Cabo Frio barrier (de Oliveira Filho 1977). Sites with temperate affinities and a *Porphyra* assemblage contrast with tropical sites and *Sargassum/Zonaria* and *Caulerpa* assemblages, while other sites have mixed communities. *Pterocladiella capillacea* develops high biomass at exposed sites as a result of the combined effect of cold, nutrient-rich upwelled water and wave action. Strong wave action of exposed rocky shores and high phytoplankton biomass also favor the development of banks of the mussel *Perna perna*, associated with rich algal communities and dense populations of cirripeds. The distribution of benthic organisms is influenced by local hydrodynamics which directly affect recruitment of the propagules and larvae, post-settling mortality, and grazing by the gastropod *Littorina ziczac* on cyanophyceans and the crab *Pachygrapsus transversus* on Ulvales and Ectocarpales (Coutinho 1995).

The impact of SACW on the structure of benthic communities of the continental shelf varies with the nature of the substrate. North of the cape (23°S), conglomerates of lithothamnioid calcareous algae and almost permanently cold waters favor recruitment and establishment of extensive *Laminaria abyssalis* banks (50–110 m depth). The conditions for the completion of the *Laminaria* life cycle and maximum growth of sporophytes (18 °C, 20 µM NO_3^-, and 35 µmol photons m^{-2} s^{-1} of blue light) are provided by SACW at depths below 50 m (Yoneshigue-Valentin 1990). *Laminaria* beds are habitat for a rich demersal fauna.

South of 23°S, unconsolidated substrates are composed of medium sand (30–45 m) and fine sand and mud (below 60 m). The action of currents, together with strong thermal gradients during frequent upwelling in spring and summer, determine the structure of benthic and demersal communities. Sea stars are an ecologically important and numerically dominant group of the soft-bottom community. The distribution of the common *Astropecten cingulatus*, followed by *Astropecten brasiliensis*, *Luidia ludwigi scotti*, *Luidia alternata*, and *Tethyaster vestitus*, seems to follow depth-related sediment characteristics. The 45-m isobath appears to represent a transitional depth for echinoderms and pelecypods, which are the principal prey of sea stars (Gomes 1989; Ventura and Fernandes 1995). The reproductive peak of *Astropecten brasiliensis* coincides with upwelling periods. The larvae benefit from abundant phytoplankton, and large quantities of prey (bivalves and crustaceans, mainly cumaceans) are ingested during the major feeding period in summer (Ventura et al. 1997). Similarly, depth-related sediment particle size seems to be the main factor for the distribution of bivalves. The common *Mactra petiti*, *Tellina petitiana*, *Tellina gibber*, and *Crassinella lunulata* are abundant in neashore areas down to 30 m depth where high turbulence and resuspension of sediment favors the dominance of sediment-feeders. The organic enrichment of inner shelf sediments by upwelling promotes the presence of detritivorous bivalves like *Nucula puelcha* and *Corbula patagonica* (below 30 m depth) and *Malletia cumingii* (below 60 m; Gomes 1989). Another important element of the inner shelf is crustaceans, like Brachyura (*Portunus spinicarpus*; Brisson 1992) and Anomoura (*Dardanus arrosor insignis*; da Gama and Fernandes 1994). Their highest abundance below 18 °C suggests a close relationship with upwelling events. Benthic octopi (*Semirossia tenera*, *Eledone massyae*), which occur below 45 m depth, migrate towards the shore during upwellings to feed and mate. They leave the shelf in late summer to spawn in the deeper waters of the shelf break (Costa and Fernandes 1993a).

7.3.3 Nekton

High abundance of neritic squids, like *Loligo sanpaulensis* (74% of the catch) and *Loligo plei* (13%), is related to recruitment and to high productivity of feeding grounds. The arrival of *Loligo sanpaulensis* recruits coincides with coastal upwelling when cold nutrient-rich waters provide ample food for juveniles (Costa and Fernandes 1993b). *Loligo sanpaulensis* is one of the upwelling's most important fisheries resources and highest catches are associated with spring-summer upwelling at depths between 45 and 60 m (Haimovici and Perez 1991).

The flow of SACW over the inner continental shelf influences the composition and abundance of benthic and demersal fishes (79 species; Netto and Gaelzer 1991). Permanent components of the demersal community are the flounder *Etropus longimanus*, the ray *Zapterix brevirostris*, the sea robin *Prionotus nudigula*, the rough-scad *Trachurus lathami*, and the flathead *Percophis brasiliensis*. Together with SACW, species typical of deep habitats (i.e. the red porgy *Pagrus pagrus*, the cusk-eel *Genypterus brasiliensis*, the silver hake *Merluccius hubbsi*, the anglerfish *Lophius gastrophysus*) reach the coastal zone (30–60 m depth). Upwelling provides favorable conditions for pelagic planktivorous fishes, such as the Brazilian sardine (*Sardinella brasiliensis*) which supports one of the most important fisheries in the region. The survival of larvae depends on the intensity of SACW intrusion during late spring and summer, thus a decline in catch from 200,000 to 50,000 tons (1973–1993) may have been caused by recruitment failure of some year-classes, owing to low frequency of upwelling events (Matsuura 1996). Other components of the pelagic fish fauna are the skipjack (*Katsuwonus pelamis*), the frigate tuna (*Auxis thazard*), the Atlantic little tuna (Euthynnus alletteratus), the chub mackerel (*Scomber japonicus*), the Atlantic moonfish (*Sele setapinnis*), the blue runner (*Caranx cryusos*), the bluefish (*Pomatomus saltator*), and the dolphin (*Coryphaena hippurus*). The abundance of nekton provides an important food source for seabirds over the inner shelf. The brown booby *Sula leucogaster* and the terns *Sterna hiriundinacea* and *Sterna eurygnatha* are top carnivores of pelagic resources and Cabo Frio Island offers habitat for reproductive colonies of numerous species of birds.

Acknowledgements. The Brazilian Council for Technological and Scientific Research (CNPq) provided financial support. I wish to thank Paulo A. Costa of Rio de Janeiro University (UNIRIO) and Eliane G. Rodriguez of the Institute of Marine Studies Admiral Paulo Moreira (IEAPM) for providing literature and information on fishery resources.

References

Boudouresque CF, Yoneshigue Y (1983) Données préliminaires sur le régime alimentaire de quelques Echinides réguliers de la région de Cabo Frio (RJ - Brésil). Symbioses (Rev Biol Hum Anim) 15:224-226

Brisson S (1992) Estudos sobre a população de *Portunus spinicarpus* (Stimpson 1871) (Crustacea Decapoda Brachyura) da região de Cabo Frio, Brasil. MSc Thesis, Universidade Federal do Rio de Janeiro, Brazil

Costa PAS, Fernandes F da C (1993a) Seasonal and spatial changes of cephalopods caught in the Cabo Frio (Brazil) upwelling system. Bull Mar Sci 52:751-759

Costa PAS, Fernandes F da C (1993b) Reproductive cycle of *Loligo sanpaulensis* (Cephalopoda: Loliginidae) in the Cabo Frio region, Brazil. Mar Ecol Prog Ser 101:91-97

Coutinho R (1995) Avaliação crítica das causas da zonação dos organismos bentônicos em costões rochosos. In: Esteves FA (ed) Estrutura, funcionamento e manejo de ecossistemas Brasileiros. Oecol Brasil 1:259-271

da Gama BAP, Fernandes F da C (1994) Distribuição de crustáceos anomuros na plataforma continental de Cabo Frio (Rio de Janeiro, Brasil). Nerítica 8:87-98

de Oliveira Filho EC (1977) Algas marinhas bentônicas do Brasil. Thesis, Universidade de São Paulo, Brazil

Druehl LD, Yoneshigue-Braga Y (1976) Growth and succession of tropical phytoplankton cultured in deep water. Publ Inst Pesqu Marinha, Rio de Janeiro 93:1-13

Gomes AS (1989) Distribuição espacial dos moluscos bivalves na região da plataforma continental do Cabo Frio, Rio de Janeiro. MSc Thesis, Universidade Federal Rio de Janeiro, Brazil

Gonzalez-Rodriguez E, Valentin JL, André DL, Jacob SA (1992) Upwelling and downwelling at Cabo Frio (Brazil): comparison of biomass and primary production responses. J Plank Res 14:289-306

Haimovici M, Perez AA (1991) Coastal cephalopod fauna of southern Brazil. Bull Mar Sci 49:221-230

Katsuragawa M, Matsuura Y, Susuki K, Dias JF, Spach HL (1993) O ictioplâncton ao largo de Ubatuba, SP: composição, distribuição e ocorrência sazonal (1985-1988). Publ Inst Oceanogr São Paulo 10:85-121

Matsuura Y (1996) A probable cause of recruitment failure of the brazilian sardine *Sardinella aurita* population during the 1974/75 spawning season. S Afr J Mar Sci 17:29-35

Moreira da Silva P de C (1973) A ressurgência em Cabo Frio. Publ Inst Pesqu Marinha, Rio de Janeiro 78:1-56

Netto EBF, Gaelzer LR (1991) Associações de peixes bentônicos e demersais na região de Cabo Frio, RJ, Brasil. Nerítica 6:139-156

Rocha GRA (1998) Modelo quantitativo das interações tróficas da plataforma continental de Ubatuba (SP), Brasil. PhD Thesis, Universidade de São Paulo, Brazil

Rodrigues RF (1977) Evolução da massa d' água durante a ressurgência em Cabo Frio. Publ Inst Pesqu Marinha, Rio de Janeiro 115:1-31

Valentin JL (1988) A dinâmica do plancton na ressurgência de Cabo Frio - RJ. In: Brandini F (ed) Mem III Enc Brasil Plancton. Univ Fed Paraná, pp 26-35

Valentin JL, Moreira AP (1978) A matéria orgânica de origem zooplanctônica nas águas de ressurgência de Cabo Frio (Brasil). An Acad Bras Ciên, Rio de Janeiro 50:103-112

Valentin JL, André DL, Monteiro-Ribas WM, Tenenbaum DR (1978) Hidrologia e plâncton da região costeira entre Cabo Frio e o estuário do Rio Paraíba (Brasil). Publ Inst Pesqu Marinha, Rio de Janeiro 127:1-14

Valentin JL, Monteiro-Ribas WM, Mureb MA, Pessotti E (1987) Sur quelques zooplanctontes abondants dans l' upwelling de Cabo Frio (Brésil). J Plank Res 9:1195-1216

Ventura CRR, Fernandes F da C (1995) Bathymetric distribution and population size structure of Paxillosid seastars (echinodermata) in the Cabo Frio upwelling ecosystem of Brazil. Bull Mar Sci 56:268-282

Ventura CRR, Falcão APC, Santos JS, Fiori CS (1997) Reproductive cycle and feeding periodicity in the starfish *Astropecten brasiliensis* in the Cabo Frio upwelling ecosystem (Brazil). Invertebrate Reproduction and Development 31:135-141

Yoneshigue Y (1985) Taxonomie et écologie des algues marines dans la région de Cabo Frio (Rio de Janeiro, Brésil). Thesis, Universite Aix-Marseille II, France

Yoneshigue-Valentin Y (1990) The life-cycle of *Laminaria abyssalis* Joly et Oliveira Filho (Laminariaceae, Laminariales, Phaeophyta) in culture. Hydrobiologia 204/205:461-466

8 Baía de Guanabara, Rio De Janeiro, Brazil

B. Kjerfve, L.D. de Lacerda, and G.T.M. Dias

8.1 Introduction

Baía de Guanabara (approx. 23°45'S 44°45'W) is a eutrophic and polluted coastal bay in Brazil, impacted by the discharge from the Rio de Janeiro metropolitan area. The shallow bay measures 28 km from west to east and 30 km from south to north, has a 131-km perimeter, a mean water volume of $1.87 \times 10^9 \text{m}^3$, and 384km^2 surface area. The entrance to Baía de Guanabara is only 1.6 km wide (Fig. 8.1). The bay is structurally controlled and has a central channel with a depth of 30 m and a sandy bottom near the entrance, reflecting wave and tidal forcing. Bottom sediments are mostly muds as a result of the Holocene transgression and rapid fluvial sedimentation, accelerated by channelization of rivers and deforestation in the drainage basin. An extensive sandbank is located seaward of the bay entrance and a flood-oriented sand wave system indicates sand transport into the bay. The mean freshwater runoff measures $125 \text{m}^3 \text{s}^{-1}$ and peaks in January. Tides are mixed and mainly semidiurnal with a range of 0.7 m, and peak spring tidal currents reach 0.5m s^{-1} inside the bay and 1.6m s^{-1} in the bay entrance. The passage of northward propagating polar fronts occasionally results in strong winds from the southwest and high-energy, long-period waves from the south. The salinity of the bay varies from 21.0 to 34.5 with a weak vertical salinity stratification. An estuarine circulation is characterized by a $900 \text{m}^3 \text{s}^{-1}$ net surface outflow. Fifty percent of the water volume in Baía de Guanabara flushes every 11.4 days. Largely untreated sewage enters the bay from the west, resulting locally in very poor water quality. The near-bottom mean dissolved oxygen concentration at the inner-most station measures 3.1mg l^{-1} and results in anoxic bottom muds. The worst water quality is indicated by average fecal coliform counts higher than $1,000 \text{ml}^{-1}$. The average chlorophyll concentration in the inner bay, in response to high nutrient loading, exceeds $130 \mu\text{g l}^{-1}$.

Fig. 8.1. Map of Baía de Guanabara, Rio de Janeiro, Brazil

8.2 Geological Setting

The Serra do Mar mountains dominate the southeast coast of Brazil. This coast-parallel mountain range was uplifted during the Tertiary period at the time of the formation of the Santos basin along the margin of the adjacent continental shelf. It consists of semi-grabens which define blocks of successive escarpments. These are oriented southwest to northeast and are aligned with the remaining, underlying Precambrian rocks.

The Baía de Guanabara basin is located in one such 30-km-wide Tertiary depression, which is referred to as Baixada Fluminense and also the Guanabara rift (Ruellan 1944; Asmus and Ferrari 1978; Ferrari 1990). Many outcrops of Precambrian origin abound throughout the basin, as in the case of the 400-m-high Sugar Loaf (Pão de Açucar) at the southwestern extreme of the present bay. Baía de Guanabara, as it currently exists, is oriented south to north. This changed physiography evolved during the Quaternary, as the drainage patterns were adjusted during one or more lower stands of sea level. The bay was further modified by the marine transgression during the Holocene. Present-day sedimentary remnants of the Tertiary are the Macacu and the pre-Macacu formations northeast of the bay (Meis and Amador 1972, 1977). Coring in the bay only revealed late Pleistocene deposits and braided fluvial sands related to the Caceribu formation (Amador 1980b, 1993). During the Holocene marine transgression 5,000 years BP, relative sea level extended 4 m above the present sea level and the bay was then more than twice the size (800 km^2) of the present bay. Elevated paleo-beaches and marine terraces (+4 m) commonly occur around the bay (Amador 1974; Amador and Ponzi 1974).

Flooding of the drainage basin during the Holocene transgression was responsible for extensive mud deposition on top of fluvial Pleistocene sands. As evidenced by high-resolution seismic profiling, mud deposits attain a thickness of more than 10 m near the Rio-Niterói bridge. Along the interior margin of the bay, the distribution of mangrove wetlands conforms with the southwest-northeast orientation of the basin and the orientation of the dominant Rio Guapimirim and Rio Macacu catchment basins.

During the last half-century, Rio Macacu and other meandering rivers have been channelized to reclaim alluvial lands for urban growth. The lengths of the rivers have in several instances been reduced to 30% of their previous lengths. This caused higher flow velocities, local scouring, and increased sediment load. River channelization combined with catchment deforestation increased sedimentation to Baía de Guanabara. The mean sedimentation rate used to be 0.24 m century^{-1} during the period 1849–1922, but increased to 0.81 m century^{-1} from 1938 to 1962 (Amador 1980a). Recent sedimentation rates vary spatially from 0.57 to 4.50 m century^{-1} within the bay (Amador 1992).

8.3 Bathymetry and Bottom Sediment

Baía de Guanabara has a complex bathymetry with a relatively flat central channel. The channel is 400 m wide, stretches from the mouth more than 5 km into the bay, and is defined by the 30-m isobath. The deepest point of the bay measures 58 m and is located within this channel. Numerous outcrops of the crystalline basement rock exist within the bay, both above and below the water level, and pose a risk to navigation. The central channel widens to 900 m inland of the main choking point, and the depth decreases to 20 m. North of the Rio de Janeiro-Niterói bridge, the channel loses its characteristics as the bay rapidly becomes more shallow. Mud covers the interior parts of the bay as a result of the high sediment load during the past century, accelerated by anthropogenic activities in the catchment. Because of the rapid mud sedimentation, the area-weighted bay depth measures only 5.7 m. Although muddy sand, sandy mud, and mud cover the bottom of most of the bay (Amador 1992), the central channel from the continental shelf to the Rio-Niterói bridge is filled with sand (Kjerfve et al. 1997). Isolated areas of relict sands are also found northeast and southwest of Ilha de Governador (Amador 1992), indicating strong tidal scouring.

A submerged sandbank, consisting of medium quartzose sands, exists 12 m below the surface seaward of the bay mouth. This asymmetrical bank appears to be entering the bay but is kept from extending further inland by tidal exchanges and the flux of freshwater. The sand shoal is similar to the adjacent sub-aerial sand barrier formations, which stretch for more than 200 km along the coast of the state of Rio de Janeiro from the entrance to Baía de Sepetiba to Arraial do Cabo. Were it not for the seaward pressure gradient due to the freshwater runoff from the bay, it is likely that the sandbank at the entrance to Baía de Guanabara would also have been transformed into an emergent barrier system.

The benthic sediment distribution reflects the hydrodynamic forcing. For example, sand waves are common along the eastern margin of the central channel at depths of 10–26 m. These sand waves have heights of 0.5–2.5 m, lengths of 18–98 m, and decrease in both height and wavelength from the ocean into the bay in response to decreasing tidal energy. The sand waves have steeper slopes facing the bay, indicating wave progression and bottom sand transport into Baía de Guanabara. The sand waves and their characteristics result from energetic ocean swells associated with meteorological frontal passages and the tidal flood-dominance of bottom currents.

8.4 Climate and Weather

The local marine climate is tropical-humid wet with wet, warm summers and dry, cool winters. The mean air temperature is 23.7 °C, and the mean relative humidity is 78%. The mean 30-year rainfall varies across the drainage basin from a maximum 291 mm month^{-1} in the mountains to a minimum 30 mm month^{-1} in the winter in low-lying areas. The mean rainfall for Baía de Guanabara is 1,173 mm annually, and the mean evaporation is 1,198 mm annually. There is a well-defined wet (December–April) and dry

(June-August) season. During the austral summer, heavy rains can produce intense runoff and flooding of low-lying fringe areas (Alcântara and Washington 1989).

During the austral winter, polar fronts propagate northward at 500 km d^{-1} from the South Atlantic and produce very intense, short-duration southerly winds. A satellite study indicated that an average of 13 cold fronts of polar origin, one per week, impacted the Baía de Guanabara coast during the austral winter (Stech and Lorenzzetti 1992). The frontal system normally passes the bay in less than 24 h, and winds from the south and southwest may occasionally exceed 25 m s^{-1} along the adjacent ocean beaches, at the same time that the air temperature may drop by 5–10 °C. Frontal passages are however not limited to the winter, but occur on a year-around basis with an average of 46 fronts annually. Ocean swells, 2–4 m high with periods of 8–12 s, at times precede the arrival of fronts. Breaking waves on the adjacent ocean beaches are translated into occasional, damaging low-frequency oscillations in the lower bay. The median significant wave height in deep water varies between 1.3 and 1.8 m and has a period of 7 s during normal periods of anticyclonic dominance (Souza 1988). The most frequent winds blow across Baía de Guanabara from the southeast (21 % of the time), from the south (17 %), and from the north (14 %). Twenty-four percent of the time winds are less than 1 m s^{-1}. The mean wind speed measures is 3 m s^{-1}, and the maximum wind in 1990 was recorded in April, blowing from the south at a speed of 14.4 m s^{-1}.

8.5 Runoff

The Baía de Guanabara drainage basin measures only 4,080 km². It is drained by 45 small rivers, although six of these are responsible for 85 % of the total runoff. The monthly freshwater discharge rate was modeled by Kjerfve et al. (1997) based on monthly measurements of temperature and rainfall form multiple stations in the Baía de Guanabara catchment. The calculations yielded varying runoff ratios, from a low 0.01 at sea level in the austral winter to 0.57 at high elevations in the austral summer. The area-weighted mean runoff ratio measured 0.20, which corresponds to a mean fresh water discharge into Baía de Guanabara of 100±59 m³ s^{-1}, with a range from 33 m³ s^{-1} in July to 186 m³ s^{-1} in January, based on the 30-year normal temperature and rainfall data. To provide sufficient quantities of potable water for metropolitan Rio de Janeiro, Companhia Estadual de Água e Esgoto (CEDAE) pumps 40 m³ s^{-1} from Rio Paraíba do Sul, the adjacent river basin to the north. An estimated 25 m³ s⁻¹ of this water becomes runoff into Baía de Guanabara. Thus, the total mean runoff of freshwater into Baía de Guanabara measures 125 m³ s⁻¹.

8.6 Tidal Variability

Baía de Guanabara experiences mixed mainly semidiurnal tides with a tidal form number of 0.33. Kjerfve et al. (1997) summarized harmonic tidal amplitude and phase data for 15 stations within Baía de Guanabara, based on water level measurements by the Departamento de Hidrografia e Navegação (DHN). The mean tidal range in Baía de Guanabara measures 0.7 m and does not vary significantly spatially. The fortnightly spring tidal range measures 1.1 m and the neap tidal range is 0.3 m. The semidiurnal M_2 and S_2 tidal amplitudes measure 0.32 and 0.18 m, respectively, and are dominant. However, the semidiurnal tidal range is strongly modulated by the lunar declinational fortnightly cycle with diurnal tides exhibiting a 0.4-m tropic tidal range and a 0.1-m equatorial tidal range. The tidal prism of Baía de Guanabara measures 2.3×10^8 m³. Tidal currents in the entrance section are flood-dominant and reach 1.6 m s⁻¹, whereas peak ebb-currents measure 1.0 m s⁻¹ in the same cross section. Further into the bay, peak tidal currents are on the order of 0.5 m s⁻¹.

8.7 Salinity, Circulation and Flushing

The water circulation in Baía de Guanabara is composed of both gravitational circulation and residual tidal circulation. Based on surface and bottom salinity measurements by FEEMA at 13 stations from 1980 to 1993, the mean salinity measures 29.5±4.8, and the mean water temperature 24.2±2.6 °C. The vertical salinity stratification is moderate to weak with a vertical net salinity difference seldom exceeding 4. Based on these net surface and bottom salinity measurements and simulated freshwater discharge, Kjerfve et al. (1997) calculated a net surface outflow of 900 m³ s⁻¹ and a net bottom inflow of 775 m³ s⁻¹. Thus, the water volume, which participates in the two-layered gravitational circulation, is on average eight times greater than the freshwater input. Baía de Guanabara, similar to many coastal plain estuaries, also has a large residual tidal circulation. Approximately half of the salt transported into the bay is due to the gravitational circulation and the remainder due to tidal dispersion. One measure of flushing in a coastal bay is the time it takes to renew 50 % of the water volume. Using a mean bay water volume of $1.87 \times 10^9 \text{m}^3$, Kjerfve et al. (1997) calculated 11.4 days for the renewal of 50 % of the bay water volume. This relatively rapid water renewal explains why Baía de Guanabara has reasonable water quality in the lower parts of the bay, in spite of input of huge quantities of untreated sewage.

8.8 Coastal Ecosystems

Besides urban and industrial areas, the drainage basin consists of agricultural fields and cattle lands in addition to high relief mountains, the Serra do Mar, with a maximum elevation of 2,263 m (Pedra do Sino) within the watershed. The catchment used to be covered by Mata Atlântica, the Atlantic rain forest, but nowadays the basin is only partially vegetated. The Mata Atlântica still extends to the shoreline in a few places to the southwest of the Baía de Guanabara. Much of the indigenous vegetation has been cut, but appears as small areas of intact rain forest on mountain slopes and as preserved mangrove wetlands along the fringes of the bay.

In spite of being encroached on by urban development, the interior margins of Baía de Guanabara are still bordered by 90 km² of fringing mangroves (Pires 1992), of which 43 km² is the Guapimirim conservation area. Other mangroves suffer from the impact of oil spills from ships and petrochemical facilities along the inner bay shores. These mangrove communities are composed of opportunistic species of the *Aizoacea* and *Amaranthacea*, which have replaced mangrove trees in many areas. However, biodiversity is decreased and it is difficult for mangroves to recolonize after a spill (Lacerda and Hay 1982). Existing fringe stands of mangroves consist of *Avicennia schaueriana*, *Rhizophora mangle*, and *Laguncularia racemosa*, which occur along the lower river courses of the northern coast of the bay. At the inner reaches of the mangrove wetlands, more than 100 taxa of plants have been recorded, and the wildlife is abundant. There are more than 68 species of birds, including local and migrating species and threatened species, such as the roseate spoonbill (*Ajaja ajaja*) and the purple gallinule (*Porphyrula martinica*). Crabs are represented by approximately 30 species, including *Callinectes danae*, *Goniopsis cruentatata*, and *Cardisoma guanhumi*, which are of major economic importance to artisinal fishermen (Araújo and Maciel 1979).

Notwithstanding the high degradation of Baía de Guanabara resources, the bay is still an important fisheries resource and supports 6,000 artisanal fishermen and their families. The annual catch measures 3,000 tons, including 2,700 tons of fish, paricularly sardine and mullet; 200 tons of mussels from hard substrates, mostly *Perna perna*, and cockles, i.e., *Anomalocardia brasiliana*, from mud flats; and 100 tons of shrimp (*Pennaeus schimidtii*). However, these catches are but 10% of the annual catches recorded prior to 1970 (Almeida 1993).

8.9 Anthropogenic Impacts

Baía de Guanabara is one of the most prominent coastal bays in Brazil. The cities of Rio de Janeiro, Duque de Caxias, São Gonçalo, Niterói, and many smaller communities are located along its margins, and the greater metropolitan area of Rio de Janeiro has a population exceeding 11 million and growing by 1.1% annually. The domestic runoff from the 8 million inhabitants who live in the bay catchment is still largely untreated. There are more than 12,000 industries in the drainage basin. Two oil refineries along the shore of the bay are responsible for the processing of 17% of the national oil. Baía de Guanabara is traversed by the 6-lane 13-km-long Rio-Niterói highway bridge, which was completed in 1975. At least 2,000 commercial ships dock in the port of Rio de Janeiro every year, which is second only to the Santos in Brazil. The bay is also the home port to two navy bases, has a shipyard, and a large number of ferries, fishing boats, and yachts.

Dredging takes place continuously in the port to a design depth of 17 m. Due to urban development, including two commercial airports, residential construction, and building of roads and bridges, coastal marginal lands have been reclaimed. The bay surface area has been reduced by 10% during the past few decades.

In spite of the pollution control plan created by FEEMA (1979), the water quality situation in the inner portion of Baía de Guanabara has become critical. Although there are six existing sewage treatment plants, only 15% of the total amount of domestic and industrial waste discharged into the bay was subjected to any form of sewage treatment in 1991. Six thousand tons of garbage are generated daily in the area surrounding the bay (Silva et al. 1990). Ferreira (1995) calculated that 18 tons of petroleum hydrocarbons enter the bay daily, 85% via urban runoff. Further, large amounts of suspended solids, organic matter, and heavy metals are discharged into Baía de Guanabara and accumulate in the bottom sediments. High inputs of nutrients have made the bay eutrophic (Rebello et al. 1988; Lavrado et al. 1991) at the same time that the fisheries yield has declined to 10% of the level three decades ago (FEEMA 1990).

The mean dissolved oxygen concentration in Baía de Guanabara measures 8.4 mg l^{-1} (124% saturation) at the surface and 5.1 mg l^{-1} (73% saturation) in the bottom layer. It is only at locations in the inner bay, near sources of significant runoff, that the dissolved oxygen concentration in the bottom layer averages 3-4 mg l^{-1}. Surface waters, in contrast, are frequently supersaturated with respect to dissolved oxygen, even in highly polluted areas. This reflects a high rate of phytoplankton primary production (Rebello et al. 1988) as indicated by chlorophyll-*a* values in the central channel of less than 25 µg l^{-1}, but mean values in the inner bay exceed 130 µg l^{-1}. Although the main part of the bay has relatively low levels of fecal coliform counts, on average 0.2-10 counts ml^{-1}, there is a serious problem at the sampling sites in the inner parts of the bay where counts on average exceed 1,000 counts ml^{-1} (Kjerfve et al. 1997).

Extensive contamination of fisheries resources by heavy metals, in particular mercury, cadmium, copper, and lead, is another problem. For example, heavy metal concentrations reach two to four times the maximum allowed concentration levels for mussels (Rezende and Lacerda 1986). However, bioconcentration of heavy metals in fish is still noncritical because of the low solubility of most heavy metals in the low-oxygen region of the inner bay as a result of sulfide precipitation (Carvalho et al. 1993). Although many studies of Baía de Guanabara have involved extensive data collection (e.g., Paranhos and Mayr 1993; Paranhos et al. 1993), there remains a great need for continued monitoring of hydrological, oceanographic, and water quality parameters for analysis, synthesis, modeling, and management of the natural resources of Baía de Guanabara.

References

Alcântara F, Washington DC (1989) An analytical synoptic-dynamic study about the severe weather event over the city of Rio de Janeiro on Jan 2, 1987. In: Magoon O, Neves C (eds) Coastlines of Brazil. American Society of Civil Engineers, New York, pp 195-204

Almeida MCM (1993) Estudo sobre a pesca do camarão realizada pelas colonias de pesca da Ilha do governador na Baía de Guanabara, Rio de Janeiro. Universidade do Estado do Rio de Janeiro, Rio de Janeiro

Amador ES (1974) Praias fósseis de recôncavo da Baía de Guanabara. An aisAcad Brasil Ciênc 46:253-262

Amador ES (1980a) Assoreamento da Baía de Guanabara - taxas de sedimentação. Anais Acad Brasil Ciênc 52(4):723-742

Amador ES (1980b) Unidades sedimentares cenózoicas do recôncavo da Baía de Guanabara (folhas Petrópolis e Itaboraí). Anais Acad Brasil Ciênc 52(4):743-761

Amador ES (1992) Sedimentos de fundo da Baía de Guanabara – uma síntese. In: Anais do III Congresso da Associação Brasileira de Estudos do Quaternário – ABEQUA. Belo Horizonte, MG, Brazil, pp 199–224

Amador ES (1993) Baía de Guanabara: um balanço histórico. In: Abreu MA (org) Natureza e sociedade no Rio de Janeiro. Prefeitura do Rio de Janeiro, RJ, Brazil, DGDIC, pp 201–258

Amador ES, Ponzi VRA (1974) Estratigrafia e sedimentação dos depósitos flúvio-marinhos da orla da Baía de Guanabara. Anais Acad Brasil Ciênc 46:253–262

Araújo DSD, Maciel NC (1979) Os manguezais do Reconcavo da Baia de Guanabara. FEEMA, Serie Tecnica 10/79, Rio de Janeiro

Asmus HE, Ferrari AL (1978) Hipótese sobre a causa do tectonismo Cenozónico na região sudeste do Brasil. Aspectos Estructurais da Margem Continental Leste e Sudeste do Brasil. Série Projeto REMAC 4:75–88

Carvalho CEV, Lacerda LD, Gomes MP (1993) Heavy metal contamination of the marine biota along the Rio de Janeiro coast, SE, Brazil. Water Air Soil Pollut 57/58:645–653

FEEMA (1979) Manual do meio ambiente: sistema de licenciamento de atividades poluidoras (SLAP). Manual de procedimentos, normas e legislação. Fundação Estadual de Engenharia do Meio Ambiente, Rio de Janeiro, RJ, Brazil

FEEMA (1990) Projeto de recuperação gradual do ecossistema da Baía de Guanabara., vol 1. Fundação Estadual de Engenharia do Meio Ambiente, Rio de Janeiro, RJ, Brazil

Ferrari AL (1990) A geologia do "Rift"da Guanabara (RJ) na sua porção centro-ocidental e sua relação com o embasamento pré-cambriano. Anais do XXXVI Congresso Brasileiro de Geologia, Natal, Brazil, pp 2858–2871

Ferreira HO (1995) Aporte de hidrocarbonetos de petróleo para a Baía de Guanabara. MSc Thesis, Universidade Federal Fluminense, Brazil

Kjerfve B, Ribeiro CHA, Dias GTM, Filippo AM, Quaresma VS (1997) Oceanographic characteristics of an impacted coastal bay: Baía de Guanabara, Rio de Janeiro, Brazil. Continental Shelf Research 17(13):1609–1643

Lacerda LD, Hay JD (1982) Evolution of a new community type after the destruction of a mangrove ecosystem. Biotropica 25:117–120

Lavrado HP, Mayr LM, Carvalho V, Paranhos R (1991) Evolution (1980–1990) of ammonia and dissolved oxygen in Guanabara Bay, RJ, Brazil. In: Magoon OT, Converse H, Tippie V (eds) Proceedings of the 7th symposium on coastal and ocean management. Coastal Zone 91, vol 4. American Society of Civil Engineers, New York, pp 3234–3245

Meis MRM, Amador ES (1972) Formação Macacu-Considerações à respeito do neocenozóico da Baía de Guanabara. Anais Acad Brasil Ciênc 44:602

Meis MRM, Amador ES (1977) Contribuição ao estudo do neocenozóico da Baixada da Guanabara. Rev Brasil Geog 7:150–174

Paranhos R, Mayr LM (1993) Seasonal patterns of temperature and salinity in Guanabara Bay, Brazil. Fresenius Environ Bull 2:647–652

Paranhos R, Mayr LM, Lavrado HP, Castilho PC (1993) Temperature and salinity trends in Guanabara Bay (Brazil) from 1980 to 1990. Arquivo Biol Tecnol 36:685–694

Pires IO (1992) Monitoramento dos manguezais da APA-Guapimirim através de correlação de dados da fitomassa e da radiância TM/Landsat. PhD Thesis, Universidade de São Paulo, Brazil

Rebello AL, Ponciano CR, Melges LH (1988) Avaliação da produtividade primária e da disponibilidade de nutrientes na Baía de Guanabara. Anais Acad Brasil Ciênc 60:419–430

Rezende CE, Lacerda LD (1986) Metais pesados em mexilhões (*Perna perna*, L.) no litoral do Estado do Rio de Janeiro. Rev Brasil Biol 46:239–247

Ruellan F (1944) A evolução geomorfológica da Baía de Guanabara e regiões vizinhas. Rev Brasil Geog 6:445–508

Silva CS, Rodrigues JCV, Câmara NL (1990) Saneamento básico e problemas ambientais na região metropolitana do Rio de Janeiro. Rev Brasil Geog 52:5–106

Souza MHS (1988) Clima de ondas ao norte do Estado do Rio de Janeiro. MSc Thesis, Universidade Federal do Rio de Janeiro, Brazil

Stech JL, Lorenzzetti JA (1992) The response of the South Brazil Bight to the passage of wintertime cold fronts. J Geophys Res 97:9507–9520

9 The Lagoon Region and Estuary Ecosystem of Cananéia, Brazil

J.G. TUNDISI and T. MATSUMURA-TUNDISI

9.1 Introduction

The lagoon region and estuary of Cananéia extends for approximately 110 km along the southern part of the coastal plain of São Paulo State, Brazil (25°S, 48°W). A distinctive feature of the estuarine-lagoon complex is the presence of several islands, Cardoso, Cananéia (27 km^2), Comprida (70 km^2), and Iguape Island. The islands are isolated from the continent by interconnected channels, the Cubatão and Cananéia Sea and the Trapandé Bay, which are 1–3 km wide and up to 20 m deep (mean depth 6 m). The backwaters and small rivers are shallower and characteristically fringed by mangrove vegetation. The system connects with the Atlantic Ocean through channels at the northern (Barra de Icaparra) and southern (Barra de Cananéia) end of Comprida Island (Tessler and Souza 1998; Fig. 9.1).

9.2 Environmental Setting

The climate is dominated by interplay of air masses of tropical fronts and cold fronts of polar origin (Fig. 9.2). Tropical air masses prevail from the end of spring (September) to the end of summer (February), while polar air masses dominate in fall and winter (March to September). Periods of frequent perturbation in February/March are related to warm air masses with stationary fronts and high precipitation and, in September/October, to the transition from predominantly cold to warm air masses with stationary fronts and intense fog. Perturbations are less frequent during common cold fronts in April/May (Garcia Occhipinti 1963; Franco da Silva 1989). Clear days occur during 10% of the year and are common in winter,

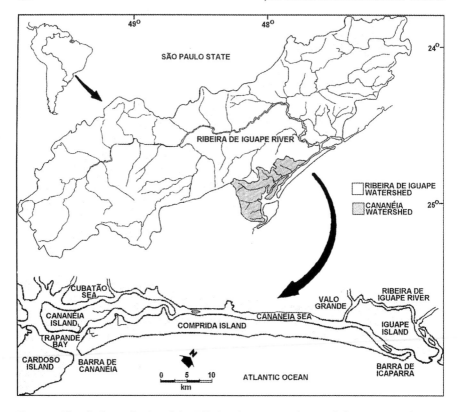

Fig. 9.1. The drainage basin of the Ribeira de Iguape River and the Cananéia lagoon region and estuarine ecosystem

while fully overcast days (36%) occur mainly during spring. The dominance of SW winds favors accumulation of coastal water inside the lagoon region and thus influences salinity and water temperature. Diurnal variations are associated with climatic forcing, daily heating and nocturnal cooling, elevated precipitation, tides, and the influence of humic substances that provide conditions for heating of surface water in rivers and channels (Teixeira et al. 1965). Wind can be a factor of turbulence in the open channels.

The geomorphology of the lagoon region is related to large-scale changes during the Quaternary period, hydrodynamics that continuously reshape the coast, and atmospheric circulation (Petri and Suguio 1971; Suguio and Petri 1973; Tessler 1982; Suguio and Martin 1987). Channels bottom sediments are predominantly sandy (76.3%), homogeneous, and flat (Tessler and Souza 1998). Where circulation is weak, fine sediments are common and clay sediments of terrestrial origin occur predominantly on

The Lagoon Region and Estuary Ecosystem of Cananéia, Brazil 121

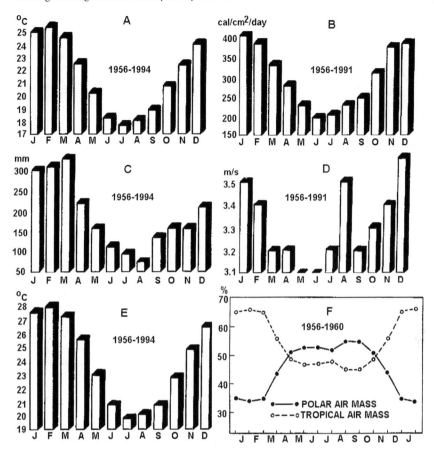

Fig. 9.2A–F. Climatic characteristics of the Cananéia lagoon region. Air temperature (A), solar radiation (B), precipitation (C), wind velocity (D), and surface water temperature (E; modified after Wainer et al. 1996). Percent frequency of polar and tropical air masses (F; modified after Garcia Occhipiati 1963)

the northern part of Cananéia Island and near the Valo Grande channel. Although submerged rocks and sandy elevations may interfere with hydrodynamic conditions, the transport and distribution of bottom sediments and the bottom morphology are a register of the hydrodynamic conditions (i.e., action of tidal currents) and the contribution of runoff and discharge of terrestrial origin (Kutner 1962).

Prior to 1828, the natural watershed of the lagoon region and estuarine system of Cananéia was restricted to an area of approximately 3,000 km², thus freshwater runoff into the system tended to be low. Between 1828 and 1830, an artificial channel (Valo Grande) was opened between the lower

Ribeira de Iguape River and the Cananéia Sea to facilitate navigation and transportation in the northern part of the estuary system. Freshwater from the Ribeira de Iguape River drainage basin (23,350 km^2) began to discharge through the Valo Grande channel into the lagoon region. The action of the river gradually widened the channel (250 m), which has become the principal tributary (mean annual input 435 m^3 s^{-1}) to the lagoon region. After the closure of the Valo Grande channel (1978–1995), the salinity in the lagoon system changed from a minimum of 0–22 and a maximum of 14–33 to 16–30 and 26–34, respectively (de Camargo 1987). As a consequence, during the last 150 years the input of freshwater and the introduction of sediments into the system have modified physiographic and hydrologic characteristics, which influence biological structure and ecological functions of the lagoon region.

The general pattern of water circulation in the main channels and small rivers (Fig. 9.1; Tessler and Souza 1998) depends on tides that enter the Barra de Cananéia and Icaparra (Myiao et al. 1986). Tides are semidiurnal (with a diurnal inequality; Mesquita 1983). The mean tidal amplitude is 0.8 m and surface current velocities ranged from 0.7 to 0.8 m s^{-1}, though during low tide surface velocities may reach 1.2 m s^{-1} in the Cananéia Sea channel near the entrance. Residual flux, probably freshwater from rivers and creeks, is dammed by inflowing coastal water (Myiao 1977). Tidal propagation is a coupling of progressive and stationary waves, thus Cananéia is a partially stratified estuary. The vertical structure is more stratified near the open channels (Trapandé Bay, Cubatão Sea) and less stratified upstream, with almost all upstream salt flux due to turbulent diffusion (Myiao et al. 1986; da Miranda et al. 1995; Fig. 9.3).

Dissolved oxygen concentrations are highest (80–90%) in the large channels, such as Trapandé Bay or the Cananéia Sea, while concentrations decrease to 20–30% in small rivers and channels, due to large amounts of mangrove-originated organic matter and decomposition processes (Teixeira et al. 1965; Kato 1966; Myiao et al. 1986). Diurnal dissolved oxygen at the surface may fluctuate between 120 and 10–20% at day and night, respectively. Dissolved inorganic nutrients tend to be low. Higher values of dissolved inorganic phosphate during high tide might be a result of nutrient addition by coastal waters or re-suspension and transport of bottom sediments by tidal action (Myiao et al. 1986). Low correlation between dissolved inorganic phosphate, nitrate, and nitrite concentrations and salinity suggests different sources of nutrients, regeneration, and distribution. The large contribution of decomposing mangrove litter (Gerlach 1958; Kato 1966) plays a fundamental role for the nutrient cycles.

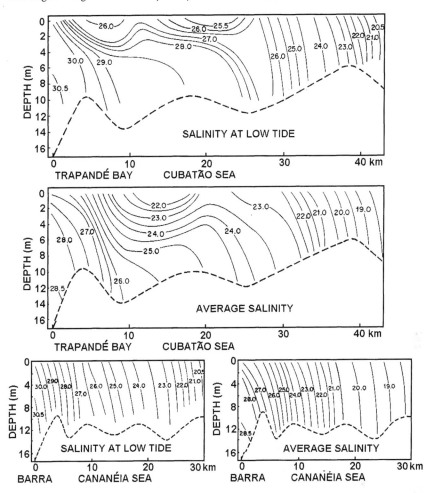

Fig. 9.3. Salinity distribution in the main channels

9.3 Biotic Components

9.3.1 Mangroves

Mangrove forests of the lagoon region of Cananéia (about 110 km^2) develop around the islands, the margin of the continent, and along small rivers and channels. Mangrove vegetation is composed of *Rhizophora mangle* (red mangrove), *Avicennia shaweriana* (black mangrove), *Laguncularia racemosa* (white mangrove), and *Conocarpus erecta*, the last species being rare. In open channel areas, 9–10 m high trees of *Rizophora mangle* are the dominant vegetation on soft sediments. River-borne sediments in front of mangrove vegetation are occupied by a fringe of *Spartina* sp. Between the margins and the inner part of the mangrove forest with more solid substrate, a mixture of *Rizophora mangle* and *Laguncularia racemosa* is associated with *Avicenia shaweriana*. Basin forests are dominated by *Laguncularia racemosa*, which grows on sandy soils to a maximum height of 2.5 m. Transitional vegetation from the mangrove to Tropical Atlantic forest is common. The structural development of the mangrove forest is a function of periodicity of inputs and nature and intensity of stress factors (Cintron and Schaeffer-Novelli 1983). The spatial and structural variability of the mangrove forest depends on the complex interaction between water level fluctuations, terrestrial nutrient input, freshwater, low temperature, high soil salinity, and drought stress (Adaime 1985).

9.3.2 Plankton Community

Phytoplankton growth and production cycles are more pronounced in inshore waters of Cananéia than in adjacent coastal waters (Teixeira and Kutner 1963; Teixeira et al. 1965, 1969; Tundisi et al. 1973, 1978), with primary production (mainly nanoplankton <50 µm) reaching approximately 1 g C m^{-2} day^{-1} and 0.1 g C m^{-2} day^{-1}, respectively (Tundisi 1970). Diatoms, with *Skeletonema costatum* as the dominant species (Kutner 1972), prevail during peak growth and primary production in the rainy summer (December to March), though up to 20% photosynthetic inhibition at the surface is common in inshore areas (Teixeira et al. 1969). The growth of *Skeletonema costatum* (Aidar Aragão 1980), as well as that of other phytoplankton, appears to be stimulated by dissolved humic substances from mangroves after mixing with coastal water (Tundisi 1970). A second peak in early winter (September/October) is dominated by *Cyclotella stylorum* (Kutner 1972). In general, the vertical distribution of phytoplankton and chlorophyll *a* is influenced by flood tides. For example, during high tide, bottom waters have denser populations of *Chaetocera* sp. and *Skeletonema costatum* than surface waters (Brandini 1982). Seasonal phytoplankton cycles and the succession of species were altered during the closing of the Valo Grande channel (1978–1995), and *Skeletonema costatum* summer blooms were absent (Kutner and Aidar Aragão 1986). The standing stock of phytoplankton decreased and diatoms, such as *Phaeodactylum tricornutum* and *Lauderia anulata*, were common. Dinoflagellates were more frequent and coastal water species (*Ceratium furca*) dominated (Kutner and Sassi 1979). The conspicuous seasonal cycle of phytoplankton in inshore waters has its origin in the mixing processes of coastal waters with nutrient-enriched waters of the lagoon region. Rainfall, solar radiation, and high summer temperatures are the main forcing functions.

Copepods, like the dominant *Oithona hebes, Acartia lilljeborghi, Pseudodiaptomus acutus, Euterpina acutifrons, Paracalamus crassirostris, Oithona oswaldocruzi, Acartia tonsa*, and *Temora turbinata*, are principal component (68–97%) of the zooplankton community (Tundisi 1972; Fonseca and Bjornberg 1976; Fonseca and Almeida Prado 1979; Almeida Prado and Lansac Toha 1984; Almeida Prado et al. 1989; Ara 1998). The distribution of several zooplankton species and larvae is strongly effected by salinity. The vertical migration of the salinity-tolerant copepod *Pseudodiaptomus acutus* towards the bottom, followed by tidal transport to upper reaches of the estuary where reproductive populations thrive under optimal conditions in backwaters of mangrove channels (Tundisi and Matsumura-Tundisi 1968), proved to be a paradigm for the behavior and distribution of estuarine zooplankton in temperate and tropical regions (Rodriguez 1975). As for phytoplankton, standing stock values of zooplankton, with a minimum in the winter (July–August), tend to be higher in the lagoon region than in coastal waters. Daily production of copepods ranges from 2.08 to 44.76 mg C m^{-3} day^{-1}.

9.3.3 Benthic Community

Among the 73 taxa, mainly Annelida, Crustacea, and Mollusca (Jorcin 1997), polychaetes dominate in number of species and individuals, *Loadalia americana, Laonice japonica, Clymene* sp., and *Clymenella* sp. being the most common species (Tommasi 1970). The distribution and abundance of macrobenthic fauna is related to the nature of the substrate and physicochemical variables, with highest density and richness occurring in the upper 5 cm of Trapandé Bay. The almost permanent influence of coastal water and elevated nutrient and organic matter concentrations are likely to favor elevated standing stock and diversity of benthic fauna (Jorcin 1997). Inside the estuarine-lagoon complex, muddy bottoms display a poorer epifauna and only on infralittoral stone bottoms a rich epifauna of hydroids (i.e., *Eudendrium carneum, Ophiotrix angulata*) and amphipods occurs. Freshwater components, such as Insecta and Naididae (Oligochaeta), are found in the upper reaches of the estuary. The seasonal changes in macrobenthic abundance and diversity are attributed to changes in salinity, redox potential, sediment granulometry, and organic matter concentrations.

9.3.4 Fish Fauna

The fish fauna (otter trawl) in Trapandé Bay (Zani Teixeira 1983) is represented by 68 species of 52 genera and 23 families. Sciaenidae (15 species), Carangidae (10), Ariidae (7), and Engraulidae (7) are the best-represented families, while Ariidae and Carangidae represent 51 and 27 % of the catch, respectively. Common species in inshore regions and near the Trapandé Bay inlet are *Arius spixii, Chloroscombrus chrysurus, Genidens genidens, Eucinostomus argenteus, Anchoa filifera, Isoposthus parvipinnis, Harengula clupeola, Stellifer rastrifer, Micropogonias furnieri, Netuma barba, Oligoplites saurus, Macrodon ancylodon, Peprilus paru, Pellona Harroweri, Menticirrhus americanus,* and *Achirus* sp. Most species are of marine origin but prefer waters of lower salinity to coastal waters. The only typically estuarine species is *Cynoscion microlepidotus*.

9.4 Nutrient Cycles, Energy Flow, and Food Chains

The nutrient dynamics of the lagoon region are influenced by inorganic nutrient runoff from the Tropical Atlantic forest, dissolved and particulate organic matter input, phytoplankton biomass, and decomposition in tidal creeks of the upper estuary (Schaeffer-Novelli et al. 1990). Mangrove litter (9.02 tons ha^{-1} year^{-1}, 65 % leaves and 23 % fruits) significantly contributes to organic matter production, which is added to rivers and channels and decomposes in sediments and waters of the lagoon region where high temperatures (36 °C at noon) of surface waters are fundamental for rapid decomposition cycles (Menezes 1994). In general, particulate organic carbon (POC) levels are similar to inshore waters elsewhere. Elevated POC (40–324 µg l^{-1}) reflects input from the land and high standing stock and productivity in the water column. The dominance of POC particle sizes between 0–4 and 10–50 µm suggests that phytoplankton does not represent a principal source of detritus-derived POC in surface waters of Cananéia. Phytoplankton POC contributions only become important during bloom periods and in January, April, June, and August when phytoplankton carbon may constitute up to 72 % of POC (Mesquita 1983).

Bacterial populations largely develop in upper estuarine regions with high organic matter concentration. Bacterial densities (mean 8.84 × 10^8 cells l^{-1}) vary between 6 × 10^9 cells l^{-1} in summer (November-January) and 4 × 10^6 cells l^{-1} in winter (June-August), with approximately 46.5 % of the bacteria attached to detritus. Approximately 10 % of gross primary production is consumed by the microbial community and metabolic activity is largely due to the presence of senescent phytoplankton cells, which show a significant correlation with free or attached bacteria (Mesquita 1994). Due to the enormous contribution of mangrove vegetation to dissolved or particulate organic matter, an organic matter-bacteria based trophic structure dominates in the upper estuary. In the lower reaches of the estuarine-lagoon system, with strong influence of coastal waters, a trophic structure based on phytoplankton and zooplankton probably prevails and, especially in winter, microbial metabolism is less important (Mesquita 1983).

9.5 Sustainable Development and Management Needs

Owing to its proximity to the São Paulo metropolitan area (16 million inhabitants), the Cananéia estuarine complex and lagoon region has economic, social, and ecological importance and its exploitation and multiple use may improve sustainable development of one of the poorest populations in São Paulo State. A large part of the system is located in the area of a national park under environmental protection, thus alternative ways for development based on a sound scientific background have to be sought. Fisheries, aquaculture, and ecological tourism seem to be promising activities for the future, giving new perspectives and promoting alternative ways of using the "ecosystem services" of the lagoon region. Of great importance is the continuous protection of the Tropical Atlantic forest in order to avoid damage to the lagoon region and to fringing mangroves by introducing potentially harmful materials from the Ribeira de Iguape River through the Valo Grande channel. The construction of mobile devices in the Valo Grande channel to regulate freshwater flow into the system would represent a man-made proposition to control ecological cycles in order to solve fundamental problems of fluxes and external energy (Jorgensen and Tundisi, unpubl.).

References

Adaime RR (1985) Produção do bosque de mangue da Gamboa Nobrega. PhD Thesis, Universidade de São Paulo, Brazil

Aidar Aragão E (1980) Alguns aspectos de autoecologia de *Sheletonema costatum* (Greville) cleve de Cananéia (25°S–48°W), com especial referência ao fator salinidade. Thesis, Universidade de São Paulo, Brazil

Almeida Prado PMS, Lansac Toha FA (1984) The distribution of freshwater Calanoida (Copepoda) along the coasts of Brazil. Hydrobiologia 113:147-150

Almeida Prado Por MS, Pompeu M, Por FD (1989) The impact of the Valo Grande Canal on the planktonic copepod populations of the Mar Pequeno – Seaway (São Paulo, Brasil). In: Spanier E, Steinberger Y, Luria M (eds) Environmental quality and ecosystem stability. Environ Qual Jerusalem ISEEQS Publ 4B:205-217

Ara K (1998) Variabilidade temporal e produção dos copepodos no complexo estramarina lagunar de Cananéia, São Paulo, Brasil. PhD Thesis, Universidade de São Paulo, Brazil

Brandini FP (1982) Diel variations of ecological factors in Cananéia Region (SP). Arq Biol Tecnol 25:313-327

Cintron G, Schaeffer-Novelli Y (1983) Metodos para la descripcion y estudio de areas de manglar. Compendio enciclopedio de los recursos naturales de Puerto Rico, vol 3

de Camargo TM (1987) Fixação e desenvolvimento de comunidades sesséis expostas a um gradiente de salinidade: influência da cor do substrato artificial. Complexo estuarino lagunar de Cananéia (25° Sul, Brasil). Brasil National Research Council, Final Rep (Proc 101742-82)

de Miranda LB, de Mesquita AR, França CAS (1995) Estudo da circulação e do processo de mistura no extremo sul do Mar de Cananéia. Condições de dezembro 1991. Bolm Inst Oceanogr S Paulo 43(2):153-164

Fonseca VL, Bjornberg TKS (1976) *Oithona oligohalina* sp.n. de Cananéia (Est. S. Paulo) e considerações sobre *O. ovalis* Herbst (Copepoda, Cyclopoida). Anais Acad Bras Ciênc 47:127-131

Fonseca VL, Almeida Prado MS (1979) Copepods of the genus Oithona from Cananéia region (Lat. 25°07'S – Long. 47°56'W). Bolm Inst Oceanogr S Paulo 28(2):1-16

Franco da Silva F (1989) Dados climatológicos de Cananéia e Ubatuba (Estado de S. Paulo). Bolm Inst Oceanogr S Paulo (6):1-21

Garcia Occhipinti A (1963) Climatologia dinâmica do litoral sul brasileiro. Bolm Inst Oceanogr S Paulo 3:1-86

Gerlach S (1958) Die Mangrove Region tropischer Küsten als Lebensraum. Morph Okol Tiere 46:636-730

Jorcin A (1997) Distribuição espacial (vertical e horizontal) do macrozoobentos na região estramarina de Cananéia (SP) e suas relações com algumas variáveis físicas e químicas. PhD Thesis, Universidade de São Paulo, Brazil

Kato K (1966) Geochemical studies on the mangrove region of Cananéia. Brazil. I. Tidal variation of water properties. Bolm Inst Oceanográfico S Paulo 15:13-20

Kutner AS (1962) Granulometria dos sedimentos de fundo da região de Cananéia, S.P. Bolm Soc Bras Geol 11(2):1-54

Kutner MBB (1972) Variação estacional e distribuição do fitoplancton na região de Cananéia. PhD Thesis, Universidade de São Paulo, Brazil

Kutner MBB, Sassi R (1979) Dinoflagellates from the Ubatuba Region (Lat. 23°30'S Long. 45°06'W), Brazil. In: Taylor DL, Seliger HA (eds) Proceedings of the 2nd International Conference on Toxic dinoflagellate blooms. Elsevier-North Holland, New York, pp 169-172

Kutner MBB, Aidar Aragão E (1986) Influência do fechamento do Valo Grande sobre a composição do fitoplancton na região lagunar de Cananéia (25°S-48°W). In: Bicudo CEM, Teixeira C, Tundisi JG (eds) Algas: a energia do amanhã. Instituto Oceanográfico, Universidade de São Paulo, pp 109-120

Menezes GV (1994) Produção e decomposição em bosques de mangue da Ilha do Cardoso, Cananéia, SP. MSc Thesis Universidade de São Paulo, Brazil

Mesquita HSL (1983) Suspended particulate organic carbon and phytoplankton in the Cananéia Estuary (25°S, 48°W). Brazil. Oceanogr Trop 18(1):55–68

Mesquita HSL (1994) Planktonic microbial community oxygen consuption rate in Cananéia waters (25°S-48°W), Brazil. Neth J Aquatic Ecol 28(3):441–451

Myiao S (1977) Contribuição ao estudo da oceanografia física da região de Cananéia (25°S, 48°W). MSc Thesis, Universidade de São Paulo, Brazil

Myiao S, Nishihara L, Sarti CC (1986) Physical and chemical characteristics of the Cananéia-Iguape lagunar region. Bolm Inst Oceanogr S Paulo 34:22–36

Petri S, Suguio K (1971) Exemplo de trabalho do Mar no litoral Sul do Brasil. Notic Geomorf 2(21):61–66

Rodriguez G (1975) Some aspects of the ecology of tropical estuaries. In: Golley FB, Medina E (eds) Tropical ecological systems. Springer, Berlin Heidelberg New York, pp 313–333

Schaeffer Novelli Y, Mesquita H, Cintron G (1990) The Cananéia lagoon estuarine system, SP, Brasil. Estuaries 13:193–203

Suguio K, Petri S (1973) Stratigraphy of the Iguape-Cananéia lagoonal region sedimentary deposits, S. Paulo State, Brazil, part I: field observations and grain size analysis. Bolm IG Inst Geoc Univ S Paulo (4):1–20

Suguio KL Martin L (1987) Classificação de costas e evolução geológica das planiceas litoraneas quaternárias do sudeste e sul do Brasil. Simpósio sobre ecossistemas da costa sul e sudeste brasileira, síntese dos conhecimentos. Acad Sci S Paulo, ACIESP 54(1):1–28

Teixeira C, Kutner B (1963) Plankton studies in a mangrove environment. I. First assessment of standing stock and principal ecological factors. Bolm Inst Oceanogr S Paulo 12:101–124

Teixeira C, Tundisi JG, Kutner MBB (1965) Plankton studies in a mangrove environment. I. First assessment of standing – stock and principal ecological factors. Bolm Inst Oceanogr S Paulo 12(3):101–124

Teixeira C, Tundisi JG, Santoro YI (1969) Plankton studies in a mangrove environment. VI. Primary production, zooplankton standing stock and some environmental factors. Int Rev Ges Hydrobiol 54:289–301

Tessler MG (1982) Sedimentação atual na região lagunar de Cananéia Iguape, Estado de São Paulo. Thesis, Universidade de São Paulo, Brazil

Tessler MG, Souza LAP (1998) Sedimentary dynamics and recognized sedimentary features in the bottom surface of Cananéia Iguape system. Braz J Oceanogr 46:69–83

Tommasi LR (1970) Observações sobre a fauna bentica do complexo estramarino lagunas de Cananéia. Bolm Ocean 19:43–56

Tundisi JG (1970) O plancton estuarino. Contr Avulsas Inst Ocean 19:1–22

Tundisi TM (1972) Aspectos ecológicos do zooplancton da Região Lagunar de Cananéia com especial referência aos Copepoda (Crustacea). PhD Thesis, Universidade de São Paulo, Brazil

Tundisi JG, Matsumura-Tundisi T (1968) Plankton studies in a mangrove environment. V. Salinity tolerances of some planktonic crustaceans. Bolm Inst Oceanogr S Paulo 17(1):57–65

Tundisi JG, Matsumura-Tundisi T, Kutner MBB (1973) Plankton studies in a mangrove environment. VIII. Further investigations on primary production, standing stock of phyto and zooplankton and some environmental factors. Int Rev Ges Hydrobiol 58:925–940

Tundisi JG, Teixeira C, Matsumura-Tundisi T, Kutner MBB, Kinoshita L (1978) Plankton studies in a mangrove environment. IX. Comparative investigations with coastal oligotrophic waters. Rev Bras Biol 38(2):301–320

Wainer IEKC, Colombo PM, Miguel AJ (1996) Boletim de monitoramento climatológico para as bases Norte e "Dr. João de Paiva Carvalho" do IOUSP. Relat Tec Inst Oceanogr S Paulo 38:1–13

Zani Texeira ML (1983) Contribuição ao conhecimento da ictofauna da baia do Trapandé, complexo estuarino lagunar de Cananéia. MSc Thesis, Universidade de São Paulo, Brazil

10 The Subtropical Estuarine Complex of Paranaguá Bay, Brazil

P.C. Lana, E. Marone, R. M. Lopes, and E. C. Machado

10.1 Introduction

Paranaguá Bay, on the coast of Paraná State in southeastern Brazil (48°25'W, 25°30'S), is part of a large interconnected subtropical estuarine system that includes the Iguape-Cananéia Bay system on the southern coast of São Paulo State. Rather than being an estuary, Paranaguá Bay (612 km²) is best defined as an estuarine system comprised of two main water bodies, the Paranaguá and Antonina bays (260 km²) and the Laranjeiras and Pinheiros bays (200 km²). The system connects to the open sea through three tidal channels, with the main entrance area around Mel Island (152 km²; Fig. 10.1). The structural properties of the bay are typical for a marine ingression environment. Forced regressions, following sea-level maxima at approx. 120,000 and 5,100 B.P., have formed an upper and a lower geomorphologic zone, represented by a drowned, narrow paleo-valley west of Paranaguá City and by wide beach ridge plains east of Paranaguá, respectively (Angulo 1992; Angulo and Lessa 1997). An extensive coastal plain surrounds Paranaguá Bay and the upper reaches of the bay originate about 50 km inland at the piedmont of the Serra do Mar mountain range. Mangrove swamps and marshes mainly fringe the interior of the system, while ocean-exposed areas adjacent to the mouth are composed of extensive sand beaches and some rocky shores (Angulo 1992).

10.2 Environmental Settings

The climate of the region depends on the north or south displacement of the semi-permanent anti-cyclone gyre in the South Atlantic and on the

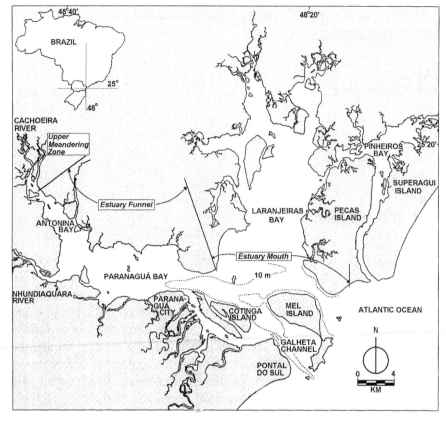

Fig. 10.1. The Paranaguá Bay system with estuarine zonation. (Modified after Lessa et al. 1998)

passage of cold polar masses in winter. The main atmospheric disturbances are cold fronts from a SW-NE direction that originate in southern South America. The Serra do Mar Mountains act as a barrier against cold fronts and cause a concentration of stationary fronts in the bay region. Owing to the intensification of the secondary tropical anti-cyclone, stationary warm fronts prevail in summer (Bigarella et al. 1978). Winds from the NE with a mean velocity of 4 m s^{-1} are most frequent, while strong storms from the southeast may reach up to 25 m s^{-1} (FUNPAR 1997). The climate of the coastal plain is classified as Cfa, with a mean annual rainfall of 2,500 mm (maximum 5,300 mm) and around 85% mean air humidity. A typical rainy season initiates in late spring and lasts during most of the summer while the dry season lasts from late autumn to late winter but is usually interrupted by a short and weak rainy period in early winter.

The mean precipitation during the rainy season is more than three times higher than that of the dry season.

10.2.1 Geomorphologic Processes

Physiographic structures of Paranaguá Bay are geologically ephemeral and depend on interactions between the upland freshwater drainage system and tidal marine conditions. The bay can be divided into a mouth zone, an estuary funnel, and a meander zone (Lessa et al. 1998; Fig. 10.1). The mouth zone corresponds to the lower part of the estuary, which is surrounded by beach ridge plains. Islands in the mouth zone (i.e., Cotinga Island) emerged from shallow estuarine areas (Angulo 1992). The main channel has a depth of more than 10 m (max. 33 m) and the bottom sediments are characterized by well-sorted fine to very fine sand, with fines (grain size <0.062 mm) increasing from sea (0%) towards land (40%). The estuary funnel represents the widest cross-section in the estuary with depths (maxima 15 m near Paranaguá harbor) varying between 4 and 10 m and decreasing upstream. Most of the bottom in the funnel zone is characterized by muddy sediments (<50% sand; Bigarella et al. 1978). In the upper funnel zone non-vegetated elongated low-tide sandbars represent a bay-head delta typical of tide-dominated environments. The upper 5 km of the estuary with shallow channels (<4 m) and strong fluvial influence represent the meander zone. The bottom sediments are characterized by poorly sorted fine to medium sand (<40% fines), though grain sizes above 2 mm are common. Several low-vegetated non-tidal islands occur near the mouth of small rivers in the lower meander zone. A larger influx of fluvial sediments into the Paranaguá Bay estuary during the last decades may have caused an expansion of tidal sandbars and the fluvial sand deposits at the head of the estuary (Fig. 10.2).

10.2.2 Physical Characteristics

The Paranaguá Bay estuarine complex has been classified as a partially mixed estuary (type B), with lateral heterogeneity (Knoppers et al. 1987), a mean depth of 5.4 m, a total water volume of $1,410^9 m^3$ and a residence time of 3.49 days (FUNPAR 1997). As circulation patterns and stratification vary between seasons, mean salinity and water temperature in summer and in winter are 12–29 and 23–30 °C and 20–34 and 18–25 °C, respectively. The structure and function of the estuarine system is influenced by different gradients. A salinity-energy gradient from freshwater to marine conditions

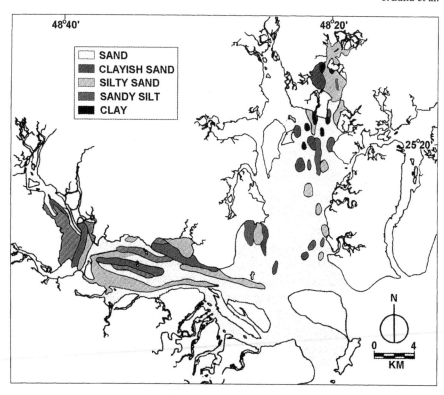

Fig. 10.2. Distribution of sediments in Paranaguá Bay. (Modified after Bigarella et al. 1978; Soares 1990)

along the east-west and north-south main axes divides the bay into a high energy, euhaline (average salinity ~30) outer region, a middle polyhaline region, and oligo- and mesohaline (average salinity 0–15) low-energy inner sectors. A lateral gradient originates from the freshwater input of rivers and tidal creeks and creates several 'micro-estuaries' in the euhaline and polyhaline sectors of the bay. A temporal gradient, with daily, seasonal and inter-annual components is super-imposed on the above gradients.

The hydrodynamics are driven by tidal forcing and river runoff (Knoppers et al. 1987; Brandini et al. 1988; Rebello and Brandini 1990; Machado et al. 2000). Waves, mainly from the southeast, are only important in the bay mouth where they range from a mean height of about 0.5 m (periods 3–7 s) to a maximum height of 2–3 m during storm events. Tides are semidiurnal with diurnal inequalities. Tidal amplitudes increase towards the head of the bay, being amplified less than twice. The tidal phase and amplitudes indicate that the tidal wave propagates in a mixed form, with a progressive form at the outer region and a standing wave form in the upper bay. During neap cycles, strong non-linear interactions allow for the formation of up to six high and low tides per day. Spring tides range from 1.7 m at the mouth to 2.7 m in the upper bay. The mean tidal range is 2.2 m, with a tidal prism of 1.34 km^3 and a tidal intrusion of 12.6 km. Following cold front forcing, storm surges elevate water levels up to 80 cm above astronomical tides (Marone and Camargo 1994). Current velocities increase upstream, with maxima of 0.8–0.85 m s^{-1} at ebb and 1–1.4 m s^{-1} at flood (FUNPAR 1997).

The average annual freshwater input from the coastal plain catchment area (about 1,918 km^2) and from the small and steep drainage basins of the Serra do Mar is close to 200 m^3 s^{-1}. The Cachoeira and Nhundiaquara rivers, with an average discharge of 21.13 and 15.88 m^3 s^{-1}, respectively, are the main tributaries (Bigarella et al. 1978), while input to the mid-sector of the bay is approximately 75 m^3 s^{-1} (Knoppers et al. 1987). Groundwater may contribute up to 10% of the total surface freshwater runoff (FUNPAR 1997). Seasonal variations of freshwater input correspond to around 30% of mean annual values during the dry period (May/October) and 170% during the rainy period (November/April). Runoff to the coastal zone ranges from approximately 7 to 28 × 10^6 m^3 day^{-1} during the dry and rainy season, respectively.

10.2.3 Chemical Characteristics

The distribution and temporal variation of physico-chemical properties in Paranaguá Bay are closely correlated to salinity and energy gradients. Annual dissolved oxygen concentrations typically range from 4 to 7.5 ml l^{-1}. The lowest oxygen concentrations are usually found in bottom waters of the inner and mid region during the rainy period, while enhanced primary productivity may cause highest saturation values in the mid bay (Brandini 1985). The distribution patterns of suspended material are less defined, probably due to the complex hydrodynamics of the estuary. Average concentrations range from 10 to 120 mg l^{-1} in rainy summers but are about 40% lower during dry winters, possibly owing to resuspension of sediments by winds and intrusion of saline bottom waters.

The nutrient and trophic status of Paranaguá Bay is the result of interactions between hydrodynamic processes and different mechanisms of sink and supply, such as biological uptake, freshwater input, sediment-water interactions and sewage discharge from the city of Paranaguá. The trophic state, based on annual rates of organic carbon supply and on chlorophyll, POC and nutrient concentrations, varies from almost oligotrophic in winter in the outer section to eutrophic during summer in the middle

Table 10.1. Chemical characteristics (1981–1995) of the water column in Paranaguá Bay. (After Knoppers et al. 1987; Rebello and Brandini 1990; Machado et al. 2000)

Chl-a (μg l^{-1})	POC (mg l^{-1})	NO$_3$ (μM)	NO$_2$ (μM)	NH$_4$ (μM)	PO$_4$ (μM)	Si(OH)$_4$ (μM)	N:P
0.4–49	0.–24.2	0.1–13	0–0.9	0.4–10	0.2–3	1.5–178	0.6–24.2

and inner sections (Table 10.1). The highest values of chlorophyll-a and dissolved inorganic nutrients occur in the middle and inner regions during the rainy period (Knoppers et al. 1987). Nitrite, nitrate, ammonia and orthophosphate behave in a non-conservative manner (Knoppers et al. 1987; Machado et al. 2000). High nitrite and ammonia concentrations are associated with warm and rainy summer periods, following the increase in freshwater input and biologically induced regenerative processes. The mid-sector of the bay is a sink for nitrogen, owing primarily to uptake by diatoms under optimum salinity and turbidity conditions for growth, and a source for orthophosphate due to sewage discharge from Paranaguá City (Brandini 1985; Brandini et al. 1988; Machado et al. 2000). When the freshwater inflow is high and the water column stratified, nitrogen follows a near-conservative behavior; however, under these conditions the residence time of surface waters is probably too short to warrant significant biological nutrient uptake. The N:P ratios vary spatially and temporally. Higher values during the rainy season in the inner bay are due to higher continental runoff. However, the prevalence of low N:P ratios (by atoms) indicates that nitrogen is a potentially limiting factor, probably as a result of denitrification of surface sediments and bottom waters. In general, silicate exhibits a conservative behavior. Although silicate assimilation by diatoms is low when compared to total silicate input to the bay, high biological demands in the mid-sector (salinity 27–33) during summer occasionally represent a minor sink (Knoppers et al. 1987; Brandini et al. 1988).

10.3 Biotic Components

Paranaguá Bay is comprised of diverse natural habitats, like dunes ("restingas"), mangrove swamps, salt marshes, sea grass meadows, rocky shores and extensive tidal flats (up to 2 km wide). Although tropical mangroves are gradually replaced by temperate salt marshes at this latitude, most intertidal areas (about 186 km^2; Martin 1992) around the bay are colonized by mangroves (*Rhizophora mangle, Avicennia schaueriana, Laguncularia racemosa, Conocarpus erecta*; Fig. 10.3). The species composition and distribution patterns of mangroves in Paranaguá Bay are similar to those of tropical mangrove areas in northern Brazil; however, local trees grow less and the mangrove swamps are structurally less complex. Nevertheless, the number of different forest types (Martin 1992; Lana 2000) and differences in litter production and export (Sessegolo 1997) suggest that local mangroves are structurally and functionally diverse. *Spartina alterniflora* marshes colonize tidal flats or creeks as monospecific, discontinuous narrow (up to 50 m wide) belts in front of the mangroves (Lana et al. 1991). In oligo- and mesohaline regions of the bay, pure stands of *Spartina* grow on muddy sediments with high organic content, in the polyhaline sector the species occurs on poorly sorted, very fine sands, and in the euhaline sector stands cover well-sorted sandy sediments. The steep physico-chemical gradients in the bay favor growth of tall-form salt marshes (up to 1 m high) in the mid- and upper bay, while shorter form marshes occur in the outer euhaline sector (Netto and Lana 1997a). Extensive sea grass meadows are absent from the bay, probably owing to high turbidity, but small patches of *Halodule wrightii* occur on shallow subtidal bottoms near Cotinga and Mel islands.

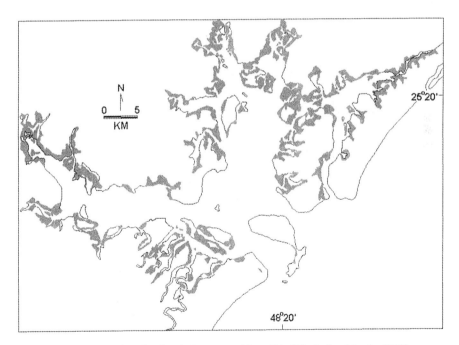

Fig. 10.3. Mangrove distribution in Paranaguá Bay. (Modified after Martin 1992)

Algal blooms in Paranaguá Bay are probably influenced by nutrient input from continental runoff during rainy periods, inflow of nutrient-rich, subtropical coastal waters during winter (Fernandes 1992), and nutrient release into the water column during advective transport associated with frontal systems (Marone and Camargo 1994; Lopes 1997). Most of the 636 pelagic and benthic diatom species identified for Paraná State occur in the bay (Brandini and Fernandes 1996) and centric diatoms and phytoflagellates are the dominant algal component. The diatom *Skeletonema costatum* is abundant throughout the year but reaches peak densities at high nutrient levels and low salinity during rainy summers while *Chaetoceros*, *Rhizosolenia* and *Leptocylindrus* are usually less abundant. Dinoflagellates and silicoflagellates are a minor component of the local phytoplankton standing-stock (Brandini 1985; Fernandes 1992; Brandini and Thamm 1994). Chlorophyll-*a* concentrations (up to 49 µg Chl-*a* l^{-1}) in the meso- and oligohaline sectors (Brandini et al. 1988; Machado et al. 2000) and the polyhaline mid bay (4–20 µg l^{-1}) depend on optimal nutrient concentration, water transparency, and salinity for diatom growth. In euhaline areas (0.1–8 µg Chl-*a* l^{-1}), concentrations are influenced by the adjacent sea (Brandini 1985; Brandini et al. 1988; Brandini and Thamm 1994; Lopes 1997) and maxima are associated with large (>200 µm) diatoms (*Coscinodiscus*, *Palmeria*; Lopes 1997). Benthic macroalgae (approximately 100 species) occur mainly on exposed rocky shores. The macroalgal community (Bostrychietum *sensu* Post) of mangrove swamps is composed of different red algae (i.e., *Bostrychia*, *Caloglossa*, *Catenella*), which dominate in different areas along the estuarine gradient. Occasionally, extensive mats of *Acanthophora spicifera* are a conspicuous feature of intertidal flats.

The zooplankton assemblages are dominated by copepods (up to 90 % of total standing stock), followed by tintinids and appendicularians as the major subdominant taxa, and by cladocerans and chaetognaths (Montú and Cordeiro 1988; Lopes 1997). Highest densities (up to 82,000 org. m^{-3}) occur in poly- and mesohaline sectors of the bay, approximately corresponding with areas of peak chlorophyll concentrations. High densities of meroplankton larvae occur during larval recruitment in tidal mangrove creeks and at the mouth of small rivers in the euhaline area. The distribution of zooplankton (i.e., copepod associations) follows a salinity gradient (Lopes 1997). The calanoid *Pseudodiaptomus richardi* represents the dominant holoplankton of oligohaline areas and is the only "true" estuarine species below salinities of 15. *Acartia tonsa* and *Oithona oswaldocruzi* occur mainly in the mesohaline sector, together with other estuarine-marine species adapted to strong salinity variations, such as *Acartia lilljeborgi*, *Pseudodiaptomus acutus* and *Oithona hebes*. Although marine-euryhaline copepods (*Temora turbinata*, *Paracalanus* spp., *Oithona simplex*, *Euterpi-*

na acutifrons) are found in salinities as low as 15, they occur mainly in euhaline areas under the influence of coastal waters. Marine stenohaline species of warm coastal and shelf waters, like *Paracalanus indicus, Clausocalanus furcatus, Temora stylifera, Oithona plumifera, Oncaea venusta, Corycaeus amazonicus* and *C. giesbrechti*, appear in the outer bay. Low densities of subtropical species (*Centropages brachiatus, Oncaea conifera*) in the outer bay (Montú and Cordeiro 1988) suggest the addition of cold nutrient-rich coastal waters during winter.

The importance of bacteria (Kolm et al. 1987), harpacticoid copepods, foraminifera (Disaró 1995) and mangrove oligochaetes and nematods (Blankensteyn 1994) in microbenthic and meiobenthic associations of Paranaguá Bay is virtually unknown. Despite similar sediment characteristics in salt marshes, mangroves and non-vegetated flats, the composition and abundance of the macrobenthic fauna can vary widely among sub- and intertidal bottoms in euhaline and polyhaline regions of the bay (Lana 1986, 1994; Blankensteyn 1994; Netto and Lana 1996, 1997b; Couto 1996; Lana et al. 1997; FUNPAR 1997). While salinity and environmental energy gradients appear to control large-scale distribution patterns in the bay, plant architecture and food availability seem to be the main source of small-scale macrofaunal variability (Lana and Guiss 1991, 1992; Lana et al. 1997). In contrast to maximum abundance of invertebrates in temperate latitudes during late spring and early summer, benthic invertebrate populations in Paranaguá Bay lack any pronounced seasonal pattern, with high abundance both in summer and winter. Seasonal oscillations have been attributed to variations in plant cover and biomass, rainfall, low temperature, or predation rates.

At least 200 species of fish (66 commercially important) and about 40 larval types have been described for Paranaguá Bay (Corrêa 1987; FUNPAR 1997). On the basis of migratory behavior, the commercially important species, which reproduce mainly in spring and summer, can be divided into four categories. Marine species, like the mullets *Mugil liza* and *M. platanus* (Mugilidae) and the sea catfish *Netuma barba* (Ariidae), visit the estuarine system for spawning while others, such as the pompanos *Trachinotus carolinus* and *T. falcatus* (Carangidae), explore the bay as a feeding and growing site. Estuarine and coastal species, like the weakfish *Cynoscion leiarchus, C. acoupa* and the white mouth croaker *Micropogonias furnieri* (Sciaenidae), migrate to oceanic waters for spawning and early development but return to the bay as juveniles and adults. Finally, the catfish *Cathorops spixii* and *Sciadeichthyes luniscutus* (Ariidae), the snook *Centropomus parallelus* (Centropomidae), the southern king croaker *Menticirrhus americanus* (Sciaenidae), and the mullets *Mugil curema* and *M. gaimardianus* are resident estuarine species (Corrêa 1987).

More than 300 bird species have been recorded for the coastal plain of Paranaguá Bay. Pelecaniformes, Ciconiiformes, Anseriformes, Falconiformes, Charadriiformes and Coraciiformes comprise about 90% and Passeriformes approximately 10% of the bird fauna. Residents comprise 52.6% of the species, the others are migratory species from the northern hemisphere or from sub-antarctic and antarctic regions of the southern coast of South America. Frigate birds (*Fregata magnificens*), brown boobies (*Sula leucogaster*) and common gulls (*Larus dominicanus*) are the most frequent and abundant forms in estuarine waters (V. Moraes and R. Krul, pers. comm.).

The gray dolphin *Sotalia fluviatilis* is the most common marine mammal in the bay, followed by scarce records of the franciscana *Pontoporia blainvillei*. Both are neritic species that appear throughout the year in coastal waters of Paraná State. Other neritic species, like *Tursiops truncatus* and *Steno bredanensis*, occur at lower densities in adjacent shelf waters. Oceanic odontocetes, including *Delphinus capensis*, *Stenella frontalis* and, less frequently, *Kogia simu*, *Ziphius cavirostris* and *Orcinus orca*, have been recorded stranded along the coast of Paraná State and in the euhaline sector of the bay (F. Rosas, pers. comm.). Several pinnipeds, like *Arctocephalus tropicalis* in winter and the rare *Arctocephalus australis* and *Mirounga leonina*, visit the bay and adjacent coastal waters, while large whales (i.e. *Balaenoptera edeni*, *Eubalaena australis*, *Megaptera novaeangliae*) have been recorded along the Paraná coast.

10.4 Trophic Structure and Energy Flow

Detrital material from phytoplankton and from riverine and terrestrial sources is associated with salinity gradients, while detrital material from marshes, mangroves and other wetlands is associated with a lateral gradient. The pronounced variability of POC:Chl-*a* ratios, with peaks in autumn/winter and low values in warm rainy periods, suggests that plant debris from mangroves and marshes as the dominant source of organic detritus (Brandini 1985) is replaced by phytoplankton in summer (Machado et al. 2000). Among others, phytoplankton productivity in Paranaguá Bay is controlled by light, nutrient and seston gradients. In contrast to nutrient-rich but turbid waters in oligohaline and low-nutrient, high-transparency waters in euhaline areas (Brandini et al. 1988), maximum productivity rates and low carotenoid:chlorophyll-*a* ratios indicate that light and nutrients provide optimal conditions for algal growth in mesohaline areas of the bay, though photoinhibition may occur at high irradiance levels during summer (Brandini 1985). In general, the grazing impact of copepods (*Acartia lilljeborgi*, *Temora turbinata*, *Paracalanus quasimodo*, *Corycaeus* spp.) on the phytoplankton standing-stock (<0.01–0.22 mg Chl-a m^{-3} day^{-1}) is low (<5% on a daily basis) and does not appear to be seasonal. However, grazing might be a major source of phytoplankton mortality in the euhaline sector, at least during part of the winter and summer (Lopes 1997).

Spartina marshes in high-energy euhaline areas of Paranaguá Bay have low biomass and productivity, probably as a result of high salinity and high sediment accretion rates, conditions that usually favor salt marsh replacement by mangroves. In general, below-ground biomass of *Spartina* is strongly seasonal, being 1.72 and 5.69 tons ha^{-1} at the end of summer and spring, respectively, and below-ground production (up to 3.58 tons ha^{-1} year^{-1}) exceeds above-ground production (1.01–1.79 tons ha^{-1} year^{-1}; Lana et al. 1991). The production of mangrove litter at euhaline sites is approximately 650 g m^{-2} year^{-1} (mainly *Rhizophora*) and litter is almost entirely (80%) composed of leaves. Litter fall rates are four to six times higher (up to 18.82 g m^{-2} day^{-1}) in rainy summers than in winter (Sessegolo 1997). Leaf litter decomposition rates are high, probably owing to grazing macrofauna; however, rates vary between 10.5 and 249 days for permanently immersed *Avicennia schaueriana* and emersed *Rhizophora mangle* leaves, respectively (Sessegolo and Lana 1991). The detritus production cycles in mangrove swamps and *Spartina* marshes seem to be seasonal; while *Spartina* produces detritus mainly in winter, mangrove litter fall is highest in summer. Since detritivorous species depend on the availability of mangrove and salt marsh detritus as a potential food source, seasonal differences in the behavior of benthic invertebrate populations are to be expected. In general, richness and abundance of benthic species is significantly lower in mangroves than in marshes and adjacent non-vegetated tidal flats. Population density of infaunal species, especially of those (i.e. the polychaetes *Isolda pulchella* and *Nereis oligohalina*) which use *Spartina* roots as substrate or refuge, correlates with below-ground plant biomass and increases with higher rhizome biomass at the end of winter and the beginning of spring (Lana and Guiss 1992).

10.5 Human Impacts and Management Needs

The estuarine complex of Paranaguá Bay is one of the least impacted coastal systems in southeastern Brazil, the present state of conservation being the result of the regional process of colonization. Contrary to most coastal regions in Brazil, the urban, agricultural and industrial development of Paraná State concentrated on the highlands ("primeiro planalto") of the Serra do Mar and even today industry and agriculture in the coastal plain are not well developed. The region's high biodiversity and some of the last, well-preserved remnants of the Atlantic Forest, extensive mangrove swamps, sand dunes ("restingas") and tidal flats might stimulate ecological tourism in coming years. However, some traditional practices, such as game hunting, exploitation of the heart-of-palm tree (*Euterpe edulis*), fisheries, slash-and-burn agriculture, and wood extraction are sources of conflict and impact (Andriguetto-Filho 1993; Andriguetto-Filho et al. 1998).

Fisheries are essentially artisanal and represent the principal activity in more than 50 fishing communities around Paranaguá Bay (Rougeulle 1993; Andriguetto-Filho 1998). Fishery landings follow a seasonal pattern, with mullets in winter and catfish and sciaenids throughout the year, though more frequent in summer. Some local fish stocks are clearly overexploited. Commercial bottom trawl and seine fisheries over the adjacent coastal shelf may be partially responsible for a decline of their landings (Andriguetto-Filho 1993). Despite evidence of the overexploitation of shrimp stocks and the mangrove crab *Ucides cordatus* and an overall decline in their landings, the shrimps *Penaeus schmitti, P. paulensis, P. brasiliensis* and *Xiphopenaeus kroyeri* still represent the most important artisanal fisheries in the region. Other resources, such as oysters (*Crassostrea* spp.), mussels (*Mytella guyanensis, Perna perna*), the pointed Venus (*Anomalocardia brasiliana*) and blue crabs (*Callinectes* spp.) are underestimated and their stocks are underexploited.

During the last three decades other environmental problems have emerged. Deforestation has increased sediment loads to the bay (Andriguetto-Filho 1993) and activities in Paranaguá harbor require dredging of the main access channel and the disposal of sediments. The population growth (200 inhabitants km^{-2}) along the southern margin of the bay, with Paranaguá City (120,000 inhabitants) as the major urban and economic center, has caused problems of domestic sewage disposal, increased metal levels in sediments and water, and stimulated the construction of marinas along the bay shores. Although federal, state and county laws legally protect natural resources of the coastal zone, including Paranaguá Bay with at least ten conservation units (Cubbage et al. 1995), and several plans for local land use have been established or proposed, they are not implemented. The failure of management to efficiently address local problems is partly due to the absence of detailed ecological knowledge, a centralized, bureaucratic model of natural resource management, lack of involvement by populations which are directly affected by the natural resources to be managed, and limited material and human resources.

Acknowledgements. We wish to thank the many researchers who often worked under adverse conditions in Paranaguá Bay for providing essential data and published information. Especial thanks go to our colleagues Rodolfo Angulo, Guilherme Lessa, Frederico Brandini, Marco Fábio Corrêa, Valéria Moraes, Fernando Rosas and José Milton Andriguetto Filho for their assistance during the preparation of the manuscript.

References

Andriguetto-Filho JM (1993) Institutional prospects in managing coastal environmental conservation units in Paraná State, Brazil. Coastal Zone '93, vol 1. Proc 8th Symp Coastal Ocean Manag 1993. Am Soc Civil Eng, New Orleans

Andriguetto-Filho JM (1998) Interações, fatores de mudança e sustentabilidade das práticas materiais e dinâmicas ambientais nos sistemas técnicos da pesca artesanal. In: Lima RE, Negrelle RR (eds) Meio ambiente e desenvolvimento no litoral do Paraná: Diagnóstico. Editora Universidade Federal do Paraná, Curitiba, pp 95–105

Andriguetto-Filho JM, Krüger AC, Lange MBR (1998) Caça, biodiversidade e gestão ambiental na Área de Proteção Ambiental de Guaraqueçaba, Paraná, Brasil. Biotemas 11(2):133–156

Angulo RJ (1992) Geologia da planície costeira do Estado do Paraná. PhD Thesis, Universidade de São Paulo, Brazil

Angulo RJ, Lessa GC (1997) The Brazilian sea level curves: a critical review with emphasis on the curves from Paranaguá and Cananéia regions. Mar Geol 140:141–166

Bigarella JJ, Becker RD, Matos DJ, Werner A (eds) (1978) A Serra do Mar e a porção oriental do Paraná, um problema de segurança ambiental e nacional. Secretaria de Estado do Planejamento do Paraná

Blankensteyn A (1994) Estrutura e análise experimental do funcionamento das associações da macrofauna bêntica do manguezal e marisma da Gamboa Perequê, Pontal do Sul, PR. DSc Thesis, Universidade Federal do Paraná, Brazil

Brandini FP (1985) Ecological studies in the Bay of Paranaguá. I. Horizontal distribution and seasonal dynamics of the phytoplankton. Bolm Inst Oceanogr S Paulo 33:139–147

Brandini FP, Fernandes LF (1996) Microalgae of the continental shelf off Paraná State, southeastern Brazil: a review of studies. Rev Bras Oceanogr S Paulo 44:69–80

Brandini FP, Thamm CAC (1994) Variações diárias e sazonais do fitoplâncton e parâmetros ambientais na Baía de Paranaguá. Nerítica 8:55–72

Brandini FP, Thamm CA, Ventura I (1988) Ecological studies in the Bay of Paranaguá. III. Seasonal and spatial variations of nutrients and chlorophyll-*a*. Nerítica 3:1–30

Corrêa MFM (1987) Levantamento e produtividade da ictiofauna da Baía de Paranaguá-Paraná-Brasil. MSc Thesis, Universidade Federal do Paraná, Brazil

Couto ECG (1996) Estrutura espaço-temporal da comunidade macrobêntica da planície intertidal do Saco do Limoeiro-Ilha do Mel (Paraná, Brasil). DSc Thesis, Universidade Federal do Paraná, Brazil

Cubbage FW, Andriguetto-Filho JM, Sills E, Motta M, Muller VY (1995) Legal and administrative frameworks for managing coastal environmental conservation units in the state of Paraná, Brazil: a review. FPEI Working Paper 56, Southeastern Center for Forest Economics Research, Research Triangle Park

Disaró ST (1995) Associações de foraminíferos da Baía das Laranjeiras. MSc Thesis, Universidade Federal do Paraná, Brazil

Fernandes LF (1992) Variação sazonal do fitoplâncton e parâmetros hidrográficos em uma estação costeira de Paranaguá-Paraná. MSc Thesis, Universidade Federal do Paraná, Brazil

FUNPAR – Fundação para o Desenvolvimento Científico e Tecnológico da Universidade Federal do Paraná (1997) Estudo de impacto ambiental (EIA) de uma usina termelétrica na Baía de Paranaguá e do Porto de desembarque, subestação e linha de transmissão associados. Companhia Paranaense de Eletricidade, technical report

Knoppers BA, Brandini FP, Thamm CA (1987) Ecological studies in the Bay of Paranaguá. II. Some physical and chemical characteristics. Nerítica 2:1–36

Kolm H, Giamberardino Filho RE, Kormann MC (1987) Spatial distribution and temporal variability of heterotrophic bacteria in sediments of Paranaguá and Antonina Bays, Paraná, Brazil. Rev Microbiol 28:230–238

Lana PC (1986) Macrofauna bêntica de fundos sublitorais não consolidados da Baía de Paranaguá (Paraná). Nerítica 1(3):79–89

Lana PC (ed) (1994) Diagnóstico ambiental oceânico e costeiro das regiões sul e sudeste do Brasil. Oceanografia biológica, vol 7. Bentos, Petrobrás, Rio de Janeiro

Lana PC (2000) Políticas públicas, legislação ambiental e conflitos de uso: subsídios para uma gestão integrada dos manguezais da Baía de Paranaguá (Paraná, Brasil). In: Lima RE, Negrelle RR (eds) O projeto PADCT-NIMAD-LITORAL: resultados finais. Editora Universidade Federal do Paraná, Curitiba (in press)

Lana PC, Guiss C (1991) Influence of *Spartina alterniflora* on structure and temporal variability of macrobenthic associations in a tidal flat of Paranaguá Bay (southeastern Brazil). Mar Ecol Prog Ser 73:231–244

Lana PC, Guiss C (1992) Macrofauna-plant biomass interactions in a euhaline salt marsh of Paranaguá Bay (SE Brazil). Mar Ecol Prog Ser 80:57–64

Lana PC, Guiss C, Disaró ST (1991) Seasonal variation of biomass and production dynamics for above- and belowground components of a *Spartina alterniflora* marsh in a euhaline sector of Paranaguá Bay (SE Brazil). Estuar Coast Shelf Sci 32:231–241

Lana PC, Couto ECG, Almeida MVO (1997) Polychaete distribution and abundance in intertidal flats of Paranaguá Bay (SE Brazil). Bull Mar Sci 60(2):433–442

Lessa GC, Meyers SR, Marone E (1998) Holocene stratigraphy in the Paranaguá Bay Estuary, Southern Brazil. J Sed Res 68:1060–1076

Lopes RM (1997) Distribuição espacial, variação temporal e atividade alimentar do zooplâncton no complexo estuarino de Paranaguá. PhD Thesis, Universidade Federal do Paraná, Brazil

Machado EC, Daniel CB, Brandini N, Queiroz RLV (2000) Temporal and spatial dynamics of nutrients and particulate suspended matter in Paranaguá Bay, PR, Brazil. Nerítica 10 (in press)

Marone E, Camargo R (1994) Marés meteorológicas no litoral do estado do Paraná: o evento de 18 de agosto de 1993. Nerítica 8(1/2):73–85

Martin F (1992) Étude de l'écosystéme mangrove de la Baie de Paranaguá (Paraná, Bresil): analyse des impacts et propositions de gestion rationnelle. DSc Thesis, University of Paris VII, France

Montú M, Cordeiro TA (1988) Zooplancton del complejo estuarial de la Bahia de Paranaguá. I. Composición, dinámica de las especies, ritmos reproductivos y acción de los factores ambientales sobre la comunidad. Nerítica 3:61–83

Netto SA, Lana PC (1996) Benthic macrofauna of *Spartina alterniflora* marshes and nearby unvegetated tidal flats of Paranaguá Bay (SE Brazil). Nerítica 10:41–56

Netto SA, Lana PC (1997a) Influence of *Spartina alterniflora* on superficial sediment characteristics of tidal flats in Paranaguá Bay (Southeastern Brazil). Estuar Coast Shelf Sci 44:641–648

Netto SA, Lana PC (1997b) Intertidal zonation of benthic macrofauna in a subtropical salt marsh and nearby unvegetated flat (SE, Brazil). Hydrobiologia 353:171-186

Rebello J, Brandini FP (1990) Variação temporal de parâmetros hidrográficos e material particulado em suspensão em dois pontos fixos da Baia de Paranaguá, Paraná (junho/87-fevereiro/88). Nerítica 5:95-111

Rougeulle MD (1993) La crise de la pêche artisanale: transformation de l'espace et destructuration de l'activité – le cas de Guaraqueçaba (Paraná, Brésil). DSc Thesis, Université de Nantes, France

Sessegolo GC (1997) Estrutura e produção de serapilheira do manguezal do Rio Baguaçu, Baía de Paranaguá, PR. MSc Thesis, Universidade Federal do Paraná, Brazil

Sessegolo GC, Lana PC (1991) Decomposition of *Rhizophora mangle, Avicennia schaueriana* and *Laguncularia racemosa* in a mangrove of Paranaguá Bay (SE Brazil). Bot Mar 34:285-289

Soares CR (1990) Natureza dos sedimentos da superfície de fundo das Baías das Laranjeiras e de Guaraqueçaba – Complexo Estuarino da Baía de Paranaguá (Estado do Paraná, Brasil). MSc thesis, Universidade Estadual Paulista, Instituto de Geociências e Ciências Exatas, 137 pp

# 11 The Convergence Ecosystem in the Southwest Atlantic

C. ODEBRECHT and J.P. CASTELLO

11.1 Introduction

The Atlantic subtropical anticyclone and the antarctic circumpolar current drive the upper oceanographic circulation in the Southwest Atlantic. The large-scale flow field comprises the south and northward flowing western boundary currents, the Brazil and Malvinas (=Falkland) currents, respectively. The Brazil Current transports warm tropical water (TW) during most of the year towards the southern Brazilian shelf (Garcia 1997), while the coastal branch of the Malvinas Current carries cool subantarctic water (SAW) northwards. The confluence of Brazil and Malvinas currents represents part of the subtropical convergence and the mixture of TW and SAW between about 25°S and 45°S forms the South Atlantic central water (SACW). The two currents outline the eastern boundary of a large marine ecosystem in the SW Atlantic between 23°S (Cabo Frio) and 55°S (Patagonia; Bisbal 1995; Fig. 11.1), separating it from the South Atlantic ocean basin. The western boundary of the confluence approaches the southern Brazilian, Uruguayan, and northeastern Argentinean coast, though latitudinal displacement occurs due to variability of the circumpolar antarctic current (White and Peterson 1996). As a consequence, high variability of physico-chemical and biological attributes occurs at the shelf and slope system between approximately 30°S and 40°S due to the influence of horizontal, vertical, and seasonal mixing processes among TW, SACW, SAW, and large continental runoff. High biological production conveys to this system importance as a nursery, feeding, and reproduction area for fishery stocks of both subtropical and antarctic origin, which utilize the western boundary currents for long distance transport (Castello et al. 1997). To comprehend the ecological variability of this warm transitional system (Boltovskoy et al. 1996), however, oceanographic processes within the larger ecosystem limits (23°S to 55°S) need to be considered.

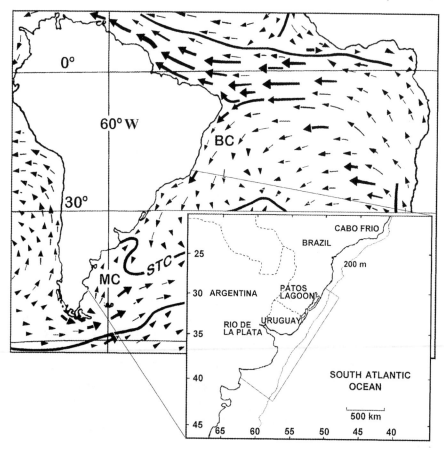

Fig. 11.1. South Atlantic oceanographic circulation with main flows of Brazil Current (*BC*), Malvinas Current (*MC*), and the Subtropical Convergence (*STC*). Detailed map of the large marine ecosystem in the SW Atlantic between 23°S and 46°S. *Shaded area* approximately corresponds to the warm transition zone ecosystem

11.2 Environmental Setting

The regional climate is largely controlled by the latitudinal migration of the Atlantic anticyclone high-pressure center and the passage of polar fronts. Monthly mean air temperatures in July (winter) and January (summer) vary between 18 and 24 °C in the north (30°S) and between 8 and 20 °C in the south (40°S; Hoffmann et al. 1997; Klein 1997). The dominant winds are NE, followed by W-SW winds; however, their path and frequency varies strongly from year to year, owing to polar front passages (6–10 day intervals). Precipitation is more frequent and higher in winter. Extreme rainfall or drought periods are associated with El Niño-southern oscillation phenomena (Seeliger et al. 1997).

The tides are amplified by SE onshore winds which increase southward (1 m at 25°S, 8 m at 45°S). Amphidromic conditions exist at 30–32°S, 41°S, and 47°S where oscillation periods of shelf water approach those of oceanic tides. Annual freshwater runoff (mean 20,000–25,000 m^3 s^{-1}) from the Río de la Plata drainage basin (3,170,000 km^2; Guerrero et al. 1997) varies considerably between seasons. During maximum runoff in fall/winter, offshore winds divide the outflowing water into a smaller, southward flowing branch and, owing to the Coriolis effect, into a larger branch along the Uruguayan coast. The northern branch mixes with freshwater runoff from the Patos Lagoon (annual mean 700–3,000 m^3 s^{-1}, maximum in winter/spring) over the southern Brazilian shelf (Möller et al. 1991). Owing to the dilution of modified SACW and SAW, the coastal waters (CW) below the low salinity surface layer are defined as subtropical shelf waters (STSW) and subantarctic shelf water (SASW) over the southern Brazilian and the Argentine shelf, respectively. STSW and SASW are separated by an oceanographic front (subtropical shelf front) at approximately 32°S. The front is an extension of the Brazil/Malvinas confluence over the shelf and runs in a north-south direction between the 50-m isobath and the shelf break (Piola et al. 2000).

11.3 Fertilization Processes

Continental runoff generates a low density and stable surface water layer, the estuarine plume (Fig. 11.2A), which extends for about 50 km (Patos Lagoon; Hartmann et al. 1986) and 240 km (Río de la Plata; Guerrero et al. 1997) offshore. The freshwater outflow influences shelf eutrophication via nutrient input and vertical stratification (sigma-t gradients up to 2.5 m^{-1}; Brandhorst and Castello 1971; Lima 1992). As a consequence, short-term and seasonal changes in freshwater volume and composition cause variability of chemical parameters over the shelf. Phosphate concentration (0.4–1.0 µM) is generally in the same range as in SAW and SACW, while average total suspended matter (>15 mg l^{-1}) and silicate (>20 µM) are highest in low density coastal water (Carreto et al. 1986; Niencheski and Fillmann 1997). Owing to in- or nearshore phytoplankton uptake (Abreu et al. 1995; Ciotti et al. 1995), outflowing freshwater tends to be poor in nitrate (<3.0 µM). Low N:P ratios (approx. 5:1) indicate that phytoplankton production is regulated by nitrogen flow, though nearshore bottom turbulence

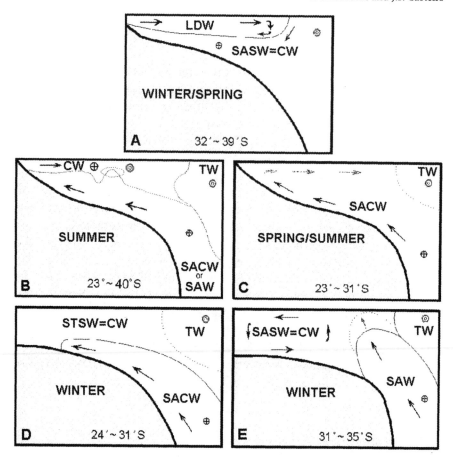

Fig. 11.2A–E. Main fertilization processes in the SW Atlantic. Low density water (*LDW*) of the winter-spring estuarine plume (**A**); summer vortex-driven shelf upwelling of SACW or SAW (**B**); spring-summer wind-driven coastal upwelling of SACW (**C**); small eddies and/or bottom topography-induced SACW (**D**) or SAW (**E**) shelf-break upwelling. (**B**, **D**, and **E** adapted from Matsuura 1996 and Lima et al. 1996)

(Odebrecht and Djurfeldt 1996) and water column nutrient recycling are additional nitrogen sources. The stabilization of the water column and concomitant nutrient enrichment of the euphotic surface layer lead to high chlorophyll-*a* concentrations and primary production rates during late winter and spring, though low incident light and high seston loads may limit phytoplankton production rates in winter. Chlorophyll-*a* concentration increases sharply at the estuarine front of the Patos Lagoon (>5 mg m^{-3}; Abreu et al. 1995) and the Río de la Plata (up to 16 mg m^{-3}; Negri et al. 1992). Since high chlorophyll-*a* concentration in shelf waters off southern Brazil coincided with large freshwater outflow in El Niño years (Ciotti et al. 1995), interannual differences seem to be controlled by continental runoff.

Upwelling of nutrient-rich SACW in northern and of SAW in southern areas between 23°S and 40°S is caused by cold cyclonic and warm anticyclonic vortices, owing to the wind regime, shelf and slope topography, and current shear between the Brazil Current and contiguous waters. In summer, dominance of NE winds causes divergence of coastal surface waters and the advection of deep SACW (10–20 °C, salinity 35–36) over the shelf between 23°S and 31°S (Fig. 11.2B,C; Ferreira da Silva et al. 1984; Matsuura 1986). SACW is the major source of dissolved nitrate (>20 µM) and N:P ratios of approximately 16:1 favor primary production. The size of production depends on the extension of SACW over the shelf and whether it reaches the euphotic zone. Ascending waters tend to reduce mixing depth in relation to euphotic depth, generating phytoplankton subsurface maxima and enhanced local primary productivity (Brandini 1990; Aidar et al. 1993; Odebrecht and Djurfeldt 1996). During strong NE winds (Fig. 11.2C), SACW enrichment reaches surface waters (Hubold 1980). Except for vortex-driven upwelling of SACW, the moderately stratified thermohaline inhibits primary production, owing to dominance of oligotrophic TW (silicate ≤5 µM, phosphate, nitrate, ammonium ≤1 µM) at the surface (Brandini 1990; Niencheski and Fillmann 1997). South of 38°S, fertilization at the shelf break front (up to 45 km wide) between shelf waters and the western edge of the Malvinas Current (Podestá 1997) sustains high phytoplankton biomass (>2 mg m^{-3}) during summer (Carreto et al. 1995). The horizontal shift of this front between outer and inner shelf regions during summer and spring/fall, respectively, depends on cyclical advance/reverse movements of the Malvinas Current.

In winter and spring, the shear between the Brazil Current and shelf waters causes wave-like meanders and vortices and leads to enhanced subsurface phytoplankton biomass along the slope (24–31°S), which is associated with nutrient-rich SACW upwelling (Fig. 11.2D; Matsuura 1986; Brandini 1990; Lima et al. 1996). Between 31°S and 35°S, the advection of northward flowing SAW at the shelf break results in a horizontal thermohaline gradient and horizontal current shear leads to the development of warm anticyclonic and cold cyclonic vortices, causing SAW upwelling (Fig. 11.2E). Under these conditions, shelf and slope waters become weakly stratified (Lima 1992). Primary production is enhanced when SAW (nitrate and silicate ~11.0 µM, ammonium 6.6 µM, phosphate 1.1 µM; Niencheski and Fillmann 1997) reaches the euphotic zone, resulting in subsurface chlorophyll-*a* nuclei and integrated chlorophyll-*a* concentrations above

100 mg m^{-2} (Ciotti et al. 1995). Further south (35–39.5°S), phytoplankton maxima off the slope are associated with cyclonic eddies and SAW upwelling (Gayoso and Podestá 1996).

11.4 The Organisms

11.4.1 Plankton

In surface waters (31–34°S; Abreu 1997) heterotrophic bacteria attached to particles (up to 5 × 10^5 cells ml^{-1}) are more abundant than free bacteria and total bacterial biomass (30–60 mg C m^{-3}) generally follows that of phytoplankton chlorophyll-*a* in coastal waters and/or under influence of SAW. Autotrophic picoplankton (<2 µm) comprises more than 50 % of low chlorophyll-*a* concentrations in oceanic surface water during spring, while highest numbers (10^8 cells l^{-1}) occur in subsurface coastal waters (Odebrecht and Djurfeldt 1996). Phytoplankters, principally diatoms (*Chaetoceros, Coscinodiscus, Lauderia, Leptocylindrus, Rhizosolenia, Skeletonema, Thalassiosira, Pseudo-nitzschia*), autotrophic (or mixotrophic) thecate dinoflagellates (*Ceratium, Dinophysis, Prorocentrum, Scrippsiella*), and naked organisms of the Gymnodiniales are the main primary producers (Odebrecht and Garcia 1997; Mendez et al. 1998). Potentially harmful and/or toxic species are dinoflagellates (*Dinophysis acuminata, Alexandrium tamarensis, Noctiluca scintillans, Gymnodinium catenatum, Gyrodinium* cf. *aureolum*), haptophycean (*Prymnesium, Chrysochromulina*) and raphidophycean flagellates (*Chatonella, Fibrocapsa, Heterosigma*) as well as diatoms (*Pseudo-nitzschia*). The autotrophic ciliate *Mesodinium rubrum* is common along the southern Brazilian and Uruguayan coast. Nanoflagellates share dominance with diatoms and dinoflagellates during high phytoplankton concentration (10^5–10^6 cells l^{-1}) in shelf and coastal waters under the influence of continental runoff, SAW and SACW. However, under the influence of TW of the Brazil Current nanoflagellates generally dominate in oceanic areas. Warm water phytoplankton species extend nearshore and abundance is usually low. Summer blooms of the cyanobacteria *Trichodesmium* occur throughout the area (23–35°S; Mendez et al. 1998; Rörig et al. 1998) and coccolithophorids are important in TW and at the wide summer shelf front south of 38°S (Brandini 1988; Podestá 1997).

Protozooplankters (heterotrophic flagellates, ciliates, and ameboids) appear to be an important component of the pelagic shelf ecosystem

(Odebrecht 1997). Cell concentration of large dinoflagellates (*Noctiluca scintillans, Polykrikos schwartzii, Protoperidinium* spp.) and of naked (*Strombidium*) and loricate (tintinnids) ciliates tend to coincide with phytoplankton concentration, higher values being common in coastal surface waters or under the influence of nutrient-rich SAW and SACW. In oligotrophic warm waters of the Brazil Current, small (<20 µm) naked ciliates are important.

Zooplankton (copepods, mysids, euphausiids, cladocerans, ostracods, chaetognaths, tunicates, medusae, siphonophores, polychaetes, mollusks) species composition and spatial and temporal distribution depend on the dominance of different water masses (Boltovskoy 1981; Montú et al. 1997). Copepods (up to 2,000 org. m^{-3}) constitute the most abundant group in neritic (*Acartia tonsa, Calanoides carinatus, Calanus australis, Centropages brachiatus, Ctenocalanus vanus, Euterpina acutifrons, Oithona nana, O. similis, Paracalanus parvus, P. quasimodo, Temora stylifera*) and oceanic (*Acartia danae, Calanus simillimus, Conaea rapax, Corycaeus furcifer, Oncaea venusta, Rhyncalanus nasutus, Sapphirina* spp.) waters (Montú et al. 1997). Mysids, like *Metamysidopsis elongata* and *Neomysis americana*, may constitute up to 90 % of the surf zone zooplankton. Adult euphausiids are more common in oceanic regions under the influence of SAW (*Euphausia longirostris, E. lucens, E. spinifera*) and TW (*Euphausia similis, E. hemigibba, E. recurva, Stylocheiron abbreviatum, S. affine, Thysanopoda* spp.; Brinton and Antezana 1981; Montú et al. 1997). Cladocerans (*Evadne nordmanni, Penilia avirostris, Pleopis polyphemoides, Podon intermedius, P. schmacheri, Pseudoevadne tergestina*) are mainly neritic with highest concentration and number of species in winter (Montú et al. 1997). Several species of chaetognaths are water masses indicators. *Sagitta tenuis* is a common coastal and shelf species, while the distribution of *S. enflata, S. serratodentata, S. minima, S. hispida*, and *Pterosagitta draco* extends over the shelf and slope (Boltovskoy 1981; Montú et al. 1997). Different species of appendicularian tunicates are abundant in nearshore (*Oikopleura dioica*), shelf and slope (*Oikopleura longicauda, O. fusiformis, O. rufescens, Fritillaria borealis, F. pellucida*), or in oceanic waters (*Fritillaria formica, Oikopleura albicans, O. cophocerca, Stegosoma magnum*; Montú et al. 1997). Thaliaceans, coelenterate medusae and siphonophores, and shelled planktonic pteropod mollusks are occasionally common.

11.4.2 Macrobenthic Invertebrates

The unstable conditions of the shallow inner shelf under influence of turbid Patos Lagoon freshwater runoff largely impede the colonization by species without pelagic life-cycle phases (Capítoli 1997). Beyond the direct influence of runoff, inner shelf sandy areas (10–15 m depth) are characterized by cumaceans, isopods, the penaeid *Artemesia longinaris*, and mollusks (*Dorsanum moniliferum, Olivancillaria urceus, Adelomelon brasiliana, Mactra isabelleana*). The mollusks *Encope emarginata, Buccinanops lamarcki, Loxopagurus loxochelis, Astropecten armatus, Adrana electa*, isopods, and polychaetes occupy deeper (15–25 m) areas with fine sand. Transitional sandy mud bottoms (25–50 m) between inner and intermediate shelf regions are typical of *Pitar rostratus, Corbula patagonica, Buccinanops gradatum, Mactra petiti, Dardanus arrosor, Persephona punctata, Hepatus pudibundus*, and ophiurids. *Phyllochaetopterus socialis, Astrangia rathbuni*, serpulid polychaetes, and *Crepidula protea* are conspicuous organisms on consolidated substrate and beach rock outcrops (Capítoli 1997). Further south (38°S), about 84 typical species occur on the inner shelf (<60 m depth) on hard (e.g., *Mytilus platensis, Lithophaga patagonica, Lumbrineris tetraura, Crepidula aculeata, Tegula patagonica, Diopatra viridis*) and soft substrate (*Nucula puelcha, Turbonilla uruguayensis, Olivella tehuelchana, Eunice argentinensis*; Roux et al. 1993).

The mollusk *Pitar rostratus*, polychaetes, amphipods, and ophiurids are abundant on muddy sand bottoms (50–90 m) of the intermediate shelf between 32°S and 34°S. The highest faunal abundance (500–1,000 ind. m^{-2}) is associated with extensive sandy mud bottoms of the intermediate shelf (80–130 m), with species like *Diopatra tridentata, Chasmocarcinus typicus, Terebellides* sp., *Onuphis tenuis, Astropecten cingulatus, Squilla brasiliensis*, and *Portunus spinicarpus*. In muddy bottoms (100–130 m) this assemblage is substituted by polychaetes (*Eupanthalis rudipalpa, Panthalis oerstedi, Spirochaetopterus* sp.) and sipunculids (*Nephasomma* sp.; Capítoli 1997). Eurybathic species (41; e.g., *Amphiura eugeniae, Idanthyrsus armatus, Philine argentina, Colpospirella algida, Hiatella solida, Limopsis hirtella, Natica isabelleana, Ninoe brasiliensis, Epicodakia falklandica, Kellia suborbicularis, Yoldia eightsii, Ophiacanta vivipara*) characterize bottoms at 38°S under the influence of SAW (Bastida et al. 1992; Roux et al. 1993).

At the outer shelf and slope (30–32°S) between 130 and 220 m depth, low abundance of gorgonians, echinoderms, isopods, barnacles, and serpulid polychaetes (100–500 ind. m^{-2}) coincides with biodetritus bottoms. Slope areas (200–500 m) have low density (<100 ind. m^{-2}) and different depths strata (200–300 m and 300–500 m) and substrate types appear to select for different species. *Eunice frauenfeldi, Zoanthidea* sp., and species of Porifera, Actinaria, and Ophiurida typically occupy consolidated substrates of upper slope regions. Lower slopes are characterized by *Acanella eburnea* and *Flabellum brasiliensis* (coral), *Ophiurolepis* sp. (ophiurid) and *Priapu-*

lus sp. are relatively constant components (Capítoli 1997). Further south (38°S), species (84) associated with SAW of the Malvinas Current (e.g., *Hiatella solida, Epicodakia falklandica, Pseudechinus magellanicus, Amphiura eugeniae, Chaetopterus variopedatus, Odontaster penicillatus, Austrocidaris canaliculata, Kenerleya patagonica, Trochodota purpurea*) characterize the outer shelf/slope (Bastida et al. 1992; Roux et al. 1993).

11.4.3 Fishes and Cephalopods

The warm- and cold-temperate elasmobranch and teleost fauna reveals a high degree of heterogeneity (López 1963; Figueiredo 1977, 1981; Figueiredo and Menezes 1978, 1980; Menezes and Figueiredo 1980, 1985; Menni and López 1984; Castello 1992; Cousseau and Perrotta 1998). Of the 57 sharks, rays, and skates (43 viviparous and 13 oviparous) over the southern Brazilian shelf and upper slope, 18 species are permanent residents (i.e., the dominant *Squatina guggenheim, Mustelus fasciatus, Rhinobatos horkelii, Zapteryx brevirostris, Raja castelnaui, R. agassizi, R. platana, R. cyclophora, Sympterigia acuta, Myliobatis* spp., *Gymnura altavela*) while another 18 species occur sporadically (Vooren 1997). Several species reproduce off Uruguay and Argentina during summer and migrate to southern Brazil in winter. Typical winter migrants are *Sympterigia bonaparte, Discopyge tschudii, Galeorhinus galeus, Mustelus schmitti, M. canis, Sphyrna zygaena, Eugomphodus taurus, Squalus megalops, S. mitskurii, S. acanthias, Notorhynchus cpedians,* and *Heptranchias perlo*, while summer migrants are represented by *Sphyrna lewini, Myliobatis freminvillei, Dasyatis say, D. centroura, Rhinoptera bonasus,* and *Narcine brasiliensis* (Vooren 1997). Among cephalopods (>40 species), *Loligo sanpaulensis* and *L. plei* are conspicuous demersal-pelagic species, the argentine shortfin squid *Illex argentinus* is common to the outer shelf and upper slope, and *Ornithoteuthis antillarum* and *Ommastrephes bartramii* are characteristic of epipelagic oceanic habitats. The octopods *Eledone gaucha, E. massyae,* and *Octopus tehuelchis* are abundant benthic species (Haimovici 1997a).

Typical migratory teleosts of the pelagic community are represented by the Argentine anchovy (*Engraulis anchoita*), mackerel (*Scomber japonicus*), Atlantic bonito (*Sarda sarda*), blue fish (*Pomatomus saltatrix*), and yellowtail amberjack (*Seriola lalandei*; Cousseau and Perrotta 1998). The community composition changes according to depth and with distance from the coast (Castello 1997). In nearshore areas to about 30 m depth, species composition (*Trachinoutus marginatus, Anchoa marinii, Lycengraulis grossidens, Brevoortia pectinata=B. aurea*, juveniles of *Engraulis anchoita*) changes seasonally. Shelf and shelf break areas (about 200 m depth) are typical of adult *Engraulis anchoita, Pomatomus saltatrix, Sarda sarda, Scomber japonicus, Trachurus lathami, Katsuwonus pelamis*, and three species of *Mugil*. Tuna fish and similar species, mostly yellowfin (*Thunnus albacares*) and swordfish (*Xiphias gladius*), pelagic sharks, and smaller oceanic fishes (lantern fish, *Maurolicus muelleri*, Micthophidae, Lampanyctinae) occupy the slope and deeper waters.

The demersal, demersal-pelagic, and demersal-benthic communities define different geographic assemblages (Angelescu and Prenski 1987; Cousseau and Perrotta 1998). Between 34°S and 41°S down to 50 m depth (Buenos Aires coastal assemblage) the white mouth croaker (*Micropogonias furnieri*), weakfisk (*Cynoscion guatucupa* = *C. striatus*) and the narrownose smooth-hound (*Mustelus schmitti*) are dominant species. Less abundant are black drum (*Pogonias cromis*), royal weakfish (*Macrodon ancylodon*), Argentine croaker (*Umbrina canosai*), "burriqueta" (*Menticirrhus americanus*), Atlantic sea robins (*Prionotus nudigula, P. punctatus*), marine catfish (*Netuma* sp.), red porgy (*Pagrus pagrus*), Brazilian codling (*Urophycis brasiliensis*), red mullet (*Mullus argentinae*), several flatfishes (*Paralichthys* spp.), and small sharks and rays. In southern Brazil this assemblage also includes Argentine conger (*Conger orbignyanus*), large head hairtail (*Trichiurus lepturus*), the sciaenids *Cynoscion jamaicensis* and *Ctenosciaena gracilicirrhus*, and horse mackerel (*Trachurus lathami*; Haimovici 1997b). Between 34°S and 48°S at 50–225 m depth (Buenos Aires and Patagonia inner and outer shelf assemblage) the dominant hake (*Merluccius hubbsi*), followed by pink cuskeel (*Genypterus blacodes*), spiny dogfish (*Squalus acanthias*), hawkfish (*Cheilodactylus bergi*), flounders, and notothenids are characteristic of SAW. The blackbelly rosefish (*Helicolenus dactilopterus lahillei*), conger (*Pseudoxenomystax albescens*), "viuda" (*Iluocoetes fimbriatus*), and six species of grenadiers (Macrouridae) are present in northern areas. In the southern part, at less than 150 m depth, the Brazilian sandperch (*Pseudopercis semifasciata*), Brazilian flathead (*Percophis brasiliensis*), elephant fish (*Callorhynchus callorynchus*), and the angel shark (*Squatina argentina*) occur. Finally, a cold water assemblage occurs along the slope between 220 and 2,300 m depth, with several species performing daily vertical feeding migrations. A gulf assemblage with reduced species number and density occurs between 42°S and 43°S, while an austral assemblage of the Patagonian-Fueguian and Malvinas shelf (up to 48°S and 200 m depth) is in the domain of SAW of the Malvinas Current.

A total of 3,193 and 3,300 species have been listed for coastal, shelf, and slope areas between 30°S and 34°S (Seeliger et al. 1997) and the adjacent Río de la Plata region (Mendez 1998), respectively. Although tropical waters tend to have higher biodiversity than cold waters, the elevated number of animal plankton species suggests high biodiversity for shelf and slope areas between 30°S and 40°S (Boltovskoy 1981). Also, diatom and silicoflagellate diversity (Shannon-Weaver index 3.4–4.6) in winter compares favorably with highly diverse marine ecosystems elsewhere (Lange and Mostajo 1985). The relative abundance and richness of demersal fishes off southern Brazil is influenced by seasonal bottom temperature due to the migration of the STC, while species richness over the shelf break seems to be related to bottom substrate, invertebrate fauna, and distinct water layers (Haimovici 1997b). Indications of increased biodiversity in this warm transition ecosystem may be partly due to long distance transport by both the coastal branch of the Malvinas Current (Stevenson et al. 1998) and the Brazil Current (Boltovskoy 1981).

11.5 Biological Production and Trophic Structure

The presence of distinct water masses over the SW Atlantic shelf and slope governs biological production and trophic interactions. Internal nutrient cycling is essential for the maintenance of biological production and trophic structure in oligotrophic TW with low chlorophyll-*a* (<0.5 mg m^{-3}) and particulate primary production (<1.0 mg C m^{-3} h^{-1}; Brandini 1990; Aidar et al. 1993; Ciotti et al. 1995) and high dissolved carbon (>20%; Vieira and Teixeira 1981) and regenerated production (ammonia, >60%; Metzler et al. 1997). Annual primary production rates are less than 50 g C m^{-2}. The importance of pico-nanoplankton (autotrophic and heterotrophic picoplankton, flagellates, and ciliates) based communities indicates that microbial processes are fundamental in structuring trophic interactions with long and complex food webs (Castello et al. 1997). Filter feeders, like the cladocera *Penilia avirostris* and Thaliaceae, are abundant and consume bacteria and picoplankton. Planktonic mollusks (pteropods) feed on small particles, phytoplankters, protozoans, and zooplankters, which adhere to their external mucus. Euphasiaceae present filter and raptorial feeding, while almost all coelenterates are carnivores and prey on plankters, small fishes, and/or nekton. Zooplanktophagous fishes (*Peprilus paru*, young *Trichiurus lepturus*, *Balistes capriscus*, *Cynoscion jamaicensis*) and the benthophagous *Ctenosciaenea gracilirrhus* arrive from the north, while other ichthyophagous and/or benthic feeding species (*Pomatomus saltatrix*, adult *Umbrina canosai*, *Cynoscion guatucupa*, *Micropogonias furnieri*, *Pagrus pagrus*, and *Mustelus schmitti*) migrate southward. In slope regions, dense schools of the skipjack tuna (*Katsuwonus pelamis*) feed heavily on macro-zooplankton (i.e., *Euphausia similis*), the mesopelagic fish *Maurolicus muelleri*, and on squid.

The input of nutrients (mainly nitrate) through fertilization processes to southern Brazilian shelf waters favors growth of large primary producers (microplankton), like single centric diatoms or chains of small-celled species, chlorophyll-*a* nuclei, and high productivity rates (160 g C m^{-2} year^{-1}; Odebrecht and Garcia 1997). Shorter and less complex food chains are characterized by the dominance of planktophagous pelagic fish, like *Engraulis anchoita*, and their predators (*Merluccius hubbsi*, *Trichiurus lepturus*, *Cynoscion guatucupa*, *Pagrus pagrus*, *Galeorhinus galeus*, and the skates *Sympterigia acuta* and *S. bonapartei*). Other zooplanktophagous species include squids (*Illex argentinus*, *Loligo sanpaulensis*) and fishes (*Anchoa marinii*, *Trachurus lathami*, juveniles of *Cynoscion guatucupa*, and *Trichiurus lepturus*). Over the outer shelf, pelagic zooplanktophagous species are represented by *Maurolicus muelleri*, mictophyds, gonastomatids, juvenile *Merluccius hubbsi*, *Brama* spp., *Loligo sanpaulensis*, *Ornithoteuthis antillarum*, and *Illex argentinus*, which also preys on small pelagic and mesopelagic fishes. Pelagic-benthic interactions are especially important over the intermediate shelf (40–60 m) where high chlorophyll-*a* concentrations under the influence of SAW in winter coincide with high demersal fish biomass (mainly sciaenids) in southern Brazilian waters (Haimovici 1997b).

11.6 Human Impacts

Demographic and industrial growth along the coast, fringing the warm transition zone ecosystem in the SW Atlantic Ocean, threatens coastal integrity. The pressure on fisheries stocks has deleterious effects on biodiversity and the sustainability of several activities. Coastal hazards, harmful algal blooms, and accidental oil and toxic spills have compounded these changes.

11.6.1 Fisheries

Fishing grounds between 23°S and 43°S represent about 75% of the total fishery production between northern Brazil and southern Argentina (FAO area 41). Today, the resources of these grounds are fully exploited (i.e., weakfish, royal weakfish, Argentine red shrimp, pink shrimp, all squids), overexploited (i.e., hake, white mouth croaker, Brazilian sardine, several demersal sharks species), or depleted (i.e., Argentine croaker, black drum, red porgy, marine catfish, school shark and guitarfish). Apart from Brazil, Uruguay, and Argentina, other countries operate under license agreement or in international waters outside the economic exclusive zone (EEZ). Catches have steadily increased at an average annual rate of 9% from less than 50×10^3 tons (1950) to a peak of $2,317 \times 10^3$ tons (1987), followed by a decline to $2,134 \times 10^3$ tons in 1994 (Fig. 11.3A). Hake (*Merluccius hubbsi*), the main demersal fishery resource, is exploited by Argentina, Uruguay and, in smaller quantities, by Brazil. After a peak of 521×10^3 tons in 1991, catches and catch per unit effort have declined considerably (Fig. 11.3B). Coastal migratory demersal resources (mainly sciaenids) are white mouth croaker (*Micropogonias furnieri*), Argentine croaker (*Umbrina canosai*), weakfishes (*Cynoscion* spp.), royal weakfish (*Macrodon ancylodon*), and redfishes and congers (Fig. 11.3C). Pelagic fisheries, mostly from southeastern and southern Brazil, contribute less than 9%. The main commer-

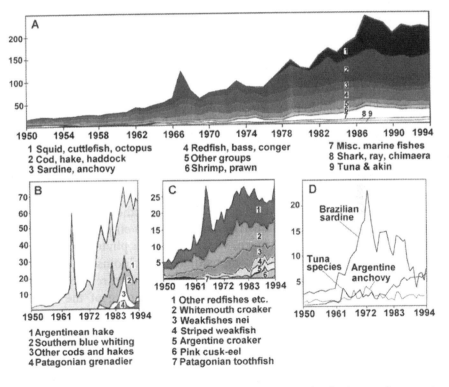

Fig. 11.3. Annual catch of A fishery resources in the SW Atlantic, B most important demersal fishery resources, C redfishes etc., and D Argentine anchovy, Brazilian sardine, and tuna species. (Adapted from FAO 1997)

cial species are the Brazilian sardine (*Sardinella brasiliensis*), skipjack (*Katsuwonus pelamis*), Argentine anchovy (*Engraulis anchoita*), albacore (*Thunnus alalunga*), bigeye tuna (*T. obesus*), swordfish (*Xiphias gladius*) and yellowfin tuna (*T. albacares*; Fig. 11.3D). Fisheries of an almost virgin stock of Argentine anchovy (average 25×10^3 tons off the northern coast of Argentina) and skipjack have potential for expansion. Smaller yields of traditional bony fish fisheries have increased the fishing pressure on elasmobranchs (bottom and pelagic sharks and rays). Today, elasmobranchs (mainly *Galeorhinus galeus*, *Mustelus schmitti*, *Prionace glauca*; Vooren 1997) represent about 10% ($49–55 \times 10^3$ tons between 1994 and 1998) of demersal catches in southern Brazil and 3.2% of the total catch in Argentina (Bonfil 1994). Squid (*Illex argentinus*, *Loligo* sp.) are fished in large quantities (1989 $>700 \times 10^3$ tons, 1995 560×10^3 tons) by Argentinean and foreign fishing boats. Shrimps (*Artemesia longinaris*, *Pleoticus muelleri*, *Farfantepenaeus paulensis*, *F. brasiliensis*) are coastal fishery resources exploited by industrial fisheries in Brazil and Argentina and artisanal fisheries in Uruguay (Fig. 11.3A).

11.6.2 Pollution and Blooms

Potentially hazardous materials reach the shelf and slope ecosystem through direct discharge from coastal urban and industrial centers, via continental freshwater runoff, or as a result of marine accidents. Owing to natural background levels (Niencheski et al. 1994; Seeliger and Costa 1997), except for copper and lead, metal contribution from the Patos Lagoon is likely to be negligible, though mercury discharge from the Río de la Plata (Kurucz et al. 1998) is of concern. Agrotoxics from extensive agricultural lands in the Río de la Plata and Patos Lagoon watershed will ultimately reach offshore waters and biota. Furthermore, large-scale agricultural activities and industrial and domestic effluents add excess nutrient loads, especially nitrogen, which increases phytoplankton production and changes the species composition. Algal blooms are a common feature off southern Brazil, Uruguay, and Argentina and red-colored surface patches have been registered since the beginning of the century. Harmful dinoflagellate (*Alexandrium tamarense*, *Gymnodinium catenatum*, *Dinophysis acuminata*) blooms cause paralythic and diarrhetic shellfish poisoning (Mendez et al. 1998). The potentially deleterious effect of increased UV radiation during the austral spring up to 30°S (Seeliger and Costa 1997) is likely to modify abundance, productivity, and trophic dynamics of the system. Red tide dinoflagellates of stable, high irradiance environments tend to be more tolerant to UV photoinhibition (Carreto et al.

1996), and thus might become more frequent in the SW Atlantic if ozone layer degradation persists. Finally, the handling of ship ballast water is likely to introduce toxic microalgae cysts and exotic species of seaweed, fish, crustaceans, polychaetes, and mollusks to this region.

11.7 Management Considerations

In southern Brazil, the annual onboard discard of small demersal fishes and elasmobranchs ($17-25 \times 10^3$ tons) during pair-, otter-, and double-ring trawling amounts to about 36% of the landed catch (Haimovici 1998). Long-lived elasmobranch species with low reproduction rates are easily impacted by growing fishing pressure, thus the absence of fishery regulations presents management and conservation problems for this resource (Bonfil 1994). In Argentina, total allowable catch (TAC) quotas have recently been applied to hake fisheries and fishing effort on the shortfin squid (*Illex argentinus*) is being controlled (FAO 1997). Closed seasons, limited boat registration and size, and control of nominal fishing effort are also being applied to other fisheries. However, an open access regime still prevails for most artisanal and small scale fisheries and poor or lack of enforcement has contributed to overexploitation and depletion of resources. To address the growing concern about the state of fish stocks and fisheries in the region, Argentina and Uruguay established a common fishing zone and a joint technical commission for the Río de la Plata Maritime Front in the early 1970s which promotes scientific meetings and joint technical reports. In Brazil, surveys of southeastern and southern pelagic and demersal stocks (Living Resources of the Exclusive Economic Zone, REVIZEE) will permit the implementation of a more realistic and scientific-based fishery policy. However, despite shared fishery resources among the three coastal countries, research cooperation and common management efforts are lacking.

The governments of Brazil, Uruguay, and Argentina have distinct coastal zone management programs and policies and the absence of comparable information on status and trends of the ecosystem has been an obstacle. In Brazil, the National Coastal Management Program (GERCO) has significantly advanced during the last decade. In Uruguay, programs for Sustainable Development of the Coastal Zone of the Río de la Plata (ECOPLATA) and Biodiversity Conservation and Sustainable Development of the Eastern Wetlands (PROBIDES) support management and restoration of coastal ecosystems. In Argentina, conservation initiatives have led to increased support for a national coastal policy. All three governments acknowledge the necessity for common marine environmental policies, standards, and incentives for compliance.

References

Abreu PC (1997) Bacterioplankton. In: Seeliger U, Odebrecht C, Castello JP (eds) Subtropical convergence environments. The coast and sea in the southwestern Atlantic. Springer, Berlin Heidelberg New York, pp 104-105

Abreu PC, Granéli HW, Odebrecht C (1995) Produção fitoplanctônica e bacteriana na região da pluma estuarina da Lagoa dos Patos, RS, Brasil. Atlantica, Rio Grande 17:35-52

Aidar E, Gaeta SA, Gianesella-Galvão SMF, Kutner MBB, Teixeira C (1993) Ecossistema costeiro subtropical: nutrientes dissolvidos, fitoplâncton e clorofila a e suas relações com as condições oceanográficas na região de Ubatuba, SP. Publ Esp Inst Oceanogr S Paulo 10:9-43

Angelescu V, Prenski B (1987) Ecología trófica de la merluza común del Mar Argentino (Merlucciidae, *Merluccius hubbsi*). 2. Dinámica de la alimentación analizada sobre la base de las condicones ambientales, la estructura y las evaluaciones de los efectivos en su área de distribución. Contr INIDEP, Mar del Plata 561:1-205

Bastida R, Roux A, Martínez DE (1992) Benthic communities of the Argentine continental shelf. Oceanol Acta 15:687-698

Bisbal GA (1995) The southeast South American shelf large marine ecosystem. Mar Policy 19:21-38

Boltovskoy D (1981) Características biológicas del Atlántico Sudoccidental. In: Boltovskoy D (ed) Atlas del zooplancton del Atlántico Sudoccidental y métodos de trabajo con el zooplancton marino. Publ Esp INIDEP Mar del Plata, pp 239-251

Boltovskoy E, Boltovskoy D, Correa N, Brandini FP (1996) Planktic foraminifera from the SW Atlantic (30o-60oS): species-specific patterns in the upper 50 m. Mar Micropaleont 28:53-72

Bonfil R (1994) Overview of world elasmobranch fisheries. FAO Fish Tec Pap 341

Brandhorst W, Castello JP (1971) Evaluación de los recursos de anchoita (*Engraulis anchoita*)frente a la Argentina y Uruguay I: las condiciones oceanográficas, sinopsis del conocimiento actual sobre la anchoita y el plan para su evaluación. Proy Des Pesq Ser Inf Tec, Mar del Plata 29:1-63

Brandini FP (1988) Composição e distribuição do fitoplâncton na região sul do Brasil e suas relações com as massas de água (julho-agosto 1982). Ciência Cult, S Paulo 40:334-341

Brandini FP (1990) Hydrography and characteristics of the phytoplankton in shelf and oceanic waters off southeastern Brazil during winter (July/August 1982) and summer (February/March 1984). Hydrobiologia 196:111-148

Brinton E, Antezana T (1981) Euphausiacea. In: Boltovskoy D (ed) Atlas del zooplancton del Atlántico Sudoccidental y métodos de trabajo con el zooplancton marino. Publ Esp INIDEP, Mar del Plata, pp 681-698

Capítoli RR (1997) Continental shelf benthos. In: Seeliger U, Odebrecht C, Castello JP (eds) Subtropical convergence environments. The coast and sea in the southwestern Atlantic. Springer, Berlin Heidelberg New York, pp 117-120

Carreto JI, Negri RM, Benavides HR (1986) Algunas caracteristicas del florecimiento del fitoplancton en el frente del Río de la Plata I: los sistemas nutritivos. Rev Invest Des Pesq Mar del Plata 5:7-29

Carreto JI, Lutz VA, Carignan MO, Colleoni ADC, De Marcos SG (1995) Hydrography and chlorophyll *a* in a transect from the coast to the shelf-break in the Argentinian Sea. Cont Shelf Res 15:315–336

Carreto JI, Benavides HR, Carignan MO, Negri RM, Akselman R (1996) Photosynthetic response of natural phytoplankton populations to environmental ultraviolet radiation. In: Yasumoto T, Oshima Y, Fukuyo Y (eds) Harmful and toxic algal blooms. Intergovernmental Oceanographic Commission of UNESCO, Paris, pp 325–328

Castello JP (1992) Necton. In: Diagnóstico ambiental costeiro e oceânico da região Sul e Sudeste do Brasil, vol 5. FUNDESPA, São Paulo

Castello JP (1997) Pelagic Teleosts. In: Seeliger U, Odebrecht C, Castello JP (eds) Subtropical convergence environments. The coast and sea in the southwestern Atlantic. Springer, Berlin Heidelberg New York, pp 123–128

Castello JP, Haimovici M, Odebrecht C, Vooren CM (1997) The continental shelf and slope. In: Seeliger U, Odebrecht C, Castello JP (eds) Subtropical convergence environments. The coast and sea in the southwestern Atlantic. Springer, Berlin Heidelberg New York, pp 171–178

Ciotti AM, Odebrecht C, Fillmann G, Moller OO Jr (1995) Freshwater outflow and subtropical convergence influence on phytoplankton biomass on the southern Brazilian continental shelf. Cont Shelf Res 15:1737–1756

Cousseau MB, Perrotta RG (1998) Peces marinos de Argentina. Biología, distribución, pesca. INIDEP, Mar del Plata

FAO (1997) Review of the state of the world fishery resources: marine fisheries. FAO Fish Cir 920

Ferreira da Silva LC, Albuquerque CAM, Cavalheiro WW, Hansen CMP (1984) Gabarito tentativo para as massas de água da costa Sudeste brasileira. Anais Hidrográgicos Niterói 41:261–312

Figueiredo JL (1977) Manual de peixes marinhos do sudeste do Brasil. I. Introdução, cações, raias e quimeras. Museu de Zoologia, S Paulo

Figueiredo JL (1981) Estudo das distribuições endêmicas de pixes da Província Zoogeográfica Argentina. PhD Thesis, Universidade de São Paulo, Brazil

Figueiredo JL, Menezes NA (1978) Manual de peixes marinhos do sudeste do Brasil. II. Teleostei (1). Museu de Zoologia, S Paulo

Figueiredo JL, Menezes NA (1980) Manual de peixes marinhos do sudeste do Brasil. III. Teleostei (2). Museu de Zoologia, S Paulo

Garcia CE (1997) Physical oceanography. In: Seeliger U, Odebrecht C, Castello JP (eds) Subtropical convergence environments. The coast and sea in the southwestern Atlantic. Springer, Berlin Heidelberg New York, pp 94–96

Gayoso AM, Podestá G (1996) Surface hydrography and phytoplankton of the Brazil-Malvinas currents confluence. J Plankt Res 18:941–951

Guerrero RA, Acha EM, Framiñan MB, Lasta CA (1997) Physical oceanography of the Río de la Plata estuary, Argentina. Cont Shelf Res 17:727–742

Haimovici M (1997a) Demersal and benthic teleosts. In: Seeliger U, Odebrecht C, Castello JP (eds) Subtropical convergence environments. The coast and sea in the southwestern Atlantic. Springer, Berlin Heidelberg New York, pp 129–136

Haimovici M (1997b) Cephalopods. In: Seeliger U, Odebrecht C, Castello JP (eds) Subtropical convergence environments. The coast and sea in the southwestern Atlantic. Springer, Berlin Heidelberg New York, pp 146–150

Haimovici M (1998) Present state and perspectives for the southern Brazil shelf demersal fisheries. Fish Manage Ecol 5:277–289

Hartmann C, Sano EE, Paz RS, Möller OO Jr (1986) Avaliação de um período de cheia (junho de 1984) na região sul da Laguna dos Patos, através de dados de sensoriamento remoto, meteorológicos e oceanográficos. Anais IV Simp Bras Sens Remoto, Gramado, pp 685-694

Hoffmann JAJ, Núñez MN, Piccolo MC (1997) Características climáticas del Océano Atlántico Sudoccidental. In: Boschi E (ed) El mar argentino y sus recursos pesqueros, vol 1. INIDEP, Mar del Plata 1:163-193

Hubold G (1980) Hydrography and plankton off southern Brazil and Rio de la Plata, Aug-Nov 1977. Atlantica, Rio Grande 4:1-21

Klein AHF (1997) Regional climate. In: Seeliger U, Odebrecht C, Castello JP (eds) Subtropical convergence environments. The coast and sea in the southwestern Atlantic. Springer, Berlin Heidelberg New York, pp 5-7

Kurucz A, Masello A, Méndez S, Cranston R, Well s P (1998) Calidad ambiental del Río de la Plata. In: Wells PG, Daborn GR (eds) The Río de la Plata, an environmental overview. An EcoPlata Project Background Report. Dalhousie University Halifax, pp 71-86

Lange CB, Mostajo EL (1985) Phytoplankton (diatoms and silicoflagellates) from the SW Atlantic Ocean. Bot Mar 28:469-476

Lima ID (1992) Distribuição e abundância de anchoita (*Engraulis anchoita*) em relação aos processos oceanográficos na plataforma continental do sul do Brasil. MSc Thesis, Universidade do Rio Grande, Brazil

Lima ID, Garcia AE, Moller OO Jr (1996) Ocean surface processes on the southern Brazilian shelf: characterization and seasonal variability. Cont Shelf Res 16:1307-1317

López RB (1963) Peces Marinos de la República Argentina. Ser Evaluación de los Recursos Naturales de la Argentina. CFI, Buenos Aires, pp 105-219

Matsuura Y (1986) Contribuição ao estudo da estrutura oceanográfica da região Sudeste entre Cabo Frio (RJ) e Cabo de Santa Marta Grande (SC). Ciência Cult S Paulo 38:1439-1450

Matsuura Y (1996) A probable cause of recruitment failure of the brazilian sardine *Sardinella aurita* population during the 1974/75 spawning season. S Afr J Mar Sci 17:29-35

Mendez S (1998) Reports from ETI branches: ETI Montevideo. ETI Partners Newsletter Amsterdam 5:4

Mendez S, Gómez M, Ferrari G (1998) Estudios planctonicos del Río de la Plata y su frente maritimo. In: Wells PG, Daborn GR (eds) The Río de la Plata, an environmental overview. An EcoPlata Project Background Report. Dalhousie University, Halifax, pp 87-115

Menezes NA, Figueiredo JL (1980) Manual de peixes marinhos do sudeste do Brasil. IV. Teleostei (3). Museu de Zoologia, S Paulo

Menezes NA, Figueiredo JL (1985) Manual de peixes marinhos do sudeste do Brasil. V. Teleostei (4). Museu de Zoologia, S Paulo

Menni RC, López HL (1984) Distribution patterns of Argentine marine fishes. Phys B Aires 42:71-85

Metzler PM, Glibert PM, Gaeta SA, Ludlam JM (1997) New and regenerated production in the South Atlantic off Brazil. Deep Sea Res 44:363-384

Möller OO Jr, Paim PSG, Soares ID (1991) Facteurs et mechanismes de la circulation des eaux dans léstuarie de la Lagune dos Patos (RS, Bresil). Bull Inst Geol Basin Aquitaine Bordeaux 49:15-21

Montú MA, Gloeden IM, Duarte AK, Resgalla C Jr (1997) Zooplankton. In: Seeliger U, Odebrecht C, Castello JP (eds) Subtropical convergence environments. The coast and sea in the southwestern Atlantic. Springer, Berlin Heidelberg New York, pp 110-114

Negri RM, Carreto JI, Benavides HR, Akselman R, Lutz VA (1992) An unusual bloom of Gyrodinium cf. aureolum in the Argentine sea: community structure and conditioning factors. J Plankt Res 14:261-269

Niencheski LF, Fillmann G (1997) Chemical characteristics. In: Seeliger U, Odebrecht C, Castello JP (eds) Subtropical convergence environments. The coast and sea in the SW Atlantic. Springer, Berlin Heidelberg New York, Berlin, pp 96-98

Niencheski LF, Windom HL, Smith R (1994) Distribution of particulate trace metal in Patos Lagoon estuary (Brazil). Mar Pollut Bull 28:96-102

Odebrecht C (1997) Protozooplankton. In: Seeliger U, Odebrecht C, Castello JP (eds) Subtropical convergence environments. The coast and sea in the SW Atlantic. Springer, Berlin Heidelberg New York, p 109

Odebrecht C, Djurfeldt L (1996) The role of nearshore mixing on the phytoplankton size structure off Cape Santa Marta Grande, southern Brazil (spring 1989). Arch Fish Mar Res 43:13-26

Odebrecht C, Garcia VM (1997) Phytoplankton. In: Seeliger U, Odebrecht C, Castello JP (eds) Subtropical convergence environments. The coast and sea in the southwestern Atlantic. Springer, Berlin Heidelberg New York, pp 105-109

Piola AR, Campos EJD, Möller OO Jr, Charo M, Martinez C (2000) The subtropical shelf front off eastern South America. J Geophys Res 105(C3):6565-6578

Podestá GP (1997) Utilizaciíon de datos satelitarios en investigaciones oceanográficas y pesqueras en el Océano Atlántico SW. In: Boschi E (ed) El Mar Argentino y sus Recursos Pesqueros I: antecedentes históricos de las exploraciones en el mar y las características ambientales. INIDEP, Mar del Plata, pp 195-222

Rörig LR, Yunes JS, Kuroshima K, Schetinni C, Pezzuto PR, Proença LA (1998) Studies on the ecology and toxicity of *Trichodesmium* spp. blooms in southern Brazilian coastal waters. In: Reguera B, Blanco J, Fernández ML, Wyatt T (eds) Harmful algae. Xunta de Galicia and Intergovernmental Oceanographic Commission of UNESCO, pp 22-25

Roux A, Bastida R, Bremec C (1993) Comunidades bentónicas de la plataforma continental argentina. Campañas transección BIP Öca Balda 1987/88/89. Bolm Inst Oceanogr S Paulo 41:81-94

Seeliger U, Costa CSB (1997) Natural and human impact. In: Seeliger U, Odebrecht C, Castello JP (eds) Subtropical convergence environments. The coast and sea in the southwestern Atlantic. Springer, Berlin Heidelberg New York, pp 197-203

Seeliger U, Odebrecht C, Castello JP (1997) Subtropical convergence environments. The coast and sea in the southwestern Atlantic. Springer, Berlin Heidelberg New York

Stevenson MR, Dias-Brito D, Stech JL, Kampel M (1998) How do cold water biota arrive in a tropical bay near Rio de Janeiro, Brazil? Cont Shelf Res 18:1595-1612

Vieira AAH, Teixeira C (1981) Excreção de matéria orgânica dissolvida por populações fitoplanctônicas da costa leste e sudeste do brasil. Bolm Inst Oceanogr S Paulo 30:9-25

Vooren CM (1997) Demersal elasmobranchs. In: Seeliger U, Odebrecht C, Castello JP (eds) Subtropical convergence environments. The coast and sea in the southwestern Atlantic. Springer, Berlin Heidelberg New York, pp 141-146

White WB, Peterson RG (1996) An Antarctic circumpolar wave in surface pressure, wind, temperature and sea ice extent. Nature 380:699-702

12 The Patos Lagoon Estuary, Brazil

U. SEELIGER

12.1 Introduction

The dominant feature of the southern Brazilian coastal plain is the Patos-Mirim Lagoon complex, the Patos Lagoon (10,360 km^2) being the world's largest choked lagoon. About 80% of the lagoon is a freshwater system but at its southern reach (32°S), where the lagoon connects to the Atlantic Ocean, brackish waters and fringing marshes comprise an estuarine ecosystem of approximately 1,000 km^2 (Fig. 12.1). The estuary performs critical interface functions between the limnic areas of the Patos Lagoon and the shelf waters of Subtropical Convergence regions in the southwestern Atlantic. The passage of continental freshwater runoff from the drainage basin (~200,000 km^2) through the estuary and the intrusion of seawater into it establish the physico-chemical characteristics which control ecological processes and vital habitats for the successful reproduction and/or growth of many commercially important marine species which visit the estuary periodically.

12.2 Environmental Setting

As is characteristic of choked lagoons, the deep (15 m) and narrow (800 m) inlet of the Patos Lagoon acts as a set-down filter and strongly attenuates the advance of tidal waves (0.47 m) into the estuary, thus estuarine water levels and currents are closely linked to regional precipitation and wind patterns. The dominant northeasterly winds (mean 5 m s^{-1}) promote flushing of the estuary, owing principally to the formation of a north-south pressure gradient inside the lagoon and to the retreat of nearshore coastal water. Especially in winter, the entire system becomes freshwater during prolonged periods, when runoff volumes and velocities in the inlet may reach 22,000 m^3 s^{-1} and 1.9 m s^{-1}, respectively. In contrast, the southerly

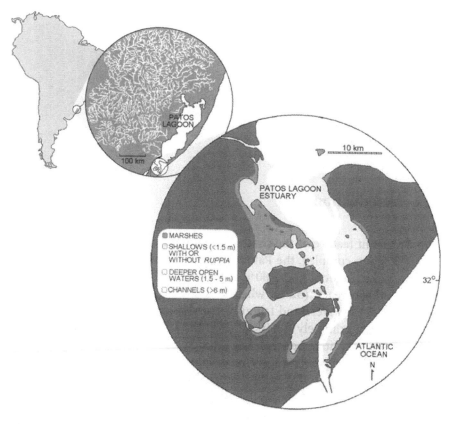

Fig. 12.1. Geographic location and major habitats of the Patos Lagoon estuary

winds (mean 8 m s^{-1}) of polar cold front passages force seawater into the estuary and into the lower lagoon, which, during drought periods in summer and fall, results in extended periods of saltwater residence (Möller et al. 1996; Garcia 1997). The average annual freshwater flow rates through the estuary are about 4,000 m^3 s^{-1} though much higher values (up to 10,000 m^3 s^{-1}) may occur during El Niño years.

As do hydrodynamic characteristics in the estuary, salinity patterns also relate to wind forcing and hydrology on scales from days to months. Most of the nutrients which are added to the upper Patos Lagoon by fluvial runoff suffer substantial reduction during transport before reaching the estuary; however, the shallow (80% <1.5 m) depth of the estuary promotes the natural deposition of large amounts of suspended matter. Sporadic elevated dissolved nitrogen (>30 μM), phosphate (1–3 μM) and silicate (~175 μM) concentrations are the result of local domestic/industrial effluents or due to remobilization of bottom sediments by inflowing seawater and strong winds (Niencheski and Baumgarten 1997).

12.3 Estuarine Habitats

The size and the physiographic heterogeneity of the Patos Lagoon estuary, with marshes, shoals (<1.5 m), deeper open waters (1.5–5 m) and channels (>6 m), provide for extensive and diverse habitats (Fig. 12.1). Many of the marine invertebrates and fishes in the southwestern Atlantic depend on estuarine habitats as nurseries during part of their life cycle while others, like marine mammals and birds, find feeding, roosting and breeding grounds in the estuary.

12.3.1 The Water Column

The meteorologic processes which control seawater intrusion are also responsible for the transport of phyto- and zooplankton and the eggs and larvae of fish into the estuary, thus marked seasonal and annual changes in the estuarine planktonic community composition are common. Organisms which are constant and often abundant components of the plankton community are typically represented by phytoplanktonic nanoflagellates (<20 µM), the copepod *Acartia tonsa*, and eggs and larvae of Atherinidae (*Odonthestes argentinensis, Atherinella brasiliensis*), Soleidae (*Achirus garmanii*), Ariidae (*Genidens genidens*), Jenynsiidae, Blenniidae, Gobiidae, and Gobiesocidae. During intense runoff in winter, oligohaline diatoms (*Skeletonema subsalsum* 10^6 cells l^{-1}), cladocerans, freshwater copepods, and eggs and larvae of some limnic fish (i.e., *Parapimelodus valencienis*) are introduced into the estuary and may become an important part of the plankton community. In contrast, after periods of seawater intrusion, euryhaline diatoms (*Skeletonema costatum, Cerataulina daemon*), marine copepods, and larvae of polychaetes, mollusks, and crustaceans become the dominant components of an increasingly diverse plankton community. Elevated water temperatures and extended periods of saltwater intrusion during summer introduce neritic marine cladocerans (i.e., ctenophores, siphonophores, medusae, chaetognats), coastal mysids, dinoflagellates, and large numbers (10^5 cells l^{-1}) of marine diatoms (*Chaetoceros, Rhizosolenia, Coscinodiscus, Odontella*), which leads to high plankton community diversity in the estuary (Montú et al. 1997; Odebrecht and Abreu 1997). Intense spawning activities of many coastal and oceanic fish species in warm nearshore waters also coincide with increased abundance of their eggs and larvae in the estuarine plankton community. Nevertheless, most of the eggs (88%) and larvae (66%) belong to the Clupeidae (i.e., *Brevoortia pectinata*), Engraulididae (*Lycengraulis grossidens*), and to the sciaenid *Micropogonias furnieri*, emphasizing the importance of the estuary as a nursery ground especially for these species (Sinque and Muelbert 1997).

The migration of more than 100 marine and some freshwater and anadromic fish species into the Patos Lagoon estuary optimizes the abundance of their populations and hence may significantly contribute to the biological production of the region. The anadromic species *Netuma barba* and *N. planifrons* arrive from the sea to reproduce in pre-limnic zones, after which the juveniles return to the sheltered conditions and abundant food of estuarine nursery grounds. Also, the juveniles and sub-adults of marine species, like the black drum *Pogonias cromis*, the white croacker *Micropogonias furnieri*, mullets (*Mugil* spp.), and the flatfish *Paralichthys orbygnianus*, are obligatory users of estuary nursery grounds; however, the early life-cycle stages of most marine species are facultative or they are opportunistic migrators or only enter the estuary occasionally (Chao et al. 1985). The juveniles of estuarine-dependent marine species and estuarine-resident Atherinidae are most common in shallow waters (<1.5 m), whilst the juveniles and/or sub-adults of bottom-oriented epibenthic and demersal species (i.e., *Netuma barba*, *Micropogonias furnieri*, *Macrodon ancylodon*, *Menticirrhus americanus*, *Paralonchurus brasiliensis*, *Cynoscion guatucupa*, *Umbrina canosai*), many of which comprise an important part of the southern Brazilian landings, are the dominant ichthyofaunal components of deeper open waters (>4 m). Pelagic species are largely represented by young individuals of *Lycengraulis grossidens*, *Anchoa marinii*, *Engraulis anchoita*, *Trichiurus lepturus*, *Peprilus paru*, *Selene setapinis*, *Brevoortia pectinata*, *Ramnogaster arcuata*, and *Platanichthys platana* (Vieira et al. 1998). Since top predators are scarce, the spatial and temporal distribution patterns of the ichthyofauna are largely a function of salinity distributions and/or food competition, though bottlenose dolphins (*Tursiops truncatus*) and sea lions (*Otaria flavescens*) frequently enter the estuary in search of food.

12.3.2 Unvegetated Subtidal Soft-Bottoms and Intertidal Flats

The soft-bottom of extensive subtidal and intertidal habitats in the estuary is comprised of unconsolidated sediments which vary with hydrodynamic energy levels from silty sand in shallow shoals to silty clay in deep channels (Calliari 1997). The benthic community changes with depth and according to environmental characteristics. The deep channels (>6 m) have low macrobenthic abundance and diversity but are an important route for many migratory organisms. The subtidal soft-bottoms between 1.5- and 5-m depths offer spatial and trophic niches for dense populations of the pelecipod *Erodona mactroides* and the gastropod *Heleobia australis*. Changes in population density of these species are related to season and depth because their predators (i.e., highly motile adult crabs and fish) tend to migrate to deeper subtidal bottoms at the end of fall/winter. *Erodona mactroides* continues as the largest macrobenthic biomass in shallower subtidal bottoms (<1.5 m), but occurs together with patches of the burrowing pelecipod *Tagelus plebeius*. These shallower areas tend to be covered by epibenthic microalgae, large quantities of detrital matter, and occasionally by aggregations of macroalgae, all of which offer protection for small sedentary peracarid crustaceans (i.e., the tanaid *Tanais stanfordi*, the amphipods *Melita mangrovi, Ampithoe ramondi, Leptocheirus* sp., the isopods *Dies fluminensis, Munna peterseni*). Furthermore, the abundant food resources constitute ideal nurseries for juvenile shrimps (*Farfantepenaeus paulensis, Palaemonetes argentinensis*), highly motile decapods (*Callinectes sapidus, Cyrtograpsus angulatus*), and estuarine-resident and estuarine-dependent fish. Owing to the dominance of burrowing species, the intertidal flats appear to be faunistically poor. However, they are an important habitat for peracarids, pelecipods, and infaunal polychaetes (*Laeonereis acuta, Nephtys fluviatilis, Heteromastus similis*). The juvenile stages of polychaetes in superficial sediment strata escape intense predation by decapods and fish in upper intertidal flats where predators have limited access or by recruitment during winter when predators abandon shallow waters, while the adults escape predation by moving to deeper sediment strata where the tube-dwelling peracarid *Kalliapseudes schubartii* becomes the dominant species (Bemvenuti 1997a, 1998).

12.3.3 Sea Grass Beds

Large areas of the shallow subtidal region (>1.5 m depth) are colonized by *Ruppia maritima* beds. The upper growth of the populations is controlled by air exposure (>20% per year), while underwater irradiance (mean monthly Secchi depth 23–78 cm) defines the lower growth limit where, owing to depth-related reproductive strategies, flowers and fruits are absent (Costa and Seeliger 1989; Seeliger 1997a). The leaves and shoots of *Ruppia* provide an important semi-permanent substrate for epiphytic and epifaunal organisms. Colonization by the diatom *Cocconeis placentula* initiates in the spring and, at the end of the summer, older leaf apices are completely covered by diatoms (i.e., *Synedra, Amphora, Nitzschia, Pleurosira laevis, Melosira, Navicula, Rhopalodia, Mastogloia*) and macroalgae (i.e., *Achrochaetium, Cladophora, Enteromorpha*; Ferreira and Seeliger 1985). *Ruppia* beds also promote the entanglement of drifting macroalgae, which may temporarily form an associated habitat of potential importance. Since drift-algae increase the drag force of waves and currents, plants become more susceptible to dislodging, which causes the thinning of beds during fall. The leaves, stems, rhizomes, and roots of *Ruppia* beds form a structurally complex habitat of calm water and stable substrate which tends to support high in- and epifaunal biomass (Garcia and Vieira 1997). *Ruppia* beds provide protection and offer food for early life-cycle stages of *Farfantepenaeus paulensis, Callinectes sapidus, Cyrtograpsus angulatus*, as well as for some species of the Sciaenidae, Mugilidae, and Atherinidae; however, the adults of these are also common in unvegetated shoals. The conditions for faunal recruitment are often suboptimal because the initiation of growth, density of plants, and the permanence and total area of *Ruppia* beds tends to vary significantly within and between years. As a consequence, the function of the beds as a nursery habitat for the development of important estuarine fisheries stock may be severely limited during some years (Seeliger 1997b).

12.3.4 Marginal Marshes

The extensive marshes of the estuary are irregularly flooded by waters of varying salinity, with oligohaline waters prevailing during the winter and spring and mesohaline conditions being more common in the summer (Costa 1997, 1998). The interstitial waters of rarely flooded upper marsh areas tend to be less saline than those of frequently inundated lower marshes, owing to rapid leaching by rain. The spatially and temporally heterogeneous marsh physiography has a pronounced influence on the diversity, abundance, and distribution of species as well as on the fate of organic matter production. About 50 % of the annual production of marsh plant litter undergoes autolysis/leaching and microbial decay in the sediments, while the larger part of the production and up to 20 % of the annual detritus pool is exported into the estuary during periods of prolonged flooding. The high plant biomass and detrital matter concentrations in the marshes attract in- and epifaunal macroinvertebrates. Detritivorous nematodes, annelids (i.e., *Heteromastus similis, Laeonereis acuta, Nephtys fluviatilis*), and gastropods (i.e., *Heleobia australis*) are common inhabitants in superficial sediments and terrestrial isopods (*Balloniscus* spp.), amphipods (*Orchestia platensis*), spiders, and insects are abundant throughout the marsh. In mid- and lower marshes the crabs *Metasesarma rubripes* and *Chasmagnathus granulata* hide between *Spartina alterniflora*, *Spartina densiflora*, and *Scirpus maritimus* plants or in burrows, respectively, and compete for space and food. Both species act upon the fragmentation and remobilization of the below-ground macrophyte biomass, thus decisively influencing the recycling of organic matter in the marshes. The dense marsh vegetation also offers potential breeding sites for birds, like the red-gartered coot (*Fulica armillata*), the black-necked swan *Cygnus melancorhyphus*, the snowy egret (*Egretta thula*), the green heron (*Butorides striatus*), as well as for shorebirds like the brown-hooded gull *Larus maculipennis* and Trudeau's tern *Sterna trudeaui* (Costa 1998).

12.3.5 Artificial Hard Substrates

The overwhelming dominance of unconsolidated sediments in the estuary reinforces the importance even small areas of permanent substrate may assume for colonization and development of encrusting organisms. Apart from estuarine salinity gradients, substrate characteristics clearly divide the local benthic algal flora into species with permanent substrate requirements growing on the rocky jetties at the inlet and those inside the estuary which tolerate unstable substrates (Seeliger 1998). The lack of permanent substrate, reduced tidal oscillation and turbid waters inside the estuary account for an encrusting community of low diversity. Here the dominant encrusting organism is the cirriped *Balanus improvisus* which settles on wooden pilings and piers together with a few tolerant species of algae (i.e., *Gomontia lignicola*, *Polysiphonia* sp., *Ulvaria oxysperma*). The dense layer of *Balanus* continuously hatches, liberating large quantities of larvae ($12{,}473\,\text{m}^{-3}$) into the water column which are food for plankton feeders or re-settle on available substrates. The density of *Balanus* eventually decreases, owing to crowding and to predation by *Stylochus* sp., and new niches open for colonization by the polychaete *Boccardia hamata*, the amphipod *Amphithoe ramondi*, the tanaiid *Tanais stanfordi*, and larvae of *Obelia*. In contrast, the rocky jetties at the inlet provide suitable conditions for a diverse encrusting community (Capítoli 1997). Furthermore, they also assume a biogeographic significance because they represent the only permanent substrate for northern tropical and southern cold temperate organisms along 700 km of sandy beaches in the southwestern Atlantic (30–34°S; Coutinho and Seeliger 1986).

12.4 Energy Flow

12.4.1 Primary Production Cycles

The unpredictable amplitude and timing and pronounced inter-annual variability of physico-chemical conditions in the estuary strongly influence primary production cycles of microalgae, benthic macroalgae, and submersed and emersed macrophytes. Phytoplankton biomass and production cycles are controlled by irradiance and temperature regimes and dinoflagellates (i.e., *Prorocentrum minimum*, *Peridinium quinquecorne*), diatoms (i.e., *Chaetoceros*, *Rhizosolenia*, *Coscinodiscus*, *Odontella*), and the autotrophic ciliate *Mesodinium rubrum* tend to reach maximum abundance (10^4–10^5 cells l^{-1}) during elevated temperatures and salinity (>20) in the summer/fall. Following frequent seawater intrusion in spring/summer, euryhaline diatoms (i.e., *Skeletonema costatum*, *Cerataulina daemon*) reach peak production (160–350 mg C m^{-3} h^{-1}) and high biomass accumulation (10^6–10^9 cells l^{-1}) during prolonged water residence. Lowest primary production (2–5 mg C m^{-3} h^{-1}) coincides with water temperatures below 20 °C. Especially in winter, when the mean water column irradiance of highly turbid waters falls below the critical light intensity (40 gcal cm^{-2} day^{-1}), net production becomes negative and heterotrophic processes prevail. The mean annual biomass values (0.07 g C m^{-2}, mean 0.03–7.0 g C m^{-2}) and production rates (72 mg C m^{-2} h^{-1}, mean 4–320 mg C m^{-2} h^{-1}) vary significantly between years, owing to differences in runoff volume (Abreu et al. 1994; Seeliger et al. 1997).

In the shoals, the frequent resuspension of benthic centric (*Melosira* spp., *Pleurosira laevis*) and pennate diatoms (*Cylindrotheca closterium*, *Bacillaria paradoxa*, *Terpsinoe americana*, *Surirella* spp.) by currents and winds leads to biomass concentrations (6–22 g C m^{-2}) which substantially exceed those of phytoplankton in deeper water. The peak abundance of cells occurs during the winter, because benthic microalgae appear to be inhibited by high incident light after resuspension in the summer. Benthic macroalgal biomass formation in the estuary is limited by the lack of suitable substrate; however, as long as suspended drift-algae remain in the photic zone of shallow waters, elevated biomass might be formed. The biomass and production peaks of the most prominent green (*Enteromorpha* sp., *Rhizoclonium riparium*, *Ulothrix flacca*) and blue-green (*Lyngbya confervoides*, *Microcoleus chthomoplastes*) algal producers and the Xantophyta *Vaucheria longicaulis* are linked to light and temperature. The highest mean monthly biomass of *Enteromorpha* sp. (34.3 g C m^{-2}), *Rhizoclonium riparium* (308 g C m^{-2}), *Ulothrix flacca* (2.2 g C m^{-2}), and *Vau-*

cheria longicaulis (70.4 g C m^{-2}) occurs in the winter and early spring, whilst biomass peaks of *Lyngbya confervoides* (55.7 g C m^{-2}) and *Microcoleus chthomoplastes* (19.8 g C m^{-2}) correspond to the summer and fall, respectively. Other blue-green algae (i.e., *Gomphosphaeria*, *Chroococuus*) are probably responsible for levels of high carbon fixation in the marshes (Coutinho and Seeliger 1986; Seeliger et al. 1997).

At optimal temperature (15 °C) and salinity the growth of *Ruppia maritima* beds commences from seed germination in spring (Koch and Seeliger 1988). Biomass peaks may attain up to 83 g C m^{-2} (30 % below-ground biomass) after the formation of reproductive shoots in the summer, though large variations between beds are common. Despite favorable light, temperature, and salinity regimes for perennial growth in the estuary, annual growth cycles of *Ruppia* with massive die-off in fall are common, owing to exposure of populations at the end of summer and/or increased sediment dynamics. Since the size and density of *Ruppia* beds and the turnover of leaves influence epiphyte colonization and growth, epiphyte biomass and production is highly variable; however, maximum epiphyte biomass may occasionally represent more than 50 % of the total *Ruppia* biomass.

The dominant primary producers in lower marshes (64 % flooding), mid-marshes (20 % flooding), and mesohaline transition zones between mid- and lower marshes are *Spartina alterniflora*, *Spartina densiflora*, and *Scirpus maritimus*, respectively. Their combined total mean net aboveground production (288–808 g C m^{-2} year^{-1}) is comparable to highly productive *Spartina alterniflora* marshes elsewhere. The net below-ground production of *Spartina alterniflora* (7289–2095 g C m^{-2} year^{-1}) and *Scirpus maritimus* (1510–2283 g C m^{-2} year^{-1}) represents more than 70 % of their total net primary production. The seasonal production cycles are likely to be a function of air temperature optima. Rapid growth of the C$_3$ plant *Scirpus maritimus* proceeds at low temperatures (<15 °C) in early spring and highest live above-ground biomass and maximum dead/total aerial biomass is reached at the end of spring and at the end of summer, respectively. In contrast, maximum growth of C$_4$ plants (*S. densiflora, S. alterniflora*) occurs above 20 °C at the end of spring, while highest dead and total aboveground biomass are reached in the early fall and during fall/winter, respectively (Costa 1997; Seeliger et al. 1997).

Conservative estimates of net primary production, based on minimum and maximum biomass difference, suggest that up to 86 % of the annual autochthonous carbon in the estuary is produced by dominant marsh plants and macrobenthic and blue-green algae. Although the total annual primary production in the Patos Lagoon estuary may considerably vary between years, owing to changes in environmental conditions, sequential pulses of carbon fixation by some producers and constant fixation by

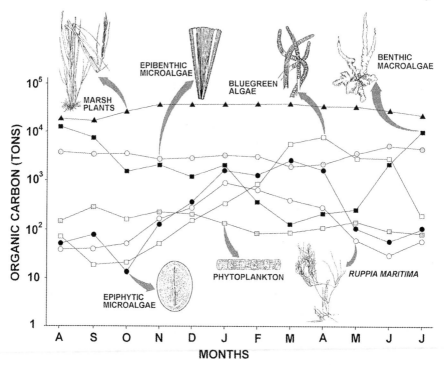

Fig. 12.2. Total monthly carbon contribution by different primary producers in the Patos Lagoon estuary. (Modified from Seeliger et al. 1997)

others (Fig. 12.2) assure continuous availability of live and detrital matter at different stages of decomposition.

12.4.2 Trophic Relations

Despite the considerable food diversity provided by primary producer organisms, only few consumers appear to be true grazers, like the red-gartered coot *Fulica armillata* and the black-necked swan *Cygnus melancorhyphus* on *Ruppia*. Most resident and migrating consumer organisms depend on a continuous supply of detrital matter as the essential energy source (Bemvenuti 1997b, 1998; Fig. 12.3).

The abundant organic detritus of soft-bottoms comprises the principal food source for deposit-feeding meiobenthic organisms, in- and epifaunal macrobenthic invertebrates, and some fish of the first consumer level. Especially meiobenthic nematodes, ostracods, and turbellarians are important intermediates of the detritus food chain, owing to their abun-

The Patos Lagoon Estuary, Brazil

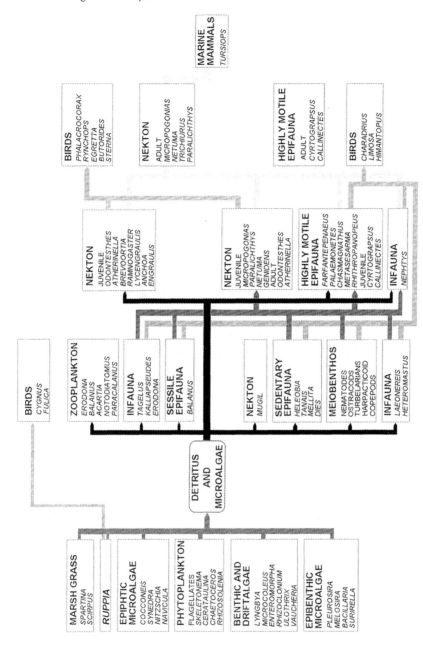

Fig. 12.3. Conceptual flow diagram of trophic relation among dominant biotic components in the Patos Lagoon estuary

dance and high population turnover rates. Most ingest organic particles in superficial sediments, but some nematodes also feed in deeper sediments and therefore facilitate energy flow to the upper aerobic layers. Macrobenthic infaunal polychaetes (i.e., *Laeonereis acuta*), the pelecipod *Tagelus plebeius*, and the peracarid *Kalliapseudes schubartii* ingest a mixture of detritus particles and epibenthic microalgae of surface layers, whilst the burrowing polychaete *Heteromastus similis* feeds on organic matter in deeper (<15 cm) sediments. In general, the low infaunal diversity accounts for abbreviated food chains though predation by the infaunal polychaete *Nephtys fluviatilis* on *Heteromastus similis* acts as an important intermediate link for epifaunal macropredators. The macrobenthic epifaunal deposit-feeding gastropods (i.e., *Heleobia australis*) and peracarids (i.e., the tanaid *Tanais stanfordii*, the amphipod *Mellita mangrovi*, the isopod *Dies fluminensis*) ingest detritus, diatoms and occasionally copepods in superficial sediments. Among the fishes, primary consumers are represented by juveniles of *Mugil curema*, *M. gaimardianus*, and monospecific schools of adult *M. platanus*, which ingest large quantities of detrital matter together with epibenthic and epiphytic diatoms (Vieira et al. 1998).

Decapods and fishes typically represent second-level consumer species. However, the omnivoruous decapods are able to act at different trophic levels because they display ontogenetic changes in diet preference and/or opportunistically exploit food sources as they become available. The early life stages of *Cyrtograpsus angulatus*, *Callinectes sapidus*, *Farfantepenaeus paulensis*, *Chasmagnathus granulata*, and *Metasesarma rubripes* largely feed on detrital matter and on epibenthic and epiphytic microalgae. The juveniles and sub-adults of *Callinectes sapidus* and *Cyrtograpsus angulatus* and the crab *Rhithropanopeus harrissi* prey on meiobenthic and small in- and epifaunal organisms, which represents an important link between detrital matter and higher trophic levels. Larger individuals of *Chasmagnathus granulata*, *Metasesarma rubripes*, as well as juveniles of *Micropogonias furnieri* (Sciaenidae) and adults of *Odontesthes argentinensis* and *Atherinella brasiliensis* (Atherinidae) selectively prey on meiobenthic ostracods, nematodes, in- and epifaunal organisms, small decapods and fishes. Especially epifaunal macrobenthic peracarids are an important food source and provide a measure of predation by secondary consumers, owing to their abundance and exposure (Bemvenuti 1997b). Many primary and secondary consumers of shallow shoals are preyed upon by wading birds, like the common stilt (*Himantopus himantopus*), the hudsonian godwit (*Limosa haemastica*), and the rufous-chested dotterel (*Charadrius modestus*) (Fig. 12.3).

In deeper waters the abundant *Erodona mactroides* represents the most important infaunal phytoplankton grazer. The larvae of both *Erodona mactroides* and the cirriped *Balanus improvisus* are the principal food source for plankton-feeding juveniles of *Atherinella brasiliensis* and *Odontesthes argentinensis* in shallow waters while demersal (*Brevoortia pectinata, Ramnogaster arcuata*) and pelagic (*Lycengraulis grossidens, Anchoa marinii, Engraulis anchoita*) fish are prominent plankton-feeders in deeper waters. Several of the deposit-, suspension-, and plankton-feeding secondary consumers are preyed upon by large individuals of *Callinectes sapidus*, adults of larger fish (i.e., *Micropogonias furnieri, Netuma barba, N. planifrons*) and by piscivorous birds, like the snowy egret (*Egretta thula*), the green heron (*Butorides striatus*), the common bigua cormorant (*Phalacrocorax olivaceus*), the common tern (*Sterna hirundo*), and the black skimmer (*Rynchops niger*). Although large predators are scarce in the estuary, *Micropogonias furnieri, Trichiurus lepturus*, and *Mugil* spp. are also preyed upon by the bottlenose dolphin *Tursiops truncatus* (Vieira et al. 1998; Fig. 12.3).

12.5 Estuary-Coast Interactions

The different estuarine habitats interface among each other and with adjacent coastal waters through processes of production, transport of matter and organisms, and migrations. Short-term and seasonal patterns of the exchange of water and organisms between the estuary and coastal nearshore areas tend to stabilize ecological processes and sustain the region's considerable secondary production. The export of inorganic nutrients and detritus at different stages of decomposition represents an important energy source for primary producers and a trophic link for coastal grazing and detrital food chains (Abreu and Castello 1997). The enrichment of coastal waters depends partly on the seeding of the estuary during saltwater intrusion with neritic diatoms (i.e., *Skeletonema costatum*) and their subsequent export as particulate organic carbon. In general, freshwater runoff favors the occurrence of dense and large plankton patches in coastal waters. Planktonic eggs and larvae (i.e., anchovy *Engraulis anchoita*) have a better chance of survival because the freshwater lens stabilizes the water column, thus inhibiting their dispersal and warranting access to food. Furthermore, the deposition of organic matter during runoff supplies food for filter- and suspension-feeding invertebrates and epibenthic and demersal fish and therefore sustains a diverse and abundant benthic community (Capítoli 1998).

Extreme events of large-scale and long-term climatic and oceanographic forcing represent a major perturbation for the stability of ecological processes in the estuary and coastal waters. Years of El Niño-Southern Oscillation (ENSO) cause extreme precipitation in the watershed. The elevated runoff extends the freshwater plume up to 50 km over the continental shelf and a concomitant increase of chlorophyll-*a* concentrations in coastal water masses is directly related to eutrophication by high fluvial nutrient input. The enhanced production processes are likely to modify the ecological dynamics of the entire coastal region. Furthermore, elevated runoff together with NE winds causes high current velocities in the inlet. Under these conditions, the recruitment of many ecologically and commercially important species into the estuary is impeded, resulting in a significant negative relationship between high spring runoff and low secondary production in summer/fall (Castello and Möller 1978). In contrast, during La Niña drought years, southerly winds may force seawater more than 200 km upstream into the Patos Lagoon. The protracted salinization (Costa et al. 1988) of brackish and freshwater areas tends to interfere with critical life-cycle stages of limnic, estuary-resident, and estuary-dependent species and may lead to salinity-dependent mortalities. The disruptive effects of extreme episodic events on ecological processes in both the estuary and coastal nearshore waters can only be assessed by long-term ecological studies.

12.6 Impact and Management

The ecological and socio-economic importance of the Patos Lagoon estuary excels among the coastal regions in the southwestern Atlantic and, despite early accounts of man's interference (von Ihering 1885), the estuary appears to have satisfied ecological and human needs without management for more than a century. However, during recent decades ever increasing demands of society have gradually altered natural runoff processes, changed the quality of water, reduced vital habitats, and exhausted the seemingly limitless natural resources. Additionally, rising sea levels and the advance of ozone-poor Antarctic air over southern Brazil (Santee et al. 1995) may affect ecological processes and pose a future hazard to the Patos Lagoon estuary ecosystem (Seeliger and Costa 1997).

Over the last 20 years, the growing demand for water for domestic and industrial use and for irrigated rice cultivation has significantly modified average annual (109 km^3) freshwater flow through the inlet. Especially during drought years with naturally reduced (75 km^3) runoff, the diversion (>13%) of freshwater resources for human use may negatively affect the balance of salinity and nutrients in the estuary and the export of organic matter to coastal waters (Abreu and Castello 1997). Natural runoff also controls the quality of water in the estuary. High nutrient loads are added to the lagoon during overflow of deficient sewage treatment facilities and from large-scale agricultural activities. The eutrophication of limnic lagoon areas induces changes in the phytoplankton composition and causes blooms (10^9 cells l^{-1}) of the toxic blue-green alga *Microcystis aeruginosa*. Sporadic increases of suspended copper and lead concentrations in the estuary may reflect metal input from industrial effluents and mining activities in the drainage basin, though metal concentrations generally correspond to natural background levels. Furthermore, natural runoff introduces large amounts of sediments into the estuary. Although dredging (approx. 30,000 m^3 month^{-1}) may solve immediate navigation problems in the inlet and main shipping channel, the accumulative effect of dredging has modified circulation patterns and deposition processes in the estuary and in coastal nearshore areas. Over the last two centuries estuarine water areas have decreased by more than 11%, as a result of increased sediment deposition after the construction of the inlet jetties in 1917 and the disposal of berth material from the expansion of the Rio Grande port. Dredging and filling has also destroyed as much as 10% of valuable estuarine salt marshes and has impacted vital *Ruppia maritima* habitats, owing to direct burial or to reduced depth distribution following increased water turbidity (Seeliger and Costa 1997).

For more than a century, trammel-, gill-, and channel net fisheries provided the socio-economic basis for many artisanal fishermen who benefited from the migration of crustaceans and fish in and out of the estuary. The introduction of synthetic fiber nets, motor-powered boats and modern means of transport and storage after 1945 significantly increased fishing efforts, and heavy exploitation during the 1970s caused the collapse of estuarine stocks of the black drum *Pogonias cromis*, the white croaker *Micropogonias furnieri* and the catfish *Netuma* spp. In contrast, annual mean landings (3,106 t) of *Farfantepenaeus paulensis* over a 20-year period indicate that the excessive catch of juvenile shrimps does not lead to the collapse of their fisheries because estuarine stocks depend principally on recruitment and growth conditions rather than on fishing pressure. Regrettably, prerequisites for the effective management of fishing activities in the estuary, such as a clear understanding of the migration of stocks and reliable landing statistics, have generally been neglected in favor of immediate financial profits and fishing efforts have not decreased (Haimovici et al. 1998).

Historically, socio-economic interests tend to collide with ecological needs. Since the present use of the estuary already conflicts with existing management recommendations, long-term research and multi-faceted management strategies must be an issue of priority in order to assess the role of natural and man-mediated processes and to predict the impact of global changes on the Patos Lagoon estuary.

References

Abreu PC, Castello JP (1997) Estuarine-marine interactions. In: Seeliger U, Odebrecht C, Castello JP (eds) Subtropical convergence environments: the coast and sea in the southwestern Atlantic. Springer, Berlin Heidelberg New York, pp 179-182

Abreu PC, Odebrecht C, González A (1994) Particulate and dissolved phytoplankton production of the Patos Lagoon estuary, southern Brazil: comparison of methods and influencing factors. J Plankt Res 16:737-753

Bemvenuti CE (1997a) Benthic invertebrates. In: Seeliger U, Odebrecht C, Castello JP (eds) Subtropical convergence environments: the coast and sea in the southwestern Atlantic. Springer, Berlin Heidelberg New York, pp 43-46

Bemvenuti CE (1997b) Trophic structure. In: Seeliger U, Odebrecht C, Castello JP (eds) Subtropical convergence environments: the coast and sea in the southwestern Atlantic. Springer, Berlin Heidelberg New York, pp 70-73

Bemvenuti CE (1998) Fundos não vegetados. In: Seeliger U, Odebrecht C, Castello JP (eds) Os ecossistemas costeiro e marinho do extremo sul do Brasil. Editora Ecoscientia, Rio Grande Brazil, pp 87-91

Calliari LJ (1997) Geologic setting. In: Seeliger U, Odebrecht C, Castello JP (eds) Subtropical convergence environments: the coast and sea in the southwestern Atlantic. Springer, Berlin Heidelberg New York, pp 13-18

Capítoli RR (1997) Rubble structures and hard substrates. In: Seeliger U, Odebrecht C, Castello JP (eds) Subtropical convergence environments: the coast and sea in the southwestern Atlantic. Springer, Berlin Heidelberg New York, pp 86-89

Capítoli RR (1998) Bentos da plataforma continental. In: Seeliger U, Odebrecht C, Castello JP (eds) Os ecossistemas costeiro e marinho do extremo sul do Brasil. Editora Ecoscientia, Rio Grande Brazil, pp 131-134

Castello JP, Möller OO Jr (1978) On the relationship between rainfall and shrimp production in the estuary of the Patos Lagoon (Rio Grande do Sul, Brazil). Atlantica (Rio Grande) 3:67-74

Chao LN, Pereira LE, Vieira JP (1985) Estuarine fish community of the dos Patos Lagoon, Brazil: a baseline study. In: Yáñez-Arancibia A (ed) Fish community ecology in estuaries and coastal lagoons: towards an ecosystem integration. Univ Nac Aut Mexico Press, Mexico, pp 429-450

Costa CSB (1997) Tidal marsh and wetland plants. In: Seeliger U, Odebrecht C, Castello JP (eds) Subtropical convergence environments: the coast and sea in the southwestern Atlantic. Springer, Berlin Heidelberg New York, pp 24-26

Costa CSB (1998) Marismas irregularmente alagadas. In: Seeliger U, Odebrecht C, Castello JP (eds) Os ecossistemas costeiro e marinho do extremo sul do Brasil. Editora Ecoscientia, Rio Grande, pp 82-86

Costa CSB, Seeliger U (1989) Vertical distribution and biomass allocation of *Ruppia maritima* L. in a southern Brazilian estuary. Aquat Bot 33:123-129

Costa CSB, Seeliger U, Kinas PG (1988) The effect of wind velocity and direction on the salinity regime in the Patos Lagoon estuary. Ciencia Cult (S Paulo) 40(9):909-912

Coutinho R, Seeliger U (1986) Seasonal occurrence and growth of benthic algae in the Patos Lagoon estuary, Brazil. Estuar Coast Shelf Sci 23:889–900

Ferreira S, Seeliger U (1985) The colonization process of algal epiphytes on *Ruppia maritima* L. Bot Mar 28:245–249

Garcia CAE (1997) Hydrographic characteristics. In: Seeliger U, Odebrecht C, Castello JP (eds) Subtropical convergence environments: the coast and sea in the southwestern Atlantic. Springer, Berlin Heidelberg New York, pp 18–20

Garcia AM, Vieira JP (1997) Abundancia e diversidade da assembleia de peixes dentro e fora de uma pradaria de *Ruppia maritima* L. no estuario da Lagoa dos Latos (RS, Brasil). Atlantica (Rio Grande) 19:161–181

Haimovici M, Castello JP, Vooren CM (1998) Pescarias. In: Seeliger U, Odebrecht C, Castello JP (eds) Os ecossistemas costeiro e marinho do extremo sul do Brasil. Editora Ecoscientia, Rio Grande Brazil, pp 205–218

Koch EW, Seeliger U (1988) Germination ecology of two *Ruppia maritima* L. populations in southern Brazil. Aquat Bot 31:321–327

Möller OO Jr, Lorenzzetti JA, Stech JL, Mata MM (1996) Patos Lagoon summertime circulation and dynamics. Cont Shelf Res 16(3):335–351

Montú M, Duarte AK, Gloeden IM (1997) Zooplankton. In: Seeliger U, Odebrecht C, Castello JP (eds) Subtropical convergence environments: the coast and sea in the southwestern Atlantic. Springer, Berlin Heidelberg New York, pp 40–43

Niencheski LF, Baumgarten MG (1997) Environmental chemistry. In: Seeliger U, Odebrecht C, Castello JP (eds) Subtropical convergence environments: the coast and sea in the southwestern Atlantic. Springer, Berlin Heidelberg New York, pp 20–23

Odebrecht C, Abreu PC (1997) Microalgae. In: Seeliger U, Odebrecht C, Castello JP (eds) Subtropical convergence environments: the coast and sea in the southwestern Atlantic. Springer, Berlin Heidelberg New York, pp 34–37

Santee ML, Read WG, Waters JW, Froidevaux GL, Manney GL, Flower DA, Jarnot RF, Harwood RS, Peckham GE (1995) Interhemispheric differences in polar stratospheric HNO_3, H_2O, ClO, and O_3. Science 267:849–852

Seeliger U (1997a) Submersed spermatophytes. In: Seeliger U, Odebrecht C, Castello JP (eds) Subtropical convergence environments: the coast and sea in the southwestern Atlantic. Springer, Berlin Heidelberg New York, pp 27–30

Seeliger U (1997b) Sea grass meadows. In: Seeliger U, Odebrecht C, Castello JP (eds) Subtropical convergence environments: the coast and sea in the southwestern Atlantic. Springer, Berlin Heidelberg New York, pp 82–85

Seeliger U (1998) Macroalgas bentônicas. In: Seeliger U, Odebrecht C, Castello JP (eds) Os ecossistemas costeiro e marinho do extremo sul do Brasil. Editora Ecoscientia, Rio Grande Brazil, pp 32–35

Seeliger U, Costa CSB (1997) Natural and human impact. In: Seeliger U, Odebrecht C, Castello JP (eds) Subtropical convergence environments: the coast and sea in the southwestern Atlantic. Springer, Berlin Heidelberg New York, pp 197–203

Seeliger U, Costa CSB, Abreu PC (1997) Primary production cycles. In: Seeliger U, Odebrecht C, Castello JP (eds) Subtropical convergence environments: the coast and sea in the southwestern Atlantic. Springer, Berlin Heidelberg New York, pp 65–70

Sinque C, Muelbert JH (1997) Ichthyoplankton. In: Seeliger U, Odebrecht C, Castello JP (eds) Subtropical convergence environments: the coast and sea in the southwestern Atlantic. Springer, Berlin Heidelberg New York, pp 51–56

Vieira JP, Castello JP, Pereira LE (1998) Ictiofauna. In: Seeliger U, Odebrecht C, Castello JP (eds) Os ecossistemas costeiro e marinho do extremo sul do Brasil. Editora Ecoscientia, Rio Grande Brazil, pp 60–67

von Ihering H (1885) Die Lagoa dos Patos. Dtsch Geogr Bl 8:182–204

13 The Río de la Plata Estuary, Argentina-Uruguay

H. Mianzan, C. Lasta, E. Acha, R. Guerrero, G. Macchi, and C. Bremec

13.1 Introduction

The Río de la Plata is an extensive and shallow coastal plain estuary on the western South Atlantic coast (35–36°S). The estuary receives freshwater from South America's second largest basin (about 3.2 million km^2) through the Paraná River, one of the longest in the world, its main tributary, the Paraguay River, and the Uruguay River. The funnel of the Río de la Plata estuary extends for over 280 km from the head (25 km wide), at the confluence of the Paraná and Uruguay rivers, to the 230-km-wide mouth between Punta Rasa and Punta del Este. A submerged shoal, the Barra del Indio, represents a geomorphological barrier and divides the Río de la Plata estuary into an inner and an outer system (Fig. 13.1). The inner fluvial system under strong tidal influence, with a depth between 1 and 5 m, is about 180 km long and up to 80 km wide and extends over 13,000 km^2. In the outer mixohaline brackish system with an area of about 22,000 km^2, both depths (5–25 m) and section width increase concomitantly (Fig. 13.1). However, the mixohaline area increases to about 38,000 km^2 if based on the mean position of the 30 isohaline (Guerrero et al. 1997a,b). The limit between mixohaline waters and continental shelf waters depends on the dynamics of the estuary. The main characteristics of the Río de la Plata estuary are its large spatial scale and the occurrence of a quasi-permanent salt wedge regime, which generates a border system (bottom and surface salinity fronts) which plays an important role in the reproductive processes of fish species and where high zooplankton biomass concentrates.

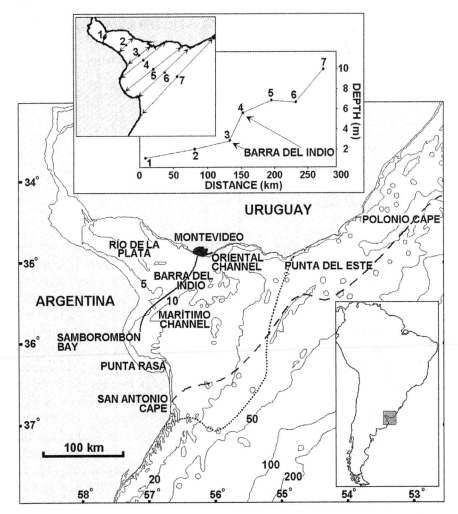

Fig. 13.1. Río de la Plata estuary with cross section areas and isobaths (m) [*solid line* bottom salinity front, *dotted line* spring-summer surface salinity front (mean 30 isohaline), *dashed line* autumn-winter surface salinity front (mean 30 isohaline)]

13.2 Environmental Setting

13.2.1 Climate

The coupling of the South Pacific and South Atlantic high-pressure systems controls the local atmospheric circulation. The frequency of on- and offshore winds is similar in the fall and winter but onshore winds dominate (about 70%) during the spring and summer. Mean monthly wind velocities in the watershed are moderate, with maxima in spring and summer. Extreme winds with velocities above 11 m s^{-1} (11–13%) are evenly distributed along the year. The large size and shallow waters make the estuary highly susceptible to atmospheric forcing. Air and water temperatures are in a quasi-permanent state of equilibrium, ranging from 10–11 °C in winter to 22–23 °C in summer (Guerrero et al. 1997a,b). Significant horizontal temperature gradients are absent. However, under the influence of NNE winds in summer, upwelling between Punta del Este and Polonio Cape may cause temperatures 4–5 °C below those of the surrounding shelf waters (Framiñan et al. 1999). Since the surrounding shelf surface waters and riverine water have a similar heat content, the effect of temperature on vertical stratification is negligible (Guerrero et al. 1997a). Mean vertical temperature differences (<1 °C) in the water column change between seasons, with cooler surface waters in winter. Maximum vertical stratification occurs during spring, when surface-bottom temperature differences vary from +2 to +5 °C.

13.2.2 Estuarine Dynamics

The freshwater discharge (annual mean 22,000 m^3 s^{-1}; Framiñan and Brown 1996) from the Paraná and Uruguay rivers into the estuary exhibits low seasonality, with a mean maximum of 26,000 m^3 s^{-1} in winter and a mean minimum of 19,000 m^3 s^{-1} in summer. The tidal wave originates on the outer shelf and enters the estuary from the southeast, with amplitudes ranging from 30 to 100 cm (Balay 1961). Tidal currents are typically below 45 cm s^{-1} (CARP 1989; Framiñan et al. 1999), thus friction energy between tidal and bottom currents is insufficient to erode vertical stratification.

Between the head of the estuary and Barra del Indio shoal a fluvial regime with vertically mixed river waters dominates. After the shoal, the cross section surface area of the estuary almost duplicates (Fig. 13.1) and denser shelf water intrudes along the bottom to maintain a gravitational balance, taking the shape of a salt wedge. The estuary therefore becomes a typically two-layer system with strong vertical stratification that gradually weakens seaward. The salt wedge regime is characterized by pronounced vertical differences in salinity, a constant halocline depth (5 m), and gradually sloping bottom topography. The boundary between the riverine regime and the salt wedge regime defines the bottom salinity front over the Barra del Indio shoal (Figs. 13.1, 13.2). Flocculation of suspended matter at the tip of the salt wedge and re-suspension of sediment due to tidal current friction at the bottom form a turbidity front (Fig. 13.2A), which extends from Montevideo along the 5-m isobath over the Barra del Indio shoal to Samborombón Bay

(Framiñan and Brown 1996). At the deeper and open Uruguayan coast, the proximity of the drainage channel causes a stratified and dynamic regime, leading to a highly variable front. In contrast, the front is relatively homogeneous in the semi-enclosed and shallow Samborombón Bay, which is distant from the main drainage channels (Lasta 1995; Guerrero et al. 1997a). Beyond the Barra del Indio shoal, the estuary discharges waters through the Oriental Channel and the gently sloping Maritime Channel (Fig. 13.1), though highly stratified salt wedge conditions may occur in one channel, while the other is not or only partially stratified coupled with the two way discharging pattern (Guerrero et al. 1997a). Further offshore, the partially stratified regime is characterized by an increasingly thicker (7–8 m) upper layer. The boundary between the partially stratified regime and the open ocean defines the surface salinity front (Fig. 13.2).

The budget between outflowing estuarine surface layer waters (82,000 $m^3 s^{-1}$) and inflowing shelf waters (60,000 $m^3 s^{-1}$) accounts for weak estuarine circulation and a residence time of 46.6 days (Guerrero, unpubl. data). The dynamics of upper layer water, discharging over the continental shelf, are mainly driven by wind stress (Guerrero et al. 1997a). During fall and winter the diluted surface waters from the Río de la Plata flow in an NNE direction along the Uruguayan coast in response to the Coriolis force, extending as far north as 28°S (Piola et al. 1999). Dominant onshore winds force surface waters in a southward direction along the Argentine coast in spring and summer. Continental shelf water, which intrudes upriver below the diluted upper layer, remains unaltered under most wind conditions, leading to an arrested position of the salt wedge regime over a long-time scale. A disruption of water column stratification and mixing of the salt wedge occurs only after several hours of strong onshore winds (>11 $m s^{-1}$), though the salt wedge at the head of the estuary becomes reestablished within 48–72 h. The dynamics of the Río de la Plata estuary is controlled by tide- and wind-driven waves and continental runoff but modified by topography and Coriolis force. The equilibrium between these forces is highly variable, depending largely on the intensity of wind stress and freshwater discharge (Guerrero et al. 1997a).

13.3 Biotic Components

Our treatment of floral and faunal assemblages is focussed at the key biotic components of the salt wedge regime and adjacent areas (Figure 2). Broad descriptions on the whole system may be found in Cousseau (1985); Boschi (1988); CARP (1989); and Wells and Daborn (1997).

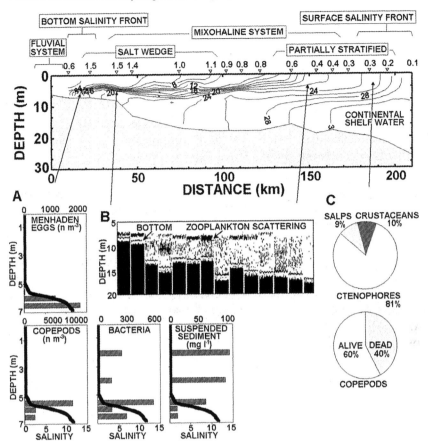

Fig. 13.2A–C. Salinity section along the major axis of the estuary under typically stratified conditions. *Numbers above upper axis* represent the stratification parameter after Hansen and Rattray (1966) (values >1 correspond to a salt wedge condition). A Vertical microdistribution of biological and physical properties at the head of the salt wedge. B Scattering layer due to zooplankton at the halocline (modified from Madirolas et al. 1997). C Mesozooplankton (copepods) and macrozooplankton aggregations at the surface salinity front. (Modified from Mianzan and Guerrero in press)

13.3.1 Freshwater Environment

The phytoplankton community of low salinity waters (0.2–5.0) is dominated by the diatom *Aulacoseira* sp. and coccal green algae, with the diatom *Stephanodiscus hantzschii* and cyanobacteria (i.e., *Microcystis aeruginosa*) occurring at polluted sites (Gómez and Bauer 1998). Clorophyll-*a* values of oligotrophic (71%) and mesotrophic (27%) environments are <4 and 4–10 mg m^{-3}, respectively (Rodolosi 1997). The species *Heleobia piscium*, *Corbicula fluminea*, *Limnoperna fortunei*, and *Chilina fluminea* are constant components of the soft bottom malacofauna, the last three species having been introduced (Darrigran 1993). Oligochaetes (*Limnodrilus claparedeianus, L. hoffmeisteri*, Tubificidae, Haplotaxidae, Naididae and Narapidae), nematodes, Hirudinea, Chironomidae, and Harpacticoidea are associated with organic matter-rich sediments (Rodrigues Capítulo et al. 1997).

All of the approximately 120 freshwater fish species (mainly Cypriniformes and Siluriformes) belong to the Paranoplatense ichtyofauna and together they constitute about 65 % of fish species in the estuary. Freshwater species sharply decrease in the mixohaline regime, where they represent about 7 % of the total species number (Cousseau 1985). *Leporinus obtusidens* and *Eigenmania virescens* frequently occur in freshwaters adjacent to the salt wedge. Catfishes like *Parapimelodus valenciennesi, Luciopimelodus pati, Pimelodus clarias, P. maculatus*, and *P. albicans* are also common species in low salinity mixohaline waters close to the salinity front. Pelagic fishes are mainly Atheriniformes (i.e., *Odonthestes bonariensis argentinensis*). Introduced species like silver carp (*Hypophthalmichthys molitrix*, Cyprinidae; García Romero et al. 1998), *Acipenser*, Acipenseriformes (Azpelicueta and Almirón in press), and *Cyprinus carpio* (Cyprinidae) have been recorded.

13.3.2 Mixohaline Environment

13.3.2.1 Plankton

Mixohaline waters with salinity between 13 and 31 and high silicate concentrations are characterized by the presence of warm-water dinoflagellates (*Ceratium candelabrum, Dinophysis caudata*) and the diatom *Thalassiosira puntigera* (Negri et al. 1988). Other marine and euryhaline diatoms include *Ditylum brightwelli, Skeletonema costatum, Asterionella glacialis*, and species of *Coscinodiscus* and *Chaetoceros* (CARP 1989). Owing to strong wave action, benthic diatoms (*Achnanthes brevipes, A. longipes, Melosira moniliforme*, and *Navicula grevillei*) are often suspended into the plankton community. The toxic dinoflagellate *Gymnodinium catenatum* is associated with the offshore limit of the surface salinity front (Carreto and Akselman 1996).

Ctenophores (*Mnemiopsis maccradyi* and *Pleurobrachia pileus*) comprised 81 % of total organic carbon at the external limit of the surface salinity (28–33) front. *Mnemiopsis* is most common in areas of thermal and saline stratification, while *Pleurobrachia* dominates less stratified areas. Gelatinous plankton tends to aggregate in areas where isolines outcrop to the surface (Mianzan and Guerrero in press). The dominant zooplankters in Samborombón Bay are *Acartia tonsa* (Copepoda), *Neomysis americana* (Mysidacea), *Mnemiopsis maccradyi*, and *Liriope tetraphylla* (Hydromedusae), with biomass peaks of *M. maccradyi* in spring and *L. tetraphylla* in summer (Sorarrain 1998). Copepods (mainly *Acartia tonsa*) and *Sagitta friderichi* are also the dominant zooplankton at the Uruguayan coast, followed by *N. americana*, the cladoceran *Pleopis polyphemoides*, chaetognats (*Sagitta*), larvae of cirripeds (*Balanus*; Méndez et al. 1997), and Scyphomedusae (*Lichnorhiza lucerna*, *Chrysaora lactea*; Mianzan et al. 1988).

Eggs and larvae of *Micropogonias furnieri* and *Brevoortia aurea* and larvae of *Gobiosoma parri* dominate the ichthyoplankton. The densities of eggs and larvae of *M. furnieri* and *B. aurea* are highest at the bottom salinity front (Acha et al. 1999; Acha 1999) and near Montevideo (Acuña et al. 1997). Both species have reproductive peaks in spring and early summer (October-December). The occurrence of larvae of *Gobiosoma parri* from March through December indicates spawning from February to April (Acha 1994). *Anchoa marinii* eggs occur mainly at the mouth of Río de la Plata and reach highest densities in front of Punta del Este (Cassia and Booman 1985; Sánchez and Ciechomski 1995; Acha 1999). The planktonic stages of most of these species (i.e., *B. aurea*, *G. parri*), as well as atherinid larvae, larvae of *Pogonias cromis* and *Cynoscion guatucupa*, and small juveniles of Syngnathidae occur in Samborombón Bay (Cassia and Booman 1985; Lasta and de Ciechomski 1988; Acha 1999).

13.3.2.2 Benthos

The muddy bottoms of the inner estuary (8–15 m depth), including Samborombón and Montevideo bays, are characterized by abundant clams (*Mactra isabelleana, Angulus gibber*), snails (*Buccinanops duartei, B. globulosum, B. gradatum, Anachis isabellei, Turbonilla uruguayensis*), worms (*Onuphis tenuis, Nephtys* sp.), and crabs (*Pinnixa brevipolex, Cyrtograpsus affinis, C. altimanus*). Most species are deposit feeders or carnivores and highest densities (*M. isabelleana*, 1,500–2,700 ind. m^{-2}) occur near Barra del Indio at 9–11 m depth. Shannon species diversity values range between 1.29 and 2.44 (A. Roux and C. Bremec, unpubl. data).

Supralittoral levels of rocky promontories between Montevideo and Cabo Polonio are dominated by lichens (*Caloplaca montevidensis, Verrucaria* sp.), cyanophytes, crustaceans, and mollusks (*Acmaea subrugosa* and *Siphonaria lessoni*). The upper midlittoral presents algae, cirripeds (*Chthamalus bisinuatus*), polychaetes (*Neanthes succinea*), and mollusks (*A. sub-*

rugosa, S. lessoni, Littorina ziczac). The lower littoral is occupied by mytilids (*Brachidontes darwinianus, B. rodriguezi, Mytella charruana, Mytilus edulis platensis, Perna perna*), crustaceans (mainly *Balanus improvisus, Cyrtograpsus angulatus*), and algae (*Ulva lactuca, Corallina officinalis, Polysiphonia* sp., *Chondria* sp., *Enteromorpha* sp.; Scarabino et al. 1975; Maytía and Scarabino 1979).

Coastal marshes of the Uruguayan coast are colonized by halophytic vegetation (*Spartina montevidensis, S. longispina, Juncus acutus, Salicornia ambigua*; Boschi 1988), while *Spartina alterniflora, S. densiflora, Scirpus maritimus, Salicornia ambigua*, and *S. virginica* are abundant along tidal creeks and mudflats in Samborombón Bay (Ringuelet 1938), which are dominated by dense populations of crabs (cangrejales). *Chasmagnatus granulata* inhabits burrows in sediments of supra- and mesolitoral zones and feeds on *Spartina* detritus or preys on small invertebrates (Botto and Irigoyen 1979). *Uca uruguayensis* occurs in upper consolidated substrates, while submerged lower levels are typical of *Cyrtograpsus angulatus* (Boschi 1979). Faunal assemblages of tidal and subtidal flats near Punta Rasa are characterized by a mixohaline, muddy-fine sand community with polychaetes (*Heteromastus similis* and *Laeonereis acuta*) and low Shannon index values (<1.60; Ieno and Bastida 1998).

13.3.2.3 Nekton

Mixohaline waters are dominated by euryhaline species of marine heritage (Boschi 1988). The Sciaenidae *Micropogonias furnieri, Macrodon ancylodon*, and *Paralonchurus brasiliensis* are common in brackish and coastal waters and some (*Micropogonias furnieri, Pogonias cromis*) have a wide distributional range. The common soles *Symphurus jennynsi* and *Paralichthys orbignyanus* (Paralichthydae) are abundant and the commercially valuable *P. orbignyanus* and *P. cromis* may exceed 20 kg and 100 cm in total length. *Conger orbignyanus* (Anguilliformes) also occurs in continental shelf areas. Other conspicuous species are Brazilian menhaden *Brevoortia aurea, Ramnogaster arcuata* (Clupeidae), *Anchoa marinii* (Engraulidae), *Austroatherina incisa* (Atherinidae), and *Netuma barbus* (Ariidae). Adults of the anadromous anchovy *Lycengraulis grossidens* (=*olidus*; Engraulidae) are common in the estuary (November to April) and breed during early spring in the Paraná and Uruguay rivers and their tributaries (Fuster de Plaza and Boschi 1961). Among the elasmobranchs *Sympterygia bonapartei, S. acuta, Raja castelnaui*, and *R. agassizi* are the most abundant species while the sting rays *Myliobatis goodei* and *Discopyge tschudii* are common, the latter mainly in spring and summer (Cousseau 1985).

13.3.2.4 Mammals and Birds

Dolphins, mainly *Pontoporia blainvillei* and *Tursiops gephyreus*, are frequent in the Río de la Plata estuary (Boschi 1988) and the sea lion *Otaria flavescens* also occurs (Vaz-Ferreira and Ponce de León 1984). Phalacrocoracidae, Ardeidae, Phoenicopteridae, Anatidae, Charadriidae, Scolopacidae, Laridae, and Passeriform represent common aquatic birds in Río de la Plata wetlands (Bonetto and Hurtado 1998). In the salt marshes of Samborombón Bay, *Calidris canutus rufa, C. fuscicoloris, Limosa haemastica, Tringa flavipes, Himantopus melanurus, Charadrius falklandicus, C. semipalmatus, C. collaris, Pluvialis dominica,* and *Arenaria interpres* feed on polychaetes (*Laeonereis acuta, Heteromastus simmili, Neanthes succinea*) and bivalves (*Tagelus plebeius*; Botto et al. 1998). *Pluvialis dominica P. squatarola, Arenaria interpres,* and *Numenius phaeopus* forage on the fiddler crab *Uca uruguayensis* (Iribarne and Martínez 1999), which constitutes an important food source.

13.3.3 Continental Shelf Environment

13.3.3.1 Plankton

The phytoplankton of shelf areas adjacent to the mouth of the Río de la Plata (depth >50 m) is richer in species than mixohaline waters (Negri et al. 1988). The community is composed of centric diatoms like *Chaetoceros curvisetus, Bacteriatrum hyalinum,* and *Rhizosolenia delicatula,* and pennate diatoms (i.e., *Asterionella glacialis, Thalassionema nitzschioides*; Elgue et al. 1991), dinoflagellates (*Ceratium tripos, C. lineatum, C. horridum*), and silicoflagellates. Generally, diatom abundance is highest in spring. Over the Uruguayan shelf dinoflagellates and coccolithophores dominate (Gayoso 1996) and blooms of toxic dinoflagellates (*Alexandrium tamarense* and *Gymnodinium catenatum*) have been reported (Carreto and Akselman 1996). Coastal areas under estuarine influence are characterized by *Acartia tonsa, Euterpina acutifrons, Paracalanus parvus, Oithona nana,* and cladocera (*Evadne nordmani, Podon polyphemoides, Penilia avirostris*; Fernández Aráoz et al. 1991), while Scyphomedusae (*Aurelia aurita, Chrysaora lactea, Lichnorhiza lucerna*) occur mainly off Punta del Este (Mianzan et al. 1988).

Eggs and larvae of *Engraulis anchoita* (Sánchez and Ciechomski 1995), which has a protracted reproductive season with a peak in late spring/early summer (November-January), dominated the ichthyoplankton of shelf waters. Eggs and larvae of *Brevoortia* sp. occur along the Uruguayan coast under strong influence of Río de la Plata waters (Hubold and Ehrlich 1981). Eggs of *Trachurus lathami* (=*picturatus*), *Prionotus* spp., *Scomber japonicus*, *Anchoa marinii*, and larvae of *Stromateus brasiliensis* have been reported for the Argentine coast (Hubold and Ehrlich 1981; Cassia and Booman 1985; Sánchez and Ciechomski 1995). Highest concentrations of eggs and larvae of *Cynoscion guatucupa* occur in shallow waters near San Antonio Cape (Cassia and Booman 1985).

13.3.3.2 Benthos

Sandy-gravel bottoms of coastal areas (34–38°S) between 15 and 40 m depth are typical of invertebrate associations dominated by filter-feeding mollusks (*Corbula patagonica*, *Nucula puelcha*, *Pitaria rostrata*) and detritivorous crustaceans (*Serolis marplatensis*, *Pagurus exilis*, *P. criniticornis*, *Ancinus* sp.). Deeper bottoms (up to 50 m) also include detritivorous and predatory echinoderms (*Amphiura eugeniae*, *Astropecten b. brasiliensis*, *Patiria stilifer*). Shannon species diversity values are 2.69 to 3.58 (C. Bremec and A. Roux, unpubl. data). The detritus-feeding echinoderm *Encope emarginata*, the isopod *Serolis marplatensis*, and coelenterate *Renilla* sp. colonies characterize sandy bottoms along the Argentine coast (20–25 m depth). The species diversity ranges from 1.32 to 2.44. Shrimps (*Artemesia longinaris* and *Pleoticus muelleri*) are common epibenthic species, with higher densities on fine sand-muddy bottoms (Boschi 1979). Beds of the commercially valuable clam *Mesodesma mactroides* occupy medium-fine sandy beaches along the coasts of Argentina and Uruguay during spring and summer (Olivier et al. 1971; Escofet et al. 1979). Associated species are mollusks (*Donax hanleyanus*, *Buccinanops* spp., *Olivancillaria* spp.), polychaetes (*Glycera americana*, *Spio gaucha*), and isopods (mainly talitrids and cirolanids). The abundance of commercial-size specimens of *M. mactroides* has decreased, possibly due to illegal exploitation, tourist activities, and commercial extraction of sand (Bastida et al. 1991); however, massive mortality in Brazil, Uruguay, and Argentina (1993–1995) has also been attributed to a natural phenomenon (J. Carreto, pers. comm.). On rocky substrates, belts of *Chtamalus bisinuatus* become denser and the *Brachydontes darwinianus-Mytella charruana* assemblage is replaced by a *Brachydontes rodriguezi-Mytilus e. platensis* assemblage as conditions become more oceanic (Maytía and Scarabino 1979; Riestra et al. 1992).

13.3.3.3 Nekton

The stripped weakfish *Cynoscion guatucupa*, with nursery grounds in coastal areas around the estuary, as well as *Umbrina canosai*, *Micropogonias furnieri*, and *Macrodon ancylodon* are abundant sciaenids in marine environments down to 50 m depth. *Paralichthys patagonicus* is the most abundant flatfish (Cousseau 1985). Other flatfishes are *Xystreuris rasile*, the small sole *Symphurus jennynsi*, and *Paralichthys isosceles*, which occurs in deeper waters (>50 m). Dominant species on rocky bottoms (<50 m), mainly along the coast of Argentina, are *Discopyge tschudii*, *Cheilodactylus bergi*, *Percophis brasiliensis* (Percophididae), *Pseudopercis semifasciata* (Lasta et al. 1998), as well as *Acanthistius brasilianus* and the red porgy *Sparus pagrus*. The feeding habits of *Sparus pagrus* and hawkfish (*Cheilodactylus bergi*) are related to the composition and structure of benthic feeding grounds (Brankevich et al. 1990; Bruno et al. in press). *Percophis brasiliensis*, *Porichthys porosissimus* (Batrachoididae), and *Prionotus punctatus* occur in marine waters (35-m depth) but may also be found in estuarine areas. The most abundant elasmobranch is *Mustelus schmitii*, followed by *M. canis*, *M. fasciatus*, and *Galeorhinus vitaminicus*. Myliobatiformes (*Myliobatis goodei*, *Dasyatis pastinaca*), Rhinobatiformes (*Zapterix brevirostris*), Hexanchiformes (*Notorhinchus pectorosus*), and Squatiniformes (*Squatina argentina*) are also common in shelf waters (Cousseau 1985). Among pelagic species, the anchovy *Engraulis anchoita* (Engraulidae) is abundant, followed by *Trachurus lathami* and *Parona signata* (Carangidae), which also enter mixohaline waters, and Stromateidae (*Stromateus brasiliensis*, *Peprilus paru*). *Pomatomus saltatrix*, *Scomber japonicus*, *Thyrsitops lepidopodea*, and *Trichiurus lepturus* reach coastal and mixohaline areas in summer, while *Balistes capriscus* (Tetraodontiformes) sporadically occurs during fall (Nion and Ríos 1991).

13.3.3.4 Mammals

The sea lion *Otaria flavescens* inhabits rocky or sandy areas and mainly hunts in coastal waters (Vaz-Ferreira and Ponce de León 1984), while the South American fur seal *Arctocephalus australis* lives on steep rocky habitats and feeds in deep shelf areas. Both species feed on fish (*Cynoscion guatucupa*, *Micropogonias furnieri*, *Macrodon ancylodon*, *Porichthys porossisimus*, *Paralonchurus brasiliensis*), squid (*Loligo* spp.), and shrimps but are preyed upon by some sharks and the killer whale *Orcinus orca*.

13.4 Biological Significance of the Salt Wedge Regime

Estuarine processes are inherently three-dimensional. As a main characteristic of the Río de la Plata estuary, the salt wedge regime plays an important role for reproductive processes of fishes and, owing to density discontinuity between two different water masses, by generating border systems, which are the surface for several ecological processes. The bottom salinity front near the head of the salt wedge is typical of pronounced density discontinuities and the retention of particulate matter and plankton coincides with the occurrence of turbidity maxima. The tip of the salt wedge represents an ecotone between the river and the estuary that few species (i.e., the anchovy *Lycengraulis grossidens*) seem to be able to cross. Density discontinuities extend for up to 200 km of the mixohaline regime along the halocline and terminate at the surface salinity front. This front represents the highly dynamic part of the salt wedge regime with a less defined salinity gradient, which delineates the boundary between estuary and continental shelf waters (Fig. 13.2).

13.4.1 Head of the Salt Wedge

In general, the upper freshwater layer at the head of the salt wedge is poorer in fish and plankton species than the bottom saline layer. During spring and summer, the whitemouth croaker (*Micropogonias furnieri*) and Brazilian menhaden (*Brevoortia aurea*) aggregate close to the bottom salinity front (Macchi et al. 1996; Acha 1999) and spawn pelagic eggs which remain below the halocline taking advantage of the retention properties at the head of the salt wedge (Fig. 13.2A). In spite of not being a common event in estuaries elsewhere, the large-scale salt wedge regime in the Río de la Plata estuary favors the spawning of pelagic eggs. The retention of eggs and larvae in the convergence zone is not complete and some are transported seaward by residual currents; however, larvae appear to have enough time to develop into a stage which permits control of their vertical position in the water column (Acha 1999; Acha et al. 1999). Vertical migrations enable the larvae to take advantage of the seaward flowing surface layer and the landward moving bottom layer, thus contributing to their permanence in the system. Although adults of *M. furnieri* are also common in continental shelf waters, the entire spawning cycle seems to occur within the estuary (Fig. 13.3). Juveniles (about 2 cm long) then migrate towards the protected waters of Samborombón Bay, which they abandon after 2–3 years. Apart from *Micropogonias furnieri* and *Brevoortia aurea*, Sam-

The Río de la Plata Estuary, Argentina-Uruguay

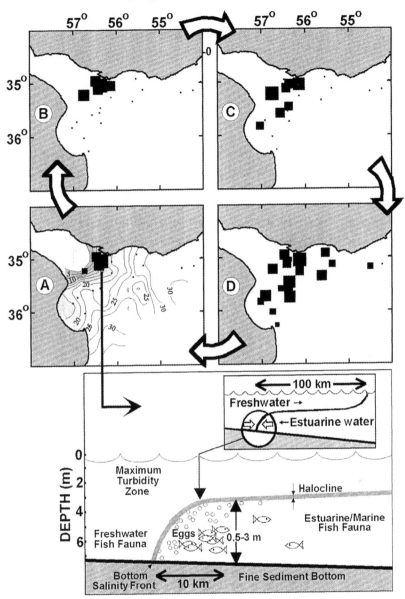

Fig. 13.3A–D. Spawning cycle of *Micropogonias furnieri* in the Río de la Plata estuary during reproductive peak (A–D modified from Macchi et al. 1996) and conceptual diagram (modified from Acha et al. 1999) with enlarged portion of salt wedge. A Spawning females with hydrated oocytes and bottom salinity field (note coincidence between spawning site and bottom salinity front). B Elapsed time from spawning <24 h (females with post-ovulatory follicles day-0). C Elapsed time from spawning 24–48 h (females with post-ovulatory follicles day-1). D Elapsed time from spawning >48 h (females with yolked oocytes and no post-ovulatory follicles). *Symbol size* is proportional to percentage of females in different stages

borombón Bay, followed by the Santa Lucia River mouth, are also the main nursery ground for *Pogonias cromis*, *Odonthestes* spp., *Mugil liza*, *Macrodon ancylodon*, *Paralichthys* spp., and other fish species that inhabit the estuary (Lasta 1995; Wells and Daborn 1997).

Zooplankton aggregations below the halocline at the head of the salt wedge are largely represented by different life stages of *Acartia tonsa* (up to 8,000 ind. m^{-3}; Fig. 13.2A). The upper freshwater layer, which may represent more than 70% of the water column, is inhabited by few freshwater copepods (<200 ind. m^{-3}; Ramírez, pers. comm.). At the thick freshwater layer, the turbidity front, and suspended sediments constrain photosynthesis, thus phytoplankton is poor. High concentrations of detrital biomass (pheopigments; CARP 1989), tintinnids, and bacteria seem to constitute the main energy supply for omnivorous species like *Acartia tonsa*. Where horizontal salinity gradients reach a maximum, beds of the deposit-feeding clam *Mactra isabelleana* concentrate along the bottom salinity front, probably taking advantage of high suspended matter deposition. Both freshwater species (Doradidae) and the estuarine *Micropogonias furnieri* prey on the clam.

13.4.2 Halocline and Surface Salinity Front

The widest range of salinity occurs in the mixohaline system and salinities control the differential penetration and spatial distribution of adult fishes in the estuary. Several species of Sciaenidae exhibit different salinity preferences like *Micropogonias furnieri* (1–33), *Paralonchurus brasiliensis* (5–33), *Macrodon ancylodon* and *Pogonias cromis* (10–25), *Menticirrhus americanus* (25–33), *Cynoscion guatucupa* (>15), and *Umbrina canosai* (about 30). At the salt wedge fish occur exclusively in the bottom layer. However, in the outer mixohaline region, where the halocline is less pronounced, pelagic species like *Trachurus lathami* and *Engraulis anchoita* occur in the increasingly saltier upper layer. Juveniles and small adults of *E. anchoita* dominate at the surface salinity (24–33) front (Hansen and Madirolas 1996) where they probably feed on plankton. *Parona signata*, *Paralichthys patagonicus*, and *Cynoscion guatucupa* also spawn near this front (Macchi 1998; Macchi and Acha 1998).

Plankton is mainly located in the lower saline layer and the highest zooplankton biomass aggregates below the halocline (Madirolas et al. 1997; Fig. 13.2B). Where the halocline intersects the surface layer, live and dead individuals (40%) of *Acartia tonsa* concentrate in the upper layer (up to 110,000 ind. m^{-3}; Ramirez and Sorarrain, pers. comm.) and gelatinous plankton (mainly *Mnemiopsis maccradyi*) aggregate at the surface salinity front (Mianzan and Guerrero in press; Fig. 13.2C). Mixing of estuarine and marine waters and enhancement of vertical nutrient flux fertilize the frontal area (Méndez et al. 1997), with maximum chlorophyll-*a* concentrations occurring at the surface salinity front. Neritic diatoms, colonies of *Phaeocystis* sp. (Nellen 1990), and aggregations of dinoflagellates and coccolithophores (Carreto and Akselman 1996; Gayoso 1996) dominate the phytoplankton, allowing for the development of dense copepod (i.e., *Acartia tonsa*) populations that constitute food for the gelatinous plankton (i.e., *Mnemiopsis maccradyi*; Mianzan and Sabatini 1985; Sorarrain 1998). Many fish species in the area feed on gelatinous plankton (Mianzan et al. 1996).

13.5 Human Impacts and Management Needs

The Río de la Plata estuary receives waters impacted by both subsistence and highly productive agriculture and cattle raising. Industrial and urban areas around the estuary, principally Buenos Aires and Montevideo with about 13 million inhabitants, affect the aquatic habitat and conflict with environmental issues. The activities of the ports in Buenos Aires and Montevideo, dredging, and subaquatic sediment disposals along the main access channels cause major environmental impacts. The estuary is the maritime access to a highly complex fluvial system communicating with the Amazon Basin and nautical accidents have caused oil spills in recent years; so, navigation activities are inadequately considered by environmental management plans.

Higher metal concentrations in the sediments and water column in areas under the influence of port activities or municipal sewage discharge are probably related to point sources. The disappearance of commercially valuable species like mussels and the red crab has been related to industrial wastes (Graña and Piñeiro 1997). Municipal and industrial wastewater of Montevideo is released close to the head of the salt wedge regime. Since some contaminants are likely to be retained, this area is particularly sensitive and, owing to the biological significance of the head of the salt wedge for the life cycles of many valuable fish species, deserves especial attention in future environmental plans.

Socio-economic activities in the Río de la Plata estuary include industrial and artisanal fisheries. During productive years, fisheries of *Micropogonias furnieri* in mixohaline waters involve about 200 vessels and 1,000 fishermen, generating approximately 30 million dollars in revenues. In the Argentine-Uruguayan Common Fishing Zone this resource is managed based on joint stock assessment, fleet monitoring, establishment of quotas, and seasonal restrictions; however, the species is clearly overexploited. In adjacent coastal shelf areas the impact of industrial fisheries, mostly bottom trawling, has still not been assessed but is partially responsible for the decline in local landings, like the disappearance of the San Antonio Cape whitemouth croaker fishery grounds (Lasta and Acha 1996). Fisheries in Samborombón Bay and the mouth of Santa Lucia River are mainly artisanal. Despite evidence of a recent decline in overall landings, sciaenids are the most important resource throughout the year, with *M. furnieri* generally being caught in winter (Lasta et al. in press). In the Santa Lucia area, *M. furnieri* is usually fished with long lines and gill nets during spawning peak from October to March and the noticeable depletion is attributed to industrial trawling fisheries (Graña and Piñeiro 1996).

Coastal resources of the Río de la Plata, including marginal areas of Samborombón Bay and Santa Lucia River, are not properly managed and regulations for estuarine or maritime waters 2 miles off the coast are lacking. The Río de la Plata Treaty (Argentina-Uruguay) establishes regulations for river water and the maritime front, though each country administers its own coastal zone. More recently, promising perspectives applying a multi-sectorial approach to face coastal problems, are being considered to produce management recommendations (international agreements, binational action plans, etc.) for the Río de la Plata estuary.

Acknowledgements. We are grateful to Dr. E. Boschi and A. Lucas for critically reading the manuscript and to our colleagues Dr. F. Ramírez, Lic. D. Sorarrain, and M.D. M. Costagliola for making unpublished data available. This is contribution no. 1124 of the Instituto Nacional de Investigación y Desarrollo Pesquero (INIDEP).

References

Acha EM (1994) Development and occurrence of larvae of the goby, *Gobiosoma parri* (Ginsburg) (Gobiidae), in the estuary of the Río de la Plata, Argentina. Sci Mar 58:337–343

Acha EM (1999) Estrategia reproductiva de la saraca, *Brevoortia aurea* (Spix y Agassiz, 1829) (Pisces: Clupeidae), en el estuario del Río de la Plata. PhD Thesis, Universidad Nacional de Mar del Plata, Argentina

Acha EM, Mianzan H, Lasta C, Guerrero R (1999) Estuarine spawning of the whitemouth croaker (*Micropogonias furnieri*) in the Río de la Plata, Argentina. Mar Freshwater Res 50:57–65

Acuña A, Arena G, Berois N, Mantero G, Masello A, Nion H, Retta S, Rodríguez M (1997) The croaker (*Micropogonias furnieri*): biological cycle and fisheries in the Río de la Plata and its oceanic front. In: Wells PG, Daborn GR (eds) The Río de la Plata. An environmental overview. An EcoPlata project background report, Dalhausie University, Halifax, Nova Scotia, pp 185-222

Azpelicueta M, Almirón A (2000) A sturgeon in the temperate waters of the río de la Plata, South America. Biogeographica (in press)

Balay MA (1961) El Río de la Plata entre la atmósfera y el mar. Servicio de Hidrografía Naval, Argentina, Publicación H-621

Bastida R, Roux A, Bremec C, Gerpe M, Sorensen M (1991) Estructura poblacional de la almeja amarilla (*Mesodesma mactroides*) durante el verano de 1989 en la provincia de Buenos Aires, Argentina. Frente Marítimo 9:83-92

Bonetto A, Hurtado S (1998) Cuenca del Plata. In: Canevari P, Blanco D, Bucher E, Castro G, Davidson I (eds) Los humedales de la Argentina. Clasificación, situación actual, conservación y legislación. Wetlands Int 46:31-72

Boschi EE (1979) Geographic distribution of Argentinian marine decapod crustaceans. Bull Biol Soc Wash 3:134-143

Boschi EE (1988) El ecosistema estuarial del Río de la Plata (Argentina y Uruguay). An Inst Cienc del Mar Limnol Univ nac autón Méx 15:159-182

Botto J, Irigoyen H (1979) Bioecología de la comunidad del cangrejal. Contribución al conocimiento biológico del cangrejo de estuario, *Chasmagnatus granulata* Dana (Crustacea, Decapoda, Grapsidae) en la desembocadura del Río Salado, Provincia de Buenos Aires. Memorias del Seminario Ecología Bentónica y Sedimentación de la Plataforma Continental del Atántico Sur (UNESCO) 9/12:161-169

Botto F, Iribarne O, Martínez M (1998) The effect of migratory shorebirds on the benthic species of three southwestern Atlantic Argentinean estuaries. Estuaries 21:700-709

Brankevich G, Roux A, Bastida R (1990) Relevamiento de un banco de pesca de besugo (*Sparus pagrus*) en la plataforma bonaerense. Características fisiográficas generales y aspectos ecológicos preliminares. Frente Marítimo 7:75-86

Bruno C, Cousseau MB, Bremec C (in press) Contributions of the polychaetous annelids to the diet of *Cheilodactylus bergi* (Pisces, Cheilodactylidae) in Argentina. Bull Mar Sci (in press)

Carreto JI, Akselman R (1996) *Gymnodinium catenatum* and autumnal toxicity in Mar del Plata. Harmful Algae News, IOC, UNESCO 15:1-3

CARP (1989) Comisión Administradora del Río de la Plata Estudio para la evaluación de la contaminación en el Río de la Plata. Informe de Avance

Cassia C, Booman C (1985) Distribución del ictioplancton en el Mar Argentino en los años 1981-1982. Physis B Aires 43:91-111

Cousseau MB (1985) Los peces del Río de la Plata y su Frente Marítimo. In: Yañez Arancibia A (ed) Fish community ecology in estuaries and coastal lagoons. Towards an ecosystem integration. Universidad Nacional Autónoma de México, México, pp 515-534

Darrigran G (1993) Los moluscos del Río de la Plata como indicadores de contaminación ambiental. In: Goin F, Goñi R (eds) Elementos de Política Ambiental. Honorable Cámara de Diputados de la Provincia de Buenos Aires, pp 309-312

Elgue JC, Bayssé C, Burone F, Parietti M (1991) Distribución y sucesión espacial del fitoplancton de superficie de la Zona Común de Pesca Argentino-Uruguaya (Invierno, 1983). Publ Com Téc Mix Fr Mar 6:67-107

Escofet A, Gianuca N, Maytía S, Scarabino V (1979) Playas arenosas del Atlántico Sudoccidental entre los 29° y 43° LS: consideraciones generales y esquema biocenológico. Memorias del Seminario Ecología Bentónica y Sedimentación de la Plataforma Continental del Atántico Sur (UNESCO) 9/12:245-258

Fernández Aráoz N, Perez Seijas M, Viñas MD, Reta R (1991) Asociaciones zooplanctónicas de la Zona Común de Pesca Argentino-Uruguaya en relación con parámetros ambientales. Frente Marítimo 8:85-99

Framiñan M, Brown O (1996) Study of the Río de la Plata turbidity front, part 1: spatial and temporal distribution. Cont Shelf Res 16 (10):1259-1282

Framiñan MB, Etala M, Acha EM, Guerrero RA, Lasta CA, Brown OB (1999) Physical characteristics and processes of the Río de la Plata estuary. In: Perillo GM, Piccolo MC, Pino-Quivira M (eds) Estuaries of South America, their geomorphology and dynamics. Springer, Berlin Heidelberg New York, pp 161-191

Fuster de Plaza ML, Boschi EE (1961) Areas de migración y ecología de la anchoa *Lycengraulis olidus* (Günther) en las aguas argentinas (Pisces, Engraulidae). Contrib Cient Fac Cienc Exact y Nat (Univ Nac B Aires) Zool 1:127-183

García Romero N, Azpelicueta M, Almiron A, Casciotta J (1998) *Hypophthalmichthys molitrix* (Cypriniformes: Cyprinidae). Other exotic cyprinid in the Río de la Plata. Biogeographica 74:189-191

Gayoso AM (1996) Phytoplankton species composition and abundance off Río de la Plata (Uruguay). Arch Fish Mar Res 44:257-265

Gómez N, Bauer DE (1998) Phytoplankton from the southern coastal fringe of the Río de la Plata (Buenos Aires, Argentina). Hydrobiologia 380:1-8

Graña F, Piñeiro D (1997) Artisanal fishing in Pajas Blancas: the fishermen's perception of the croaker and its environment. In: Wells PG, Daborn GR (eds) The Río de la Plata. An environmental overview. An EcoPlata Project Background Report, Dalhausie University, Halifax, Nova Scotia, pp 223-238

Guerrero RA, Acha ME, Framiñan M, Lasta C (1997a) Physical oceanography of the Río de la Plata Estuary. Cont Shelf Res 17(7):727-742

Guerrero RA, Lasta C, Acha M, Mianzan H, Framiñan M (1997b) Atlas Hidrográfico del Río de la Plata. CARP, INIDEP

Hansen DV, Rattray M Jr (1966) New dimensions in estuary classification. Limnol Oceanogr 11:319-326

Hansen JE, Madirolas A (1996) Distribución, evaluación acústica y estructura poblacional de la anchoíta, resultados de las campañas del año 1993. Rev Invest Des Pesq 10:5-21

Hubold G, Ehrlich M (1981) Distribution of eggs and larvae of five clupeoid fish species in the Southwest Atlantic between 25°S and 40°S. Meeresforschung (Rep Mar Res) 29:17-29

Ieno E, Bastida R (1998) Spatial and temporal patterns in coastal macrobenthos of Samborombón Bay, Argentina: a case study of very low diversity. Estuaries 21:690-699

Iribarne O, Martínez M (1999) Predation on the Southwestern Atlantic fiddler crab (*Uca uruguayensis*) by migratory shorebirds (*Pluvialis dominica, P. squatarola, Arenaria interpres* and *Numenius phaeopus*). Estuaries 22:47-54

Lasta C (1995) La Bahía Samborombón: zona de desove y cría de peces. PhD Thesis, Universidad Nacional La Plata, Argentina

Lasta CA, Ciechomski JD de (1988) Primeros resultados de los estudios sobre la distribución de huevos y larvas de peces en Bahía Samborombón en relación a temperatura y salinidad. Publ Com Téc Mix Fr Mar 4:133-141

Lasta C, Acha EM (1996) Cabo San Antonio: su importancia en el patrón reproductivo de peces marinos. Frente Marítimo 16:39-46

Lasta C, Bremec C, Mianzan H (1998) Areas ícticas costeras en la Zona Común de Pesca Argentino-Uruguaya y en el litoral de la provincia de Buenos Aires. In: Lasta CA (ed) Resultados de una campaña de evaluación de recursos demersales costeros de la provincia de Buenos Aires y del Litoral Uruguayo, Noviembre, 1994. Informe Técnico INIDEP 21:91-101

Lasta C, Carozza C, Suquelle P, Bremec C, Errasti E, Perrotta R, Bertelo C, Bocanfusso J (in press) Característica y dinámica de la explotación de corvina rubia (*Micropogo-*

nias furnieri) durante la zafra invernal. Años 1995 y 1996. Informe Técnico del INIDEP (in press)

Macchi GJ (1998) Preliminary estimate of spawning frequency and batch fecundity of striped weakfish, *Cynoscion striatus*, in the coastal waters off Buenos Aires province. Fishery Bull Natn Mar Fish Serv US 96 (2):375-381

Macchi GJ, Acha EM (1998) Aspectos reproductivos de las principales especies de peces en la Zona Común de Pesca Argentino-Uruguaya y en El Rincón, Noviembre, 1994. In: Lasta CA (ed) Resultados de una campaña de evaluación de recursos demersales costeros de la provincia de Buenos Aires y del Litoral Uruguayo, Noviembre, 1994. Informe Técnico INIDEP 21:67-89

Macchi GJ, Acha EM, Lasta CA (1996) Desove y fecundidad de la corvina rubia *Micropogonias furnieri* Desmarest, 1823 del estuario del Río de la Plata, Argentina. Boln Inst Esp Oceanogr 12(2):99-113

Madirolas A, Acha EM, Guerrero RA (1997) Sources of acoustic scattering near a halocline in an estuarine frontal system. Sci Mar 61(4):431-438

Maytía S, Scarabino V (1979) Las comunidades del litoral rocoso del Uruguay: zonación, distribución local y consideraciones biogeográficas. Memorias del Seminario Ecología Bentónica y Sedimentación de la Plataforma Continental del Atántico Sur (UNESCO) 9/12:149-160

Méndez S, Gómez M, Ferrari G (1997) Planktonic studies of the Río de la Plata and its oceanic front. In: Wells PG, Daborn GR (eds) The Río de la Plata. An environmental overview. An EcoPlata project background report, Dalhausie University, Halifax, Nova Scotia, pp 85-112

Mianzan H, Olagüe G, Montero R (1988) Scyphomedusae de las aguas uruguayas. Spheniscus 6:1-9

Mianzan H, Guerrero R (in press) Environmental patterns and biomass distribution of gelatinous macrozooplankton. Three study cases in the Southwestern Atlantic. Sci Mar (in press)

Mianzan HW, Sabatini ME (1985) Estudio preliminar sobre distribución y abundancia de *Mnemiopsis maccradyi* en el estuario de Bahía Blanca (Ctenophora). Spheniscus 1:53-68

Mianzan H, Marí N, Prenski B, Sanchez F (1996) Fish predation on neritic ctenophores from the Argentine continental shelf: a neglected food resource? Fish Res 27:69-79

Negri R, Benavídez H, Carreto J (1988) Algunas características del florecimiento del fitoplancton en el frente del Río de la Plata. II. Las asociaciones fitoplanctónicas. Publ Com Téc Mix Fr Mar 4:151-161

Nellen W (1990) Working report on Cruise N° 11, Leg 3 of RV METEOR. Inst Hydrobiol Fischereiwissenschaft 9

Nion H, Ríos C (1991) Los recursos pelágicos del Uruguay. Atlantica 13:201-214

Olivier SR, Capezzani D, Carreto JI, Christiansen H, Moreno V, Aizpun de Moreno J, Penchaszadeh P (1971) Estructura de la comunidad, dinámica de la población y biología de la almeja amarilla (*Mesodesma mactroides* Deshayes 1854) en Mar Azul (Pdo. de General Madariaga, Bs. As., Argentina). Proy Desarr Pesq FAO, Inf Téc 27:1-90

Piola AR, Campos EJD, Möller OO, Charo M, Martinez C (1999) Continental shelf water masses off eastern South America 20° to 40°S. 10th Symposium on Global Change Studies, American Meteorological Society, Boston, pp 9-12

Riestra G, Gimenez JL, Scarabino V (1992) Análisis de la comunidad macrobentónica infralitoral de fondo rocoso en Isla Gorriti e Isla de Lobos (Maldonado, Uruguay). Frente Marítimo 11:123-127

Ringuelet RA (1938) Estudio fitogeográfico del Rincón de Viedma (B. de Samborombón). Rev Fac Agronom (Univ Nac La Plata) 21:15-186

Rodolosi A (1997) Clorofila "A". In: Consejo Permanente para el Monitoreo de la Calidad de Aguas de la Franja Costera Sur del Río de la Plata (ed) Calidad de Aguas de la Franja Costera Sur del Río de la Plata (San Fernando – Magdalena). Contartese Gráfica S.R.L., Buenos Aires, pp 109–112

Rodrigues Capítulo A, César I, Tassara M, Paggi A, Remes Lenicov M (1997) Zoobentos. In: Consejo Permanente para el Monitoreo de la Calidad de Aguas de la Franja Costera Sur del Río de la Plata (ed) Calidad de Aguas de la Franja Costera Sur del Río de la Plata (San Fernando – Magdalena). Contartese Gráfica S.R.L., Buenos Aires, pp 131–142

Sánchez R, Ciechomski J (1995) Spawning and nursery grounds of pelagic fish species in the sea-shelf off Argentina and adjacent areas. Sci Mar 59:455–478

Scarabino V, Maytía S, Cachés M (1975) Carta bionómica litoral del departamento de Montevideo. I. Niveles superiores del sistema litoral. Com Soc Malac Uruguay 4:117–130

Sorarrain D (1998) Cambios estacionales en la biomasa de organismos gelatinosos en relación con otros zoopláncteres en la Bahía Samborombón. Thesis, Universidad Nacional de Mar del Plata, Argentina

Vaz-Ferreira R, Ponce de León A (1984) Estudios sobre *Arctocephalus australis* (Zimmermann 1783), lobo de dos pelos sudamericano, en el Uruguay. Contrib Depto Oceanogr (Fac Hum y Cienc Montevideo) 1:1–18

Wells PG, Daborn GR (eds) (1997) The Rio de la Plata. An environmental overview. An EcoPlata project background report, Dalhausie University, Halifax, Nova Scotia

14 The Bahía Blanca Estuary, Argentina

G.M.E. Perillo, M.C. Piccolo, E. Parodi, and R.H. Freije

14.1 Introduction

Bahía Blanca is a mesotidal coastal plain estuary in the southwest of the Buenos Aires Province, Argentina. The estuary represents the remaining part of a complex delta, formed by the Desaguadero river (Perillo 1989), which once extended over almost 200 km from the northern coast of Bahía Blanca to Bahía San Blas (Fig. 14.1). The Principal Channel of the modern estuary approximately corresponds to the principal valley of the late Pleistocene-early Holocene delta complex, which was formed by a now vanished river and its tributary, the Colorado River (or one of its arms). The southward migration of the Colorado River subsequently carved the other major channels (i.e., Falsa, Verde, and Brightman bays) of the estuary (Perillo 1989). The modern Bahía Blanca Estuary extends over about 2,300 km^2 and is formed by several tidal channels (740 km^2), extensive tidal flats (1,150 km^2) with patches of low salt marshes, and islands (410 km^2; Montesarchio and Lizasoain 1981). Owing to marked differences in surface morphology, the area can be divided along the northern shore of Falsa Bay into a funnel-shaped northern part, characterized by the Principal Channel and many small tidal creeks, and a southern part, dominated by the Falsa, Verde, and Brightman bays (Perillo and Piccolo 1999; Fig. 14.1).

14.2 Geomorphology

The Principal Channel of the Bahía Blanca Estuary has a total length of 60 km and varies in width from about 3–4 km at the mouth (22-m depth) to 200 m at the head (3-m depth); both depth and width increase almost exponentially from head to mouth. Like other major channels (bays) that flow towards the inner shelf, the Principal Channel is partly closed by a

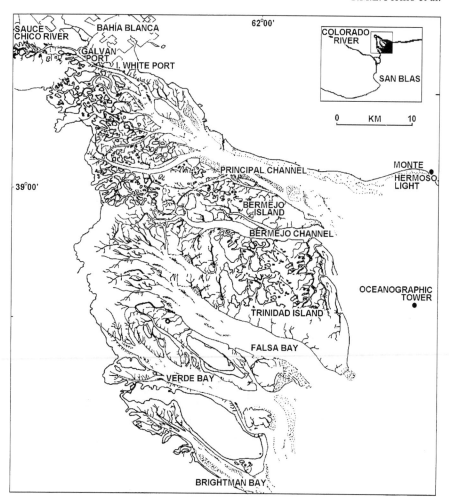

Fig. 14.1. General map of the Bahía Blanca estuary, Argentina

modified ebb delta (Cuadrado and Perillo 1997). The channel cross section is steep on the sides, with a U-shaped bottom having a small asymmetry to the right. Upstream of Puerto Galván (Fig. 14.1) the channel narrows and becomes more V-shaped with the asymmetry following the meandering pattern headward (Gómez et al. 1997). At the confluence with the Principal Channel, the funnel-shaped mouth of tributary channels is turned seawards, due to ebb dominance (Perillo et al. 1996), and up to 25-m-deep scour holes may develop (Ginsberg and Perillo 2000).

Most ebb deltas at the mouth of major channels have undergone changes (Gómez and Perillo 1992); however, the original delta shape and southward orientation of the ebb channel and associated shoals in the Principal Channel are still preserved. Despite strong marine dynamics, which constantly modify the tidal delta channels and shoals, the position and shape of the delta has been conserved, due to its stable connection to the northern shore and its position on top of a sill (Chasicó Formation), which has reduced sediment transport and served as an anchor (Cuadrado and Perillo 1997). The mobility of delta shoals depends on the approach angle of tidal currents (Perillo and Cuadrado 1991), while 3D dunes (sand waves) in the Principal Channel and tributary channels are formed due to high current velocity and geomorphologic "traps" which favor sedimentation (Aliotta and Perillo 1987; Gómez et al. 1997). Except for bed forms on flood-dominated channels, all 3D dunes and shoals in the Bahía Blanca Estuary have ebb dominance. Since sediment runoff from rivers is virtually absent and ebb delta characteristics impede sediment input from the shelf, the high concentration of suspended sediments in the estuary is due to erosion of tidal flats and island shores (Ginsberg and Perillo 1990). The southern coast of the Principal Channel across Ingeniero White Port has retreated up to 50 m between 1980 and 1986 and 1.5 million m^3 were exported from an 8-km stretch at the mid-reach of the channel (Perillo and Sequeira 1989). Because most sediments in the estuary are silt and clay, strong currents and short slack water intervals impede their deposition in channels and on the tidal flats, and waves erode old sediments and prevent any accumulation of new ones. These conditions explain the erosional stage of most of the estuary and the prevalence of sediments from the deltaic deposition period.

14.3 Physical Processes

14.3.1 Freshwater Input

Two freshwater tributaries enter the estuary from the northern shore. The Sauce Chico River (drainage area of 1,600 km^2) discharges into the Principal Channel about 3 km downstream from the head of the estuary and the Naposta Grande Creek (drainage area of 920 km^2) reaches the estuary about 1 km downstream of Ingeniero White Port (Fig. 14.1). Both tributaries behave similarly in spring and summer during maximum mean rainfall but they are out of phase in autumn when the Sauce Chico River has a secondary mode. Although mean annual runoff for the Sauce Chico River and the Naposta Grande Creek are low (1.9 and 0.8 m^3 s^{-1}, respectively), runoff from the Sauce Chico River may peak between 10 and 50 m³ s⁻¹, with a recorded maximum of 106 m³ s⁻¹ in 1977. Freshwater inflow from other, smaller tributaries into the estuary is intermittent and only significant during periods of high local precipitation.

14.3.2 Tides

The principal energy input into the Bahía Blanca system is produced by a standing, semidiurnal tidal wave. The propagation of the wave is affected by the geometry of channels. Reflection on the channel flanks and the head convert the original progressive form into a standing wave, while bottom and wall friction drain energy from the wave. The amplitude of the tidal wave increases with a decrease in depth and/or breadth of the channel; therefore, the actual amplitude of the tide along the estuary is directly related to a balance between friction and convergence. Bahía Blanca is a hypersynchronous-type estuary (Fig. 14.2A), where the amplitude increases steadily from the mouth to the head, implying that the convergence effect on the tidal wave is larger that the friction effect. A strong reduction occurs along the Sauce Chico River, due to a frictional effect associated with dampening produced by river flow at one side and spreading over salt flats at the other side. The average rate of tidal energy dissipation per unit mass of fluid due to friction is 0.0017 m² s⁻³ (Perillo and Piccolo 1991; Fig. 14.2B). Large negative dissipation rates at the outer reach of the estuary are due to differences in amplitude between the mouth (Bermejo Island) and the head. Differences in duration between flood and ebb indicate that depth

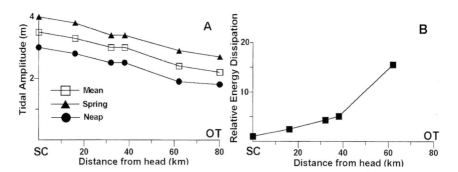

Fig. 14.2A,B. Tidal characteristics in the Principal Channel of Bahía Blanca. Distribution of the average mean, spring, and neap tide range between Sauce Chico River (*SC*) and Oceanographic Tower (*OT*; A) and energy dissipation of the tidal wave due to bottom and border friction (**B**). (Modified after Perillo and Piccolo 1999)

mean longitudinal current velocities in the Principal Channel and other major channels are asymmetrical and largely caused by the extensive tidal flats bordering the Principal Channel (Piccolo and Perillo 1990). The peak ebb current is about twice the maximum flood current in the upper reaches; however, flood usually has a longer duration. The residual circulation shows a significant difference in the direction of mass transport, causing salt concentrations in the inner portion of the estuary often exceeding those of the inner continental shelf.

14.3.3 Winds

Mid-latitude westerly winds and the influence of the Subtropical South Atlantic High dominate the typical weather pattern of the region. The resulting circulation induces strong NW and N winds with a mean velocity of 24 km h^{-1} during 40% of the year and gusts of over 100 km h^{-1}. Winds generate waves, storm surges, and subtidal sea level variations in the estuary. Wind waves (about 5–10 cm high and 1–3 m long) occur in channels and on tidal flats when covered by water. The incoming tide, together with N and NW winds, forms interaction waves, which are steep and up to 1.5 m high and 10–20 m long. However, the main effect of the predominant NW and N winds is advancing the time of low water, delaying the time of high water, and reducing predicted water levels. Large deviations between predicted astronomical tides and actual tide levels occur at Ingeniero White Port (–4.01 and 2.39 m) and the Oceanographic Tower (–1.51 and 1.87 m; Perillo and Piccolo 1991). The maximum negative values coincide with NW winds and maximum positive values with winds from the SW. At the same stations, the low-frequency sea level response to wind indicates that time scales of more than 10 days prevail while short-scale energy peaks correspond to approximately 3-day periods (Piccolo and Perillo 1989).

14.3.4 Salinity and Temperature

Mean annual (13 °C), summer (21.6 °C), and winter (8.5 °C) surface water temperatures in the Principal Channel are always slightly higher at the head of the estuary (Piccolo et al. 1987; Fig. 14.3A), while mean surface salinity increases exponentially from the head to mid-reaches of the estuary (Fig. 14.3B). Longitudinal temperature distributions vary between rainy periods in spring/summer and low runoff in winter, when the vertical thermal structure of the estuary is homogeneous and longitudinal variations

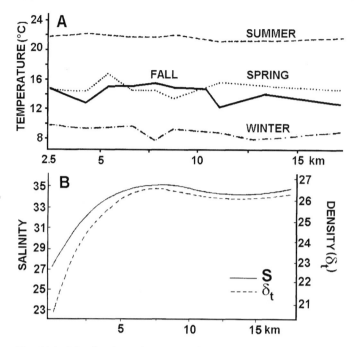

Fig. 14.3. Distribution of mean surface temperature (**A**) and salinity/density (**B**) in inner and mid-reaches of Bahía Blanca estuary (1967–1984). (Modified after Perillo and Piccolo 1999)

are less than 3 °C (Fig. 14.4A,C). Depending on runoff conditions, salinity differences between the mouth and head of the estuary may reach 17 and more than 4 between surface and bottom. At high water, stratification (halocline 1–3 m) may occur in areas of freshwater inflow, while salinity tends to be homogeneous in the outer estuary (Fig. 14.4B,D). Salinity and temperature distribution therefore characterize the Bahía Blanca Estuary as type 1a (Hansen and Rattray 1966). During normal runoff, the partially mixed inner region between the mouth of the Sauce Chico River and Ingeniero White Port tends to become homogeneous at low runoff, while salinity patterns in the outer homogeneous region are similar to the adjacent continental shelf. The boundary between these regions depends on river discharge (Martos and Piccolo 1988; Piccolo and Perillo 1990).

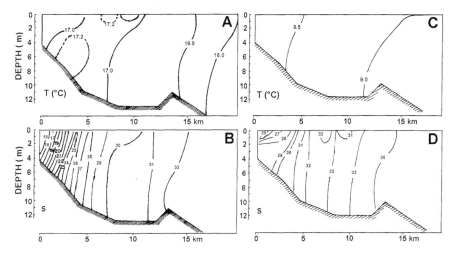

Fig. 14.4. Longitudinal distribution of temperature (*T*) and salinity (*S*) in inner and mid-reaches of the Principal Channel during spring-summer with high freshwater discharge (**A,B**) and during winter with low freshwater discharge (**C,D**). (Modified after Perillo and Piccolo 1999)

14.4 Biological Communities

14.4.1 Benthos

The distribution of benthic microalgae (mainly diatoms) on the extensive intertidal mudflats in the Bahía Blanca Estuary is a result of tidal influence, sedimentation, and physico-chemical factors. Apart from benthic freshwater and saltwater species (i.e., *Nitzschia*), several planktonic species (i.e., *Amphriprora alata, Cylindroteca closterium, Gyrosigma fasciola, Nitzschia sigma, Surirella gema*) are part of the microalgal flora and contribute to the first level of the trophic web (Cicerone 1987). Owing to generally unstable substrate characteristics, the establishment of a benthic macroalgal flora is limited to artificial substrates. Among the dominant freshwater species are *Enteromorpha flexuosa* and *Cladophora surera* while marine species are represented by *Ulva* sp., *Chaetomorpha aerea, Gracilaria verrucosa, Polysiphonia* sp., *Ceramium* sp., *Ectocarpus siliculosus, Hincksia hincksiae,* and *Punctaria* sp. Large salt marsh areas of the Bahía Blanca Estuary are covered by different species of the cordgrass *Spartina* ("espartillar"). *Spartina alterniflora* is exclusive to flooded littoral zones while *S. densiflora* coexists with *Salicornia ambigua* on salty substrata with permanent air exposure.

Three different macrobenthic faunal associations occur in the estuary (Elías 1985). An association of heterogeneous species is found in different vertical substrates. The lower and middle mesolittoral is occupied by *Laeonereis acuta* and *Eteone* sp., which are occasionally replaced by *C. altimanus* or bivalves. *Littoridina australis*, *Cyrtograpsus altimanus*, and *Tagelus gibbus* inhabit the middle and upper mesolittoral and may be substituted by *Scolecolepides* sp. An association composed of tellinids, thracids, Buccinidae, amphipods, nemerteans, and polychaetes, which tend to be deposit feeders, occupies unvegetated mudflats with *Venus* and *Tellina* in the lower littoral of the southern estuary. Dense populations of the burrowing crab *Chasmagnatus granulata* represent a third association on sandy substrate in the upper intertidal of salt marshes and mudflats ("cangrejal"). In *Spartina*-dominated marshes *Chasmagnatus* acts as a herbivore and on mudflats it largely feeds on organic deposits (Iribarne et al. 1997). The extensive burrows of the crabs tend to retain detritus, and thus control organic matter flow between marshes and the open estuary (Botto and Iribarne 2000).

14.4.2 Plankton

Diatoms are the principal components of the phytoplankton community, which appears to have low specific diversity since only 36 taxa have been cited for the estuary (Popovich 1996). *Thalassiosira curviseriata*, *T. eccentrica*, *T. minima*, *Rhizosolenia delicatula*, *Skeletonema costatum*, and *Cyclotella* sp. are the most frequent (>50%) species. Other representative groups are Prasinophyceae, Crytophyceae, Euglenids, Cyanophyceae, and Sillicoflagellates, while species like *Ophiocytium* sp. (Xanthophyceae) and dinoflagellates (*Protoperidinium* sp., *Prorocentrum micans*) are occasionally responsible for red tides (Gayoso 1988). Neritic warm water forms and oceanic elements (i.e., *Actinoptychus*, *Coscinodiscus*) may occasionally become dominant. Phytoplankton succession initiates in winter (June/September) with species like *Skeletonema costatum*, *Thalassiosira hibernalis*, *T. anguste-lineata*, and *Chaetoceros debilis*, followed by *Rhizosolenia delicatula* and *Ditylum brightwelli* in spring (September/October), and by "pulses" of flagellates (i.e., *Pyramimonas* sp.) in December and January (Gayoso 1988). High nutrient levels and low zooplankton grazing pressure in winter and spring favor blooms of *Tallassiosira anguste-lineata*, *T. rotula*, *T. pacifica*, and *Chaetoceros* sp. (Freije et al. 1980). However, *Tallassiosira curviseriata* dominates because the species is well adapted to low temperatures and high turbidity in the inner estuary (Popovich 1996). The shallow euphotic zone of the estuary is compensated for by active mixing of the water column by winds; thus, chlorophyll-*a* concentrations (40–54 µg l^{-1}) and net primary production values (>200 µg C l^{-1} h^{-1}) reflect high photosynthetic activity and, at the end of blooms, nitrate, phosphate, and silicate are almost completely depleted.

In general, composition and spatial distribution of zooplankton in the Bahía Blanca Estuary is influenced by highly variable environmental parameters in the inner estuary and more stable conditions of the coastal and marine environment. Coastal ciliates without a lorica, like the abundant *Strombidium sulcatum*, *Strombidium* sp., and *Cyrtostrombidium* sp., are among the principal components of the microzooplankton community. *Tintinnidium balechi* is most abundant and, together with *Tintinnopsis* sp., serves as an indicator of water quality in the Bahía Blanca Estuary (Barría de Cao 1992). The presence of typical warm water species (i. e., *Tintinnopsis radix*) and *Stylicauda platensis* of the Brazil Current suggests mixture of warm and cold waters. The meso- and macrozooplankton community (0–4 m depth) is composed of Cnidaria, Ctenophora, Lophochordata, Mollusca, Chaetognata, Annelida, Arthropoda (Copepoda), and Chordata (Hoffmeyer 1994). During the entire year, the calanoid copepod *Acartia tonsa* is the most frequent holoplanktonic species, followed by *Paracalanus parvus* and *Euterpina acutifrons*. Meroplankton (35 taxa) and freshwater elements (*Daphnia* sp., *Boeckella poopensis*) dominate the estuarine zooplankton community. The eggs of the cupleid *Brevoortia aurea* and larval stages of *Conopeum* sp. and *Chasmagnatus granulata* are frequent in summer (20 °C) and fall (18.2 °C). The zoea larvae and megalopa forms of *Chasmagnatus granulata* are important elements in spring (October 78 %) and fall (March 51 %). Most zoea larvae of the Grapsidae (*Pachysteles haigae*), Caridae (*Pagurus* sp.), and Majidae (*Pilumnus reticulatus*, *Peisos petrunkevitchi*) are widely distributed in the estuary and well adapted to temperature changes (Cervellini 1987). Other species, like Caridea (*Betaeus lilianae* and *Latreutes parvulus*) have coastal preference or are stenothermal (i. e., *Corystoides chilensis*, *Artemesia longinaris*, *Pleoticus muelleri*). Stenohaline forms of decapods tend to remain near the estuary entrance.

14.4.3 Nekton

Important components of the estuarine ichthiofauna are *Brevoortia aurea*, the whitemouth croaker *Micropogonas furneri*, the striped weakfish *Cynoscion striatus*, *Odontesthes bonaerensis*, and *Mustellus schmitti*. The adults of *Brevoortia aurea* migrate into the estuary during spawning in spring and summer (October-March) and leave as juveniles at the end of summer. Juveniles and adults of *Micropogonas furneri* and *Cynoscion striatus* coexist in the estuary during fall (López Cazorla 1996) when they prey on *Peisos petrunkevitchi*. The rest of the year they prey on other fishes, such as *Ramnogaster arcuata*, but also on *Micropogonas furneri* and *Cynoscion striatus*. Fishing activities inside the estuary are reduced and only about 600 tons of *Micropogonias furnieri, Cynoscion striatus*, and shrimps are landed each year.

Large shallow areas of the estuary are visited by the cetacean "Franciscana" (*Pontoporia blainvillei*). A small colony of sea lions (*Otaria flavescens*) at the outer part of the Principal Channel and at the tip of Trinidad Island is occasionally preyed upon by killer whales. Several migratory bird species of the Charadriiformes use Bahía Blanca Estuary as a resting and feeding place and are abundant in channels, on muddy shores, and on the islands (Belenguer et al. 1992). The abundant *Limosa haemastica* is present during the whole year, *Calidris fuscicolis* migrates during the winter, *Calidris alba* occurs during summer and fall, and *Charadrius falklandicus* migrates during spring. Rare breeding colonies of the endemic crab or Otrog seagull (*Larus attruticus*) are found on some islands where *Chasmagnatus granulata* is their exclusive diet.

14.5 Impact and Management

During the last 15 years, the city of Bahía Blanca has rapidly expanded to a total of 350,000 inhabitants. The demographic growth has been fueled by a large petrochemical park, fertilizer and thermoelectric plants, as well as expanding port activities since much of Argentina's export moves through Ingeniero White Port (and the subsidiaries Puerto Rosales, Base Naval Puerto Belgrano, Puerto Galván). As a consequence, the discharge of industrial wastes (oil derivatives, pesticides, heavy metals, etc.) and untreated domestic sewage (estimated at $10\,m^3\,s^{-1}$) has generated increasing problems of contamination. Furthermore, the Bahía Blanca region represents the southern border of the wet "pampas" ($680\,mm\ year^{-1}$) with extensive agricultural production and cattle breeding, thus pesticide and fertilizer addition to the estuary might be a potential hazard. However, concentrations of Cu ($7.30\,\mu g\ g^{-1}$), Cd ($0.44\,\mu g\ g^{-1}$), and Zn ($35.5\,\mu g\ g^{-1}$) in sediments and total dissolved Cu and Zn levels in water (Villa 1988) tend to be below contamination levels. Cadmium concentration ($0.9-8.8\,\mu g\ l^{-1}$) is comparable to that in estuarine mixing areas (Sericano and Pucci 1982) and tributaries (Zubillaga and Pucci 1986) while mercury concentration tends to be highest in top predators (Marcovecchio et al. 1988, 1991; Marcovecchio 1994). Finally, dredging of the Principal Channel from 9.5- to 13.5-m depths has generated over 35 × 10⁶ m³ of sediment, which was dumped on tidal flats and in offshore locations. Dredging and deposition has introduced major changes of circulation patterns in the estuary. It appears, however, that the dynamics of Bahía Blanca Estuary and the residual export of water and dissolved and suspended elements favor auto-depuration of the system with the concentration of pollutants generally being lower than international standards.

Acknowledgements. The Consejo Nacional de Investigaciones Científicas y Técnicas (CONICET) de la República Argentina (1983-1998) and Universidad Nacional del Sur (1994-1998) provided financial support. The authors thank Lic. Walter D. Melo for drawing some figures.

References

Aliotta S, Perillo GME (1987) A sand wave field in the entrance to Bahia Blanca Estuary, Argentina. Mar Geol 76:1-14

Barría de Cao MS (1992) Abundance and species composition of Tintinnina (Ciliophora) in Bahía Blanca Estuary, Argentina. Estuar Coast Shelf Sci 34:295-303

Belenguer C, Delhey K, Di Martino S, Petracci P, Scorolli A (1992) Observaciones de aves playeras migratorias de Bahía Blanca. Bol Inform Grupo Arg Limícolas 10:2-4

Botto F, Iribarne O (2000) The effect of the burrowing crab *Chasmagnathus granulata* on the benthic community of a SW Atlantic coastal lagoon. J Exp Mar Biol Ecol (in press)

Cervellini M (1987) Las larvas y postlarvas de Crustaceos Decapoda en el estuario y aguas marinas de Bahía Blanca (Pcia. de Buenos Aires). Distribución espacial, variación estacional y su relación con los factores ambientales. PhD Thesis, Univ Nac del Sur, Bahía Blanca, Argentina

Cicerone D (1987) Estimación de la biomasa de diatomeas bentónicas en áreas mesolitorales de la porción interna del Estuario de Bahía Blanca, a partir del análisis de pigmentos fotosintéticos. Lic Thesis, Univ Nac del Sur, Bahía Blanca, Argentina

Cuadrado DG, Perillo GME (1997) Migration of intertidal sand banks at the entrance of the Bahía Blanca Estuary, Argentina. J Coastal Res 13:139-147

Elías R (1985) Macrobentos del estuario de la Bahía Blanca (Argentina) I. Mesolitoral. Spheniscus 1:1-33

Freije RH, Zavatti JR, Gayoso AM, Asteasuain RO (1980) Pigmentos, producción primaria y fitoplancton del estuario de Bahía Blanca. IADO Contribución Científica 46:1-13

Gayoso AM (1988) Variación estacional del fitoplancton en la zona más interna del estuario de Bahía Blanca (Argentina). Gayana Bot 45:241-248

Ginsberg SS, Perillo GME (1990) Channel bank recession in the Bahía Blanca Estuary, Argentina. J Coastal Res 6:999-1010

Ginsberg SS, Perillo GME (1999) Deep scour holes at the confluence of tidal channels in the Bahía Blanca Estuary, Argentina. Mar Geol 160:171-182

Gómez EA, Perillo GME (1992) Largo Bank: a shoreface-connected linear shoal at the Bahía Blanca Estuary entrance, Argentina. Mar Geol 104:193-204

Gómez EA, Ginsberg SS, Perillo GME (1997) Geomorfología y sedimentología de la zona interior del Canal Principal del Estuario de Bahía Blanca. Rev Asoc Arg Sediment 3:55-61

Hansen DV, Rattray M (1966) New dimensions in estuary classification. Limnol Oceangr 11:319-326

Hoffmeyer M (1994) Seasonal succession of Copepoda in the Bahía Blanca Estuary. Hydrobiologia 292/293:303-308

Iribarne O, Bortolus A, Botto F (1997) Between-habitats differences in burrow characteristics and trophic modes in the southwestern Atlantic burrowing crab *Chasmagnathus granulata*. Mar Ecol Prog Ser 155:132-145

López Cazorla A (1996) The food of *Cynoscion striatus* (Cuvier) (Pisces: Sciaenidae) in the Bahía Blanca area, Argentina. Fish Res 28:371-379

Marcovecchio JE (1994) Trace metal residues in tissues of two crustacean species from the Bahía Blanca Estuary, Argentina. Environ Monit Assess 29:65-73

Marcovecchio JE, Moreno VJ, Perez A (1988) Determination of heavy metal concentrations in biota of Bahía Blanca, Argentina. Sci Total Environ 75:181-190

Marcovecchio JE, Moreno VJ, Perez A (1991) Metal accumulation in tissues of sharks from the Bahía Blanca Estuary, Argentina. Mar Environ Res 31:263-274

Martos P, Piccolo MC (1988) Hydrography of the argentine continental shelf between 38 and 42°S. Continent Shelf Res 8:1043-1056

Montesarchio L, Lizasoain W (1981) Dinámica sedimentaria en la denominada ría de Bahía Blanca. Instituto Argentino de Oceanografía Contrib Cient 58:1-202

Perillo GME (1989) Estuario de Bahía Blanca: definición y posible origen. Bol Centro Naval 107:333-344

Perillo GME, Cuadrado DG (1991) Geomorphologic evolution of El Toro Channel, Bahía Blanca Estuary (Argentina), prior its dredging. Mar Geol 97:405-412

Perillo GME, Piccolo MC (1991) Tidal response in the Bahía Blanca Estuary. J Coastal Res 7:437

Perillo GME, Piccolo MC (1999) Bahía Blanca Estuary: a review of its geomorphologic and physical characteristics. In: Perillo GME, Piccolo MC, Pino Quivira M (eds) Estuaries of South America: their geomorphology and dynamics. Springer, Berlin Heidelberg New York (in press)

Perillo GME, Sequeira ME (1989) Geomorphologic and sediment transport characteristics of the middle reach of the Bahia Blanca Estuary (Argentina). J Geophys Res 94:14351-14362

Perillo GME, Garcia Martinez MB, Piccolo MC (1996) Geomorfología de canales de marea: análisis de fractales y espectral. Actas VI Reunión Arg Sediment 155-160

Piccolo MC, Perillo GME (1989) Subtidal sea level response to atmospheric forcing in Bahía. Proceedings of the 3rd International Congress on Southern hemisphere meteorology and oceanography, pp 323-324

Piccolo MC, Perillo GME (1990) Physical characteristics of the Bahía Blanca Estuary (Argentina). Estuar Coast Shelf Sci 31:303-317

Piccolo MC, Perillo GME, Arango JM (1987) Hidrografía del estuario de Bahía Blanca, Argentina. Rev Geof 26:75-89

Popovich C (1996) Autoecología de *Thalassiosira curviseriata* (Bacillariophyceae) y su importancia en el entendimiento de la floración anual de diatomeas en el estuario de Bahía Blanca (Pcia. Bs.As.), Argentina. PhD Thesis, Univ Nac del Sur, Bahía Blanca, Argentina

Sericano JL, Pucci AE (1982) Cu, Cd and Zn in Blanca Bay surface sediments, Argentina. Mar Pollut Bull 13:369-371

Villa N (1988) Spatial distribution of heavy metals in seawater and sediments from coastal areas of the Southeastern Buenos Aires Province, Argentina. In: Seeliger U, Lacerda LD, Patchineelam SR (eds) Metals in coastal environments of Latin America. Springer, Berlin Heidelberg New York, pp 30-44

Zubillaga HV, Pucci AE (1986) Cu, Cd, Pb and Zn in tributaries to Blanca Bay, Argentina. Mar Pollut Bull 17:230-232

15 The Sand Beach Ecosystem of Chile

E. JARAMILLO

15.1 Introduction

The Chilean coast extends for about 4,200 km. South of Chiloé Island the coast (42–56°S) is characterized by archipelagos, fjords, channels, and islands, owing to the sinking of the Chilean longitudinal Central Valley. Exposed sandy beaches of different morphodynamic types alternate with intertidal sand flats at the mouth of rivers in south-central Chile between 38°S and 42°S and with rocky peninsulas in the extreme north (19–30°S; Fig. 15.1). Seasonal cycles of beach sand erosion and accretion are typical. During erosion, beaches have concave profiles with coarser grained sand than during accretion periods, when profiles are convex.

15.2 Environmental Setting

The influence of different ocean currents causes a gradual northward increase of seawater temperatures (Viviani 1979; Brattström and Johanssen 1983). Spring-tide winter and summer water temperatures in the surf zone at Mar Brava (42°S) and Aguila (21°S) range from 12 to 14.5°C and from 17 to 18.8°C, respectively (Fig. 15.1). The northern coast (19–30°S) is characteristic of a dry desert climate with low and irregular precipitation (Di Castri and Hajek 1976) and mean monthly air temperatures between 18 and 22°C in the summer and 12 to 17°C in the winter. Further south, along a warm-temperate zone (30–38°S), annual precipitation and mean monthly temperature are 110 mm and 15–22°C at 30°S and 760 mm and 10–13°C at 38°S (Brattström and Johanssen 1983), respectively. In the rainy Patagonia-Tierra del Fuego zone (38–56°S) mean annual temperatures decrease from 12.5°C (38°S) to 5.4°C at Cape Horn (56°S).

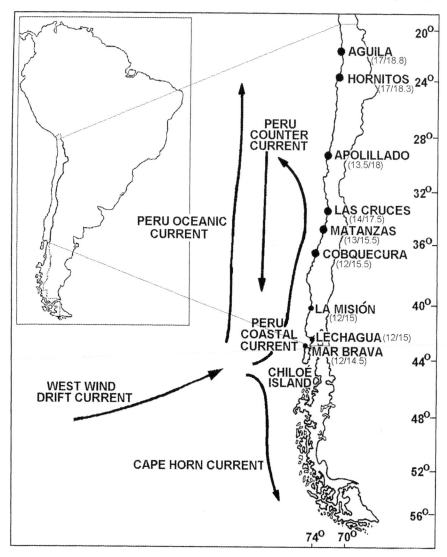

Fig. 15.1. Major ocean currents and nine sandy beaches along the Chilean coast with spring-tide winter (1998) and summer (1999) water temperature (°C) in the surf zone

15.3 Biological Components

The most common species of the sandy beach macroinfauna are the cirolanid isopod *Excirolana braziliensis*, the anomuran crab *Emerita analoga*, the insect *Phalerisida maculata*, the bivalve *Mesodesma donacium*, and the polychaete *Nephtys impressa*, all of which occur along the Chilean coast

The Sand Beach Ecosystem of Chile

Table 15.1. Sandy beach fauna from the northern (*N*, 20–23°S), north-central (*NC*, 29–30°S), central (*C*, ca. 33°S), and south-central (*SC*, 40–42°S) coast of Chile. Insecta Coleoptera (*IC*), Crustacea Amphipoda (*CA*), Crustacea Isopoda (*CI*), Crustacea Anomura (*CAN*), Crustacea Brachyura (*CB*), Crustacea Macrura (*CM*), Crustacea Stomatopoda (*CS*), Mollusca Bivalvia (*MB*), Annelida Polychaeta (*AP*) (based on Castilla et al. 1977; Sanchez et al. 1982; Jaramillo 1987b; Jaramillo et al. 1993, 1998; Brazeiro et al. 1998)

Taxa	N	NC	C	SC
Phalerisida maculata Kulser IC	x	x	x	x
Orchestoidea tuberculata Nicolet CA		x	x	x
Bathyporeiapus magellanicus Schellenberg CA			x	x
Phoxocephalopsis mehuinensis Varela CA				x
Huarpe sp. Barnard and Clark CA				x
Lysianassidae Dana CA				x
Tylos spinulosus Dana CI		x		
Excirolana braziliensis Richardson CI	x	x	x	x
Excirolana hirsuticauda Menzies CI		x	x	x
Excirolana monodi Carvacho CI			x	x
Macrochiridothea setifer Menzies CI			x	x
Macrochiridothea mehuinensis Jaramillo CI				x
Macrochiridothea aff. *lilianae* Moreira CI				x
Chaetilia paucidens Menzies CI			x	x
Emerita analoga (Stimpson) CAN	x	x	x	x
Lepidopa chilensis Lenz CAN	x	x	x	x
Blepharipoda spinimana Philippi CAN	x	x	x	x
Ogyrides tarazonai Wicksten and Méndez CM	x			
Ocypode gaudichaudii Milne Edwards CB	x			
Ovalipes punctatus (De Haan) CB				x
Pseudocorystes sicarius (Poeppig) CB				x
Bellia picta Milne Edwards CB		x	x	x
Nannosquilla chilensis Dahl CS				x
Mesodesma donacium (Lamark) MB	x	x	x	x
Donax peruvianus Deshayes MB	x			
Nephtys impressa Baird AP	x	x	x	x
Euzonus heterocirrus Rozbaczylo and Zamorano AP		x	x	x
Scololepis chilensis (Hartmann-Schröder) AP	x		x	
Leitoscoloplos sp. Monro AP	x	x		x
Lumbrinereis sp. Blainville AP	x			x
Spionidae undet. Grube AP		x		
Gliceridae undet. Grube AP	x	x		x
Onuphidae undet. Kinberg AP				x
Nemertina undet.	x	x	x	x

between 20°S and 42°S. Crustaceans (22 taxa), mainly peracarids, with three species of *Excirolana* and the idotheid isopods *Chaetilia paucidens* and *Machrochiridothea* (three species), are the most diverse group followed by polychaetes (eight taxa) (Table 15.1).

Emerita analoga and *Excirolana braziliensis* represent 90–99 % of the macroinfauna in sandy beaches at Aguila and Hornitos in northern Chile (20–23°S) while the same crab and *Excirolana hirsuticauda* represent about 72 % of the macroinfauna at Apolillado beach in north-central Chile (29°S). *Emerita analoga, Excirolana hirsuticauda*, and the talitrid amphipod *Orchestoidea tuberculata* are the dominant taxa at Cobquecura and Mar Brava beaches in south-central Chile (40–42°S), with a similar abundance of the polychaete *Euzonus heterocirrus* at Mar Brava. Most of these species are recruited during spring/early summer and in the fall (*Emerita analoga*; Osorio et al. 1967; Nuñez et al. 1974; Conan et al. 1975; Contreras et al. 1998) or during summer (*Excirolana braziliensis, E. hirsuticauda, O. tuberculata*; Zuñiga et al. 1985; Jaramillo 1987a). In general, macroinfaunal abundance varies significantly between seasons (Table 15.2). Low abundance in winter appears to be a result of habitat loss, owing to the erosion of beaches by storms that leads to steep beach-face slopes with coarse-grained sediments (Jaramillo 1987a).

The composition of macroinfauna changes along the shore. The brachyuran crabs *Ocypode gaudichaudii* and *Excirolana braziliensis* are commonly found in the drift line and dry zone of upper shores in northern Chile (20–23°S). The tylid isopod *Tylos spinulosus, Orchestoidea tuberculata*, and *Excirolana braziliensis* occupy similar levels in sandy beaches of north-central Chile (29–30°S). In contrast, only *Orchestoidea tuberculata* and *Excirolana braziliensis* are found in the upper shores of central (32–33°S) and south-central Chile (39–42°S; Fig. 15.2), probably due to their high desiccation tolerance (Jaramillo 1987a). Changes in the species composition of the upper shore levels have been related to latitudinal gradients of rainfall and sediment temperature (Jaramillo 1987b). The number of cirolanid isopods occupying the retention zone of mid-shore levels increases from lower to higher latitudes. However, in contrast to *Excirolana braziliensis*, the inter-

Table 15.2. Seasonal variability (1997–1998) of macro-infaunal abundance (ind. m^{-1}) at five sandy beaches of the Chilean coast (see Fig. 15.1 for approximate location of beaches)

Beaches	Spring/summer 1997	Winter 1998
Aguila	597,700	3,730
Hornitos	153,637	6,377
Apolillado	50,124	145,362
Cobquecura	28,660	1,280
Mar Brava	134,225	54,568

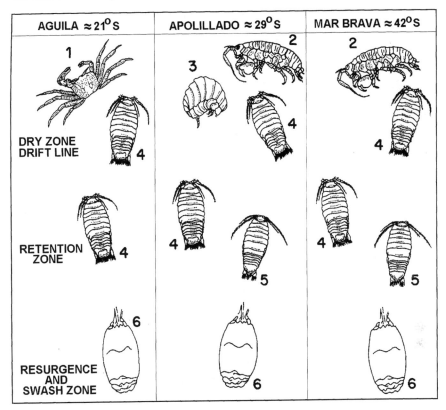

Fig. 15.2. Across-shore zonation of common sandy beach crustaceans at three selected beaches of the Chilean coast. Zonation scheme after Salvat (1969). *Ocypode gaudichaudii* (1), *Orchestoidea tuberculata* (2), *Tylos spinulosus* (3), *Excirolana braziliensis* (4), *Excirolana hirsuticauda* (5), and *Emerita analoga* (6)

tidal distribution of *Excirolana hirsuticauda*, which occupies lower beach levels (resurgence and swash zone), appears to be more affected by wave turbulence during tides (Jaramillo and Fuentealba 1993).

As elsewhere (Defeo et al. 1992), the structure of the macroinfauna community is closely related to different beach types, with richness and abundance tending to increase from reflective to dissipative beaches (McLachlan and Jaramillo 1995; McLachlan et al. 1993, 1996). Over a wide zoogeographic range (20–42°S), however, species number and abundance appear to be highest at beaches with intermediate characteristics (Las Cruces, Apolillado, Matanzas), and intermediate beaches at Matanzas (Dean=3.3) and Hornitos (Dean=3.5) have the same number of species as a dissipative beach at Mar Brava (Dean=10.5; Fig. 15.3). Furthermore, the morphology (presence of bays and horns) of intermediate beaches appears

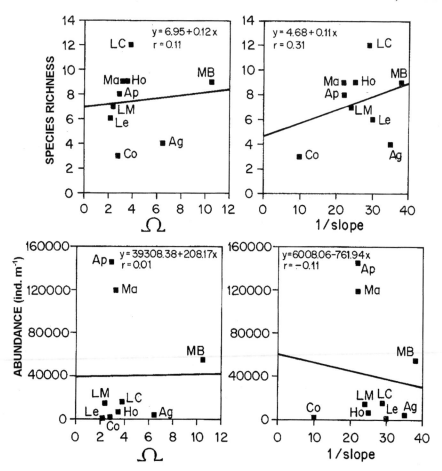

Fig. 15.3. Species richness and macrofaunal abundance in nine beaches of the Chilean coast in relation to Dean's parameter (Ω=wave height (cm)/wave period (s) × sand fall velocity; Short and Wright 1983) and beach face slope. Aguila (*Ag*), Hornitos (*Ho*), Apolillado (*Ap*), Las Cruces (*LC*), Matanzas (*Ma*), Cobquecura (*Co*), La Misión (*LM*), Lechagua (*Le*), and Mar Brava (*MB*)

to influence macroinfaunal abundance and zonation in central Chile (Brazeiro et al. 1998). It therefore seems that abundance is more affected by beach sand dynamics than by a single physical factor like mean particle size (Jaramillo 1987a). Physical factors related to El Niño events and large-scale oceanographic processes such as upwelling do not appear to induce differences in the macroinfaunal community structure of Chilean sand beaches (Jaramillo et al. 1998).

Competitive interactions affect the macrofaunal community organization elsewhere (Croker and Hatfield 1980); however, they appear to be absent from Chilean sandy beaches, even between closely related species like *Excirolana braziliensis* and *E. hirsuticauda* (Jaramillo 1987a). Fishes (*Eleginops maclovinus*) and birds, like the whimbrels (*Numenius phaeopus*) and sanderlings (*Calidris* spp.), prey on *Emerita analoga* in the swash zone of exposed beaches (Pequeño 1979), though predation does not appear to regulate the abundance of the population. The adults of upper shore species like *O. tuberculata* and the tenebrionid beetle *Phalerisida maculata* display similar patterns of nocturnal surface locomotor activity (Jaramillo et al. 1980), probably to escape predation by visually feeding shorebirds as well as to avoid desiccation at high temperatures and low humidity during daylight. The slope and water content of the sandy substrate probably influence daily migrations between different beach zones (Kennedy 1997).

15.4 Human Impacts

The surf clam *Mesodesma donacium* is the only invertebrate species widely harvested in the surf zone and the shallow subtidal area of exposed sandy beaches between 19°S and 42°S (Tarifeño 1980). In some areas clams ("machas") are collected by divers behind the surf zone. However, clams are mostly harvested during low tide by the fishermen ("macheros") who use their feet and body weight to excavate the sand until clams come up to the sediment-water interface. In north-central Chile (30°S), surf clams reach commercial size (70 mm) after 4–5 years. The maximum size of 109 mm at an age of 15 years decreases to 86 mm from north-central Chile to southern beaches at 37–38°S (Alarcón 1979; Tarifeño 1980, 1984). Landing data suggest a significant decrease in clam fisheries from 12,000–17,000 tons in 1987/89 to about 6,000 tons during 1995–1997 due to overexploitation.

The disposal of liquid and solid wastes along the Chilean coast has increased over the years. Accidental oil spills and the deposition of mine tailings have changed the zonation patterns of the intertidal macroinfauna and caused a decrease in macroinfaunal species richness and abundance (Castilla et al. 1977; Castilla 1983). Furthermore, Chilean sand beaches represent an important recreational resource due to their proximity to major urban centers. The mechanical disturbance of sands in the intertidal area by beach visitors is common during the summer holiday season, though it seems to have little effect on intertidal macroinfauna (Jaramillo et al. 1996).

Acknowledgements. I appreciate the collaboration of C. Añazco, M. Avellanal, H. Contreras, C. Duarte, L. Filún, M. González, F. Kennedy, M. Lastra, A. McLachlan, J. Muñoz, E. Naylor, M. Pino, P. Quijón, S. Silva and R. Stead during the last 10 years of research. Financial support was given by CONICYT (Chile) FONDECYT Projects 904-88, 191-92, 950001 and FONDAP O&BM, Programa Mayor no. 3.

References

Alarcón RT (1979) Informe sobre el recurso de machas (*Mesodesma donacium*) de la IV Región. Informe del Departamento de Oceanografía Biológica, Universidad del Norte, Coquimbo, Chile

Brattström C, Johanssen A (1983) Ecological and regional zoogeography of the marine benthic fauna of Chile. Sarsia 68:289–389

Brazeiro A, Rozbaczylo N, Fariña JM (1998) Distribución espacial de la macrofauna en una playa expuesta de Chile central: efectos de la morfodinámica intermareal. Investnes Mar Valparaíso 26:119–126

Castilla JC (1983) Environmental impact in sandy beaches of copper mine tailings at Chañaral, Chile. Mar Pollut Bull 14:459–464

Castilla JC, Sánchez M, Mena O (1977) Estudios ecológicos en la zona costera afectada por contaminación del Northern Breeze. I. Introducción general y comunidades de playas de arena. Medio Ambiente 2:53–64

Conan G, Melo C, Yany G (1975) Evaluation de la production d'una population littorale du crabe Hippidae *Emerita analoga* (Stimpson) par intégration des paramétres de croissance et de mortalité. 10th European Symposium on Marine Biology, vol 2, pp 129–150

Contreras H, Defeo O, Jaramillo E (1998) Life history of *Emerita analoga* (Stimpson) (Anomura, Hippidae) in a sandy beach of south central Chile. Estuar Coast Shelf Sci 48:101–112

Croker RA, Hatfiel EB (1980) Space partitioning and interactions in an intertidal sand-burrowing amphipod guild. Mar Biol 61:79–88

Defeo O, Jaramillo E, Lyonnet A (1992) Community structure and zonation of the macroinfauna on the Atlantic coast of Uruguay. J Coast Res 8:830–839

Di Castri F, Hajek (1976) Bioclimatología de Chile. Editora Universidad Católica de Chile, Chile

Jaramillo E (1987a) Community ecology of Chilean sandy beaches. PhD Thesis, University of New Hampshire, USA

Jaramillo E (1987b) Sandy beach macroinfauna from the Chilean coast: zonation patterns and zoogeography. Vie Milieu 37:165–174

Jaramillo E, Fuentealba S (1993) Down-shore zonation of two cirolanid isopods during two spring-neap tidal cycles in a sandy beach of south central Chile. Rev Chil Hist Nat 66:439–454

Jaramillo E, McLachlan A (1993) Community and population responses of the macrofauna to physical factors over a range of exposed sandy beaches in south-central Chile. Estuar Coast Shelf Sci 37:615–624

Jaramillo E, Stotz W, Bertrán C, Navarro J, Román C, Varela C (1980) Actividad locomotriz de *Orchestoidea tuberculata* (Amphipoda, Talitridae) sobre la superficie de una playa arenosa del sur de Chile (Mehuín, Provincia de Valdivia). Stud Neotrop Fauna Environ 15:9–33

Jaramillo E, McLachlan A, Coetzee P (1993) Intertidal zonation patterns of macroinfauna over a range of exposed sandy beaches in south-central Chile. Mar Ecol Prog Ser 101:105-118

Jaramillo E, Contreras H, Quijón P (1996) Macroinfauna and human disturbance in a sandy beach of south-central Chile. Rev Chil Hist Nat 69:655-663

Jaramillo E, Carrasco F, Quijón P, Pino M, Contreras H (1998) Distribución y estructura comunitaria de la macroinfauna bentónica en la costa del norte de Chile. Rev Chil Hist Nat 71:459-478

Kennedy F (1997) The locomotor activity of peracarid crustaceans on wave-exposed sandy beaches of Chile. PhD Thesis, University of Wales, UK

McLachlan A, Jaramillo E (1995) Zonation on sandy beaches. Oceanogr Mar Biol Rev 33:305-335

McLachlan A, Jaramillo E, Donn TE, Wessels F (1993) Sandy beach macroinfauna communities and their control by the physical environment: a geographical comparison. J Coast Res 15:27-38

McLachlan A, De Ruyck A, Hacking N (1996) Community structure on sandy beaches: patterns of richness and zonation in relation to tide range and latitude. Rev Chil Hist Nat 69:451-467

Nuñez J, Aracena O, López MT (1974) *Emerita analoga* en Lico, provincia de Curicó (Crustacea, Decapoda, Hippidae). Boln Soc Biol Concepción 48:11-22

Osorio C, Bahamonde N, López MT (1967) El limanche [*Emerita analoga* (Stimpson)] en Chile. Boln Mus Nac Hist Nat Chile 29:60-116

Pequeño G (1979) Antedentes alimentarios de *Eleginops maclovinus* (Valenciennes, 1830) (Teleostomi: Notothenidae), en Mehuín, Chile. Acta Zool Lilloana (Argentina) 35:207-230

Salvat B (1969) Les conditions hydrodynamiques interstitielles des sediments meubles intertidaux et la repartition verticale de la faune endogée. CR Acad Sci Paris 259:1575-1579

Sanchez M, Castilla JC, Mena O (1982) Variaciones verano-invierno de la macrofauna de arena en playa Morrillos (Norte Chico, Chile). Stud Neotrop Fauna Environ 17:31-49

Short A, Wright L (1983) Variability of sandy shores. In: McLachlan A, Erasmus T (eds) Sandy beaches as ecosystems. Junk Publishers, The Hague, pp 133-144

Tarifeño E (1980) Studies on the biology of the surf clam *Mesodesma donacium* (Lamarck, 1818) (Bivalvia: Mesodesmatidae) from Chilean sandy beaches. PhD Thesis, University of California, USA

Tarifeño E (1984) Manejo y evaluación de la macha (*Mesodesma donacium*) en la Provincia de Arauco, VIII Región. Documentos de Difusión, Intendencia Regional, SERPLAC VIII Región. Pontificia Universidad Católica de Chile, Talcahuano

Viviani CA (1979) Ecogeografía del litoral chileno. Stud Neotrop Fauna Environ 14:65-123

Zuñiga O, Peña R, Clarke M (1985) Historia de vida y producción de *Excirolana braziliensis* Richardson, 1912 (Isopoda, Cirolanidae). Estud Oceanol 4:9-19

16 The Peruvian Coastal Upwelling System

J. TARAZONA and W. ARNTZ

16.1 Introduction

The Peruvian coastal upwelling ecosystem is part of an upwelling area that extends between 4°S and about 40°S along the western coast of South America (Guillén 1983), with high productivity originating from equatorial upwelling south of the Galapagos Islands. Peruvian coastal upwelling is peculiar because winds sustain the upwelling process throughout the year (Zuta and Guillén 1970) and the El Niño-Southern Oscillation (ENSO) cycle induces distinct interannual ecosystem variability. The coastal upwelling system, which hardly comprises 0.02% of the total ocean surface, is of great significance because it determines the enormous productivity of Peruvian coastal waters, representing almost 20% of the world's landings of industrial fish.

16.2 Environmental Characteristics

16.2.1 Currents and Winds

The complex current system off Peru is associated with the Equatorial Current system (Fig. 16.1A). The superficial currents flow predominantly northward. The Peru (or Humboldt) Coastal Current (4–15 cm s^{-1}) is closest to the coast and confined to the uppermost 200 m depth, transporting cold water (14–16 °C) in summer. The Peru Oceanic Current (down to 700 m depth) reaches higher velocities than the Peru Coastal Current. Between these currents the weak, irregular southward flow of the Peru (or Humboldt) Subsurface Countercurrent is usually subsuperficial but occasionally reaches the surface. The subsurface currents constitute an extension of the Equatorial Undercurrent. The principal current, the Peru-

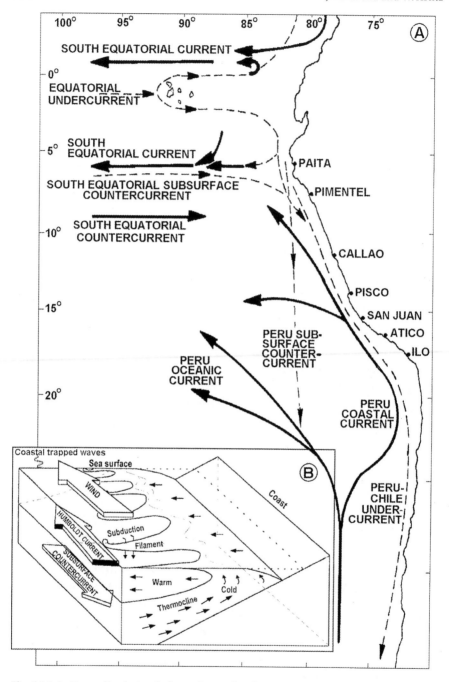

Fig. 16.1.A Generalized circulation scheme for the eastern South Pacific Ocean with sites of strong coastal upwelling. B Two-layer coastal upwelling system for the eastern South Pacific Ocean with the superficial layer constituted by the Peru (Humboldt) Coastal Current and the subsuperficial layer constituted by the Peru Subsurface Countercurrent. (After Arntz and Fahrbach 1991)

Chile Undercurrent, transports oxygen-deficient waters (4–10 cm s^{-1}) southward, which seriously impacts demersal and benthic species distribution. This countercurrent is more intense along the Peruvian than the Chilean coast.

Wind intensity and persistence are highest in winter and lowest in summer and wind stress, and consequently upwelling strength, has intensified over the last decades (Bakun 1990). Owing to anomalous warming of the sea during the 1983 El Niño event, which led to a steep thermal gradient between ocean and continent, a positive correlation between surface temperature and wind stress caused winds up to 10 m s^{-1} along the central Peruvian coast (up to 5°S).

16.2.2 Coastal Upwelling

Southeastern trade winds blow northward and parallel to the Peruvian coast. The Coriolis force together with friction forces in the water column generate a deflection towards the left that causes Ekman transport perpendicular to the coast, forcing deeper water to rise and forming an upwelling cell (Fig. 16.1B). In general, coastal upwelling along the Peruvian coast is superficial, with cold, low-oxygen, nutrient-rich water (salinity 34.6–35.1) being drawn from the lower border of the permanent thermocline (50–150 m depth). Upwelling is more intense in winter. Prominent upwelling plumes (4–6°S, 7–9°S, 11–13°S, 14–16°S) receive water from the southern extension of the Cromwell Current (4–6°S and north of 9°S), Peru Subsurface Countercurrent (north of 12°S), and Subantarctic Temperate Water (south of 14°S; Zuta and Guillén 1970). Filaments (cold water tongues), which extend 25–50 km offshore into the oceanic zone, generally persist less than 1 month. Depending on wind permanence, the width of the continental shelf, and the type of upwelling cell, some upwelling events may last only 3–10 days (Hill et al. 1998). Since filaments are deeper than the mixed layer, they enhance water exchange between the neritic and oceanic zones and may considerably impact the benthic organisms of deeper shelf and continental slope regions. The upwelled water off Peru totals approximately 3×10^{12} m^3 s^{-1}, with an ascending velocity of $5-30 \times 10^{-5}$ cm s^{-1} (mean 20×10^{-5} cm s^{-1}). Upwelling events off Chile (i.e., 18°30'S, 19°30'S, 23–29°S, 35–38°S) are less intense during winter, though filaments may extend from 60 to 125 km offshore (Strub et al. 1998).

16.2.3 Oceanographic Features

Interannual sea levels off Peru fluctuate by more than 40 cm (seasonal differences <15 cm), with even higher amplitudes during El Niño events. Tidal amplitude along the Peruvian coast does not exceed 1.2 m and semidiurnal tides predominate north of 5°S while diurnal tides occur towards the south. Sea surface temperatures in Peru and Chile tend to increase towards the west and north, thus creating both a zonal and latitudinal gradient, respectively, the latter being mainly a consequence of coastal upwelling. The average seasonal variation of surface water temperatures off Peru (winter 13–17 °C, summer 17–27 °C) and Chile (winter 11–17 °C, summer 13–24 °C) is 5–7 °C and 2–8 °C, respectively. Temperatures increase in El Niño years (1982–1983, 1986–1987, 1991–1992, 1997–1998), with an increment of almost 10 °C above the long-term mean during the 1983 event. Salinity ranges from 34.8 to 35.1 in nearshore upwelling zones off Peru and between 33.8 and 35.2 north of 34°S off Chile. During El Niño events, equatorial waters with lower salinity are displaced towards the south and oceanic water (salinity >35.1) approaches the coast (Tarazona et al. 1985).

Dissolved superficial oxygen concentration off the Peruvian and Chilean coast vary between 2–7 and 4–8 ml l^{-1}, respectively. Off the central Peruvian coast, oxygen minima (<0.5 ml l^{-1}) and frequent anoxia occur mostly between 50 and 700 m depth, owing to high organic matter accumulation and microbial activity (Rowe 1985). However, low oxygen values may also be associated with shallow depths (<20 m) in protected bays (Rosenberg et al. 1983; Tarazona et al. 1991). Although oxygen concentrations at the sea floor tend to diminish from north to south, below the minimum oxygen zone they rapidly increase with depth. Owing to an influx of equatorial water, dissolved oxygen values increase strongly during El Niño events (Arntz et al. 1991) and, on the whole, dissolved oxygen concentrations seem to be the principal selective factor for the benthic, demersal, and pelagic invertebrate and fish fauna on the continental shelf off Peru.

Coastal waters of the upwelling system off Peru are characterized by high phosphate (maximum >2.5 µg l^{-1}) and silicate (up to 20 µg l^{-1}) concentrations, while nitrite (0.1–1.6 µg l^{-1}) and nitrate (0.5–20 µg l^{-1}) concentrations tend to decrease towards offshore areas (approx. 90 km). High nitrate concentration corresponds to upwelling centers and denitrification is an important process because it is restricted to oxygen-deficient situations that characterize the coastal upwelling ecosystem. High rates of organic matter deposition are typical of sediments in upwelling areas. The magnitude of organic matter loss is controlled by the rate of heterotrophic activity and euphotic production. The remineralized nitrogen can be returned as ammonium and urea by the near-bottom flow towards near-shore areas (Barber and Smith 1981). Surface nutrients values tend to be lower off the Chilean coast and fluctuation of nitrate concentrations are associated with El Niño events (Chávez et al. 1989). In shallow areas along the central Peruvian coast, smelly "milky waters" or "white tides" are associated with abundant hydrogen sulfide and dissolved nitrite in the water column, due to lack of incorporation of elevated primary production into the pelagic food chains rather than to organic pollution.

16.3 Community Structure and Dynamics

As a result of a thermal barrier to the north and the distance of the northern limit of the Magellan Province, marine biota of the Peruvian Province shows a high degree of endemism. In general, coastal upwelling communities are characterized by low diversity. However, strong El Niño events introduce numerous species from the open ocean and the Panamanian Province, thus species richness can be considerable (Tarazona and Valle 1999).

16.3.1 Phytoplankton

The phytoplankton community (168 diatoms, 209 dinoflagellates) is dominated by cosmopolitan species, though some species are indicators of specific water masses. In shallow waters pennate diatoms are most conspicuous (i.e., *Cocconeis* sp., *Licmophora abbreviata*, *Grammatophora marina*, *Navicula membranacea*, *Pleurosigma elongatum*). In the Peru Coastal Current small diatoms (i.e., *Skeletonema costatum*, *Actinocyclus octonarius*, *Lithodesmium undulatum*, *Chaetoceros affinis*, *Thalassiosira subtilis*) and elevated reproduction rates prevail. Subtropical surface water of the Oceanic Province is dominated by dinoflagellates (*Ceratocorys horrida*, *C. trichoceros*, *C. hexacanthum* f. *sirale*, *C. gravidum*) and some large diatoms (*Chaetoceros radicans*, *Rhizosolenia temperei*, *Planktoniella sol*). Coastal upwelling areas are characterized by the dominance of chain-forming, colonial diatoms (5–30 μm) with high division rates, which use recently upwelled nutrients more effectively than pico- and nanoplankton. Phytoplankton primary production (3.84 mg C m^{-2} day^{-1}) and bacterioplankton biomass (up to 4.05 g C m^{-2}) of the Peruvian coastal upwelling ecosystem are fueled by the supply of additional nutrients, though silicates may be a limiting factor.

The success of pioneer species is a result of both rapid growth rates in nutrient-rich and turbulent environments and the composition and abundance of seed populations. Seeding of upwelling water may be a result of the presence of bloom-forming diatoms with resistant spores that quickly settle from the euphotic zone to the bottom, heteromorphic dinoflagellates with planktonic and non-mobile benthic stages, and active migration of the photosynthetic ciliate *Mesodinium rubrum* (Barber and Smith 1981). Species dominance and succession of the phytoplankton community largely depend on the frequency and intensity of upwelling and are associated with spatial and temporal transitions from a turbulent to a stratified water column. The stability of the water column increases along the axis of an upwelling plume with increasing distance from the coast. Since in each upwelling cell succession proceeds from small to large diatoms and dinoflagellates and from a dependence on new nitrogen to a dependence on regenerated nitrogen, an upwelling area represents a mosaic of water masses with different stages of succession.

16.3.2 Zooplankton

The dominant species of the zooplankton community in coastal water are copepods (*Acartia tonsa, Centropages brachiatus*) and larvae of polychaetes, brachiopods, cirripedes, and other crustaceans. At the margin of the continental shelf and above the slope the dominant or conspicuous taxa are represented by copepods (*Paracalanus parvus, Calanus* spp., *Oncaea* spp.) and large holoplankters such as siphonophores, chaetognaths, and euphausiids. Due to hypoxic conditions at greater depths, most of the zooplankton is restricted to the uppermost 30 m of the water column; however, some species (i.e., the copepod *Eucalanus inermis*) migrate to greater depths with low (<0.1 ml l^{-1}) oxygen concentration. The volume of zooplankton decreases from north to south, though seasonal variability is similar along the coast, with a marked increase in spring.

A fundamental characteristic of the Peruvian upwelling system is the low (5%) transfer efficiency from primary to secondary production. Nevertheless, high primary production provides abundant food for grazers and sustains large populations of copepods, euphausiids, anchovies, and also myctophiids at greater depth on the slope. Protozoans, principally ciliates and zooflagellates, also attain high biomass (up to 0.97 g C m^{-2}). Bacterial and protozoan plankton seem to constitute the primary diet of herbivorous mesozooplankton and the abundance of foraminiferans, coccolithophorids, and siliceous radiolarians in the sediments suggests their importance in the benthic subsystem. Copepods and euphausiids quickly respond to an increase in food by intensifying egg production, although their response in terms of growth rate is much slower. Juveniles (nauplii and first copepodites) remain close to the sea surface, whereas older stages undertake downward migrations into the subsurface layer and are transported back to the coast. Vertical migration is impeded where hypoxic conditions prevail in the subsurface layer. Zooplankton grazers have developed behavioral responses that maximize grazing time on phytoplankton blooms and transport them further offshore with surface layer waters, although blooms of primary producers and small crustacean grazers do not necessarily coincide. Different phyto-zooplankton communities, one close to upwelling centers and the other further offshore, appear to constitute an early and a mature stage of succession in the Peruvian upwelling cycle, respectively (Vinogradov and Shushkina 1984). Re-circulation in a poleward direction, favored by the Peru Countercurrent and by eddies of the upwelling centers, prevents the loss of juvenile stages towards the equator.

16.3.3 Benthic Organisms

Intertidal and shallow subtidal zones of Peru and northern Chile display high species diversity (432 decapods and stomatopods, 348 polychaetes, 972 mollusks). Several commercially important filter-feeding bivalves (*Mesodesma donacium*, *Donax marincovichi*, *Gari solida*, *Semele solida*, *Protothaca thaca*, *Aulacomya ater*, *Argopecten purpuratus*, *Venus antiqua*, *Tagelus dombeii*, *Ensis macha*) form large populations. The community of *Aulacomya ater* (up to 40,000 g m^{-2}) off Pisco hosts 186 species. Populations of the mussel *Argopecten purpuratus*, the gastropod *Thais chocolata*, the decapod crustaceans *Cancer setosus*, *Cancer porteri*, and *Platyxanthus orbignyi*, the sea urchin *Loxechinus albus*, and the kelps *Lessonia* spp. and *Macrocystis pyrifera* develop high density and biomass. Kelp forests cover rocky bottoms down to about 15 m and provide niches and refuges for numerous associated species, as do intertidal mussel beds where *Perumytilus purpuratus* and *Semimytilus algosus* communities harbor about 87 and 77 species, respectively. The structure of the floral and faunal communities tends to be controlled by grazers (sea urchins, chitons, limpets, and other gastropods) and predators (sea stars, brachyurans, crabs, and fish), respectively. Macroalgal (i.e., *Macrocystis pyrifera*) and bivalve (i.e., *Aulacomya ater*, *Argopecten purpuratus*) population dynamics are influenced by the success of larval recruitment and by warmer and cooler conditions, which are determined by upwelling periodicity and the ENSO cycle (Wolff 1988; Arntz and Tarazona 1990). Dominant macrobenthic organisms of sandy beaches are large-sized species such as the ghost crab *Ocypode gaudichaudii*, the mole crab *Emerita analoga*, and the surf clams *Mesodesma donacium* and *Donax marincovichi*. The exposed sandy beaches of the Peruvian central coast reveal exceptionally high faunal density (up to 10,000 ind. m^{-2}) and biomass (up to 35,500 g wet weight m^{-2}) as a consequence of the extremely high water column productivity (Rowe 1985).

Elevated organic matter accumulation and high H$_2$S content (Rowe 1985; Arntz et al. 1991) of the Peruvian coastal upwelling ecosystem beyond 50 m depth induce hypoxic (<0.5 ml l^{-1}) or anoxic conditions and lead to a benthic community with low diversity and high dominance of only a few species. These conditions favor high biomass (>1,000 g wet weight m^{-2}) of giant filamentous bacteria (*Thioploca araucae*, *T. chileae*; Gallardo 1977) which link nitrogen and sulfur cycles in the sediment (Fossing et al. 1995). At lower depths and closer to the coast sulfide-rich sediments (>1,200 M) are dominated by different species of the giant filamentous bacteria *Beggiatoa*. Since benthic diversity follows an oxygen-related gradient, the species number and density of deeper benthic communities tend to increase along the continental shelf from the south to the north and during El Niño events, when the benthic system is flushed with high dissolved oxygen concentrations (Fig. 16.2; Arntz et al. 1991).

16.3.4 Fishes and Other Vertebrates

The high productivity of the Peruvian upwelling causes low diversity among pelagic and demersal ichthyofauna with elevated densities of some epipelagic and mesopelagic species. The dominant coastal water fishes are pilchard (*Sardinops sagax*) and anchovy (*Engraulis ringens*), of which the latter has been the dominant species in this ecosystem for at least 12,000 years (DeVries and Pearcy 1982). The jack mackerel (*Trachurus picturatus murphii*) and Pacific mackerel (*Scomber japonicus*) are common along the shelf margin and mesopelagic species of the outer upwelling system are largely represented by lantern fishes such as *Vinciguerria lucetia*. The Peruvian pelagic anchovy fishery initiated in the early 1950s reached annual landings of about 12.5 million tons in the early 1970s, followed by a sharp decline as a result of overfishing and due to the 1972–1973 El Niño impact. Although fisheries recovered, landings are much lower than in the 1970s (Pauly and Tsukayama 1987; Pauly et al. 1989). During the last decades, anchovy and pilchard have experienced considerable fluctuations and alternating patterns off Peru, with dominance of anchovy in cold La Niña and dominance of pilchard in warm El Niño years (Ñiquen et al. 1999). Oceanic species such as yellowfin tuna (*Thunnus albacares*) and giant squid (*Dosidicus gigas*) also sustain important fisheries.

The Peruvian Coastal Upwelling System 237

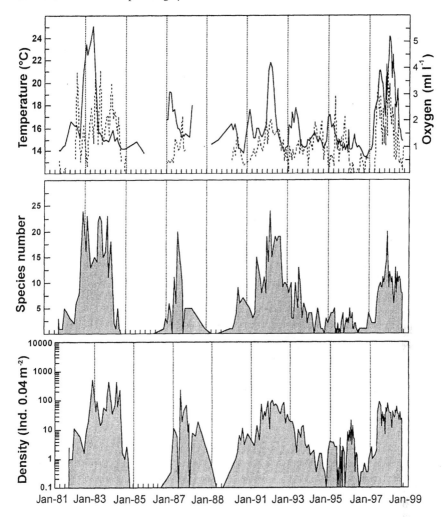

Fig. 16.2. Near bottom temperature (*full line*) and oxygen (*broken line*) concentrations at 34 m depth and species number and density of the soft bottom macrobenthic community

Demersal fisheries use trammel and drift nets as well as hook and line in well-oxygenated nearshore waters, while trawl fishery is limited to some areas off northern Peru (5–9°S), with Peruvian hake (*Merluccius gayi peruanus*) representing 70% (56,000 tons in 1993, 192,000 tons in 1996) of the landings. Other common species include gurnard (*Prionotus stephanophrys*), congrios (*Genypterus maculatus, Lepophidium negropinna*), flatfish (*Paralichthys adspersus, Hippoglossina macrops*), mullets (i.e., *Mugil cephalus, M. liza*), sea bass (*Paralabrax humeralis*), dogfish (*Mustelus* spp.), rays (*Myliobatis* spp.), and several sciaenids (*Cynoscion analis, Paralonchurus peruanus, Isacia conceptionis, Sciaena gilberti, S. deliciosa*).

The trophic flow of the upwelling ecosystem (4–14°S) off Peru is subject to intense fluctuations as a result of upwelling and fishery impacts. The length of food chains is subject to changes caused by alterations in stratification, upwelling, physico-chemical properties of the water column, and the distribution of fish and their food (Fig. 16.3). Lantern fishes are zoo-

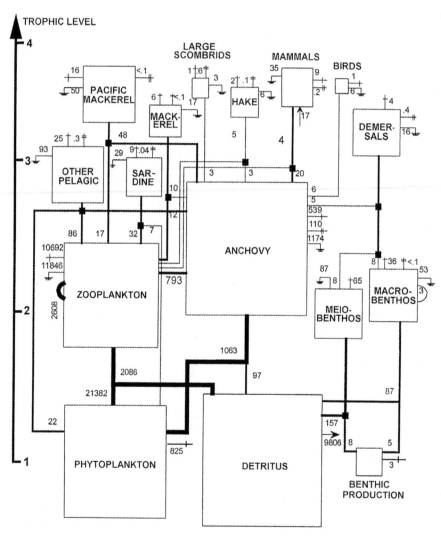

Fig. 16.3. Average trophic flows (10^3 kg km^{-2} year^{-1}) between 1973 and 1981 in the Peruvian coastal upwelling ecosystem. (After Jarre-Teichmann 1998)

plankton feeders while pilchards and anchovies feed on neritic and epipelagic resources such as diatoms and zooplankton. Although zooplankton (i.e., large copepods and euphausiids) is the principal diet of anchovy, during phytoplankton dominance direct consumption of microalgae may also occur (Pauly et al. 1989). In the absence of anchovies and pilchards, most of the plankton production is likely to be used further offshore by euphausiids and mesopelagic fishes, including hake (*Merluccius gayi*). Both jack and Spanish mackerels prey on anchovy and after the collapse of anchovy stocks (1964-1971), the trophic flow via pilchard increased sevenfold (Jarre-Teichmann 1998). Anchovy and, to a lesser degree, pilchard, mackerels, and hake sustain dense populations of guano birds, like cormorant (*Phalacrocorax bougainvillii*), booby (*Sula variegata*), and pelican (*Pelecanus thagus*), as well as sea lions (*Otaria byronia*) and fur seals (*Arctocephalus australis*; Muck 1989).

16.4 El Niño Impact on the Ecosystem

The oceanographic conditions along the Peruvian coast are influenced by an interannual variability known as El Niño-Southern Oscillation (ENSO). The ENSO cycle consists of a cold period (La Niña) and a warm period (El Niño) with anomalous warming of the eastern Pacific during about 3 months. El Niño occurs every 2-7 years (mean, 4 years), causing a deepening of the pycnocline and thus rendering the upwelling process inefficient (Barber et al. 1985). Other profound ecological alterations of the Peruvian coastal upwelling system during El Niño events are due to the arrival of Kelvin waves, temperature rise and deeper thermocline, altered currents, reduced nutrient transport, dissolved oxygen increase and increase in H_2S reduction in bottom sediments, etc. (Wyrtki 1982; Arntz 1986; Arntz and Tarazona 1990; Arntz and Fahrbach 1991; Jaimes 1999).

The severity of El Niño depends on the month of occurrence, relative strength and duration, and distance from the major (i.e., equatorial) zone of impact. Among the principal manifestations of El Niño is the "tropicalization" of the Oceanic Province off central and southern Peru. Under these circumstances the distribution of many species in the upwelling system changes due to immigration of species from the Panamanian Province and emigration of species to northern Chile. Some of the long-term changes, however, appear to be associated with global climatic change since simultaneous oscillations of pelagic shoaling fish occur in different upwelling systems.

Although upwelling may increase during El Niño events, primary production in the euphotic zone decreases by about 50% (1.03–1.68 g C m^{-2} d^{-1}) because the deeper pycnocline moves nutrient-rich waters out of upwelling reach (Chávez et al. 1989). During the 1982–1983 El Niño event, most autochthonous phytoplankton species of the Peru Coastal Current disappeared and were replaced by equatorial and oceanic warm water species. The replacement of native species and reduction of phytoplankton density negatively affected higher food web levels, in particular zooplankton grazers and pelagic fish. Furthermore, zooplankton and pelagic fish species suffered mortality due to the inflow of warm water and/or were replaced mainly by tropical predators, such as chaetognaths, salps, jellyfish, and siphonophores. The reduction of local holoplankton and meroplankton was accompanied by an increase of tropical holoplanktonic organisms and the range of tropical meroplankton distribution extended 10° further south (Tarazona et al. 1985).

Several warm water invertebrates immigrated from tropical areas, such as shrimps (mostly *Xiphopenaeus riveti* and *Penaeus californiensis*), swimming crabs (*Euphylax robustus, Euphylax dovii, Portunus acuminatus, Callinectes arcuatus, Arenaeus mexicanus*), spiny lobster (*Panulirus gracilis*), stalked barnacle (*Pollicipes elegans*), and mollusks (i.e., *Pteria sterna, Atrina maura, Malea ringens*; Arntz and Tarazona 1990; Paredes et al. 1998). The high temperatures and the increased swell during the 1982–1983 El Niño event together with warm water predatory swimming crabs from tropical areas caused mass mortality of key species like the mussel *Semimytilus algosus* and brown algae (*Macrocystis pyrifera, Lessonia* spp.) along Peruvian rocky shores (Tarazona et al. 1988). The subsequent decrease in density, biomass, and species numbers opened new space in the intertidal zone that was colonized by green algae (e.g., *Ulva lactuca*), followed by a red algal stage until the community stabilized with brown algal dominance and the return of invertebrates. On sandy beaches, reduced latitudinal distribution (7°) and mass mortality of the dominant surf clam *Mesodesma donacium* induced drastically lower density and biomass of the community. Mussel beds (*Aulacomya ater*) died and detached from the substrate and commercially important bivalves (*Semele* spp., *Gari solida, Tagelus dombeii*) and decapods (*Cancer setosus, Platyxanthus orbignyi*) suffered almost complete mortality. In contrast, the gastropod *Thais chocolata* and the octopus *Octopus mimus* benefited from El Niño conditions and landings of the warm water-tolerant scallop (*Argopecten purpuratus*) increased by 60- to 100-fold. Soft bottoms at the seafloor are favored by a pronounced increase in dissolved oxygen concentrations during El Niño events, leading to increased species richness and densities (Fig. 16.2).

El Niño events have both negative and positive effects on many commercially exploited species, with substantial economic implications for the economies of the region. During El Niño events (1972–1973, 1976, 1997–1998) with up to 8 °C temperature increase, anchovy changed from a wide distribution along the entire Peruvian coast to a dispersed, asymmetric distribution at the southern Peruvian littoral, whereas pilchards migrated from 5°S (Paita) to the central-southern coast (Ñiquen et al. 1999). Pelagic fisheries, which are mostly monospecific for anchovy, became multispecific because a large number of tropical and oceanic fish migrated into the pelagic zone of Peruvian-Chilean upwelling areas. Dolphinfish (*Coryphaena hippurus*), yellowfin tuna (*Thunnus albacares*), picuda (*Sphyraena ensis*), and Pacific mackerel (*Scomber japonicus*) were the first to invade the area, followed by samasa (*Anchoa nasus*), machete de hebra (*Opisthonema libertate*), penaeid shrimps, skipjack (*Katsuwonus pelamis*), and Myctophidae (i.e., *Bregmaceros bathymaster*). Other invading species were ayamarca (*Cetengraulis mysticetus*), fine jack mackerel (*Decapterus macrosoma*), big-eye jack mackerel (*Caranx hippos*), ribbonfish (*Desmodema polysticta*), and swimming crabs, mainly *Euphylax* sp. Some, like dolphinfish, frigate tuna (*Auxis rochei*, *A. thazard*), flying fish (*Hirundichthys rondeletii*, *Cheilopogon heterurus*, *Exocoetus volitans*), and pampanito (*Stromateus stellatus*) are reliable indicators of El Niño conditions.

During the strong 1982–1983 El Niño event, part of the anchovy and pilchard shoals migrated from the warmer epipelagic zone to remaining nearshore upwelling cells with cold, nutrient-rich water (Arntz 1986). The increase in density led to exceptional catches (170,000 tons of anchovy in 1 day) at the beginning of El Niño (1972). Another part of anchovy and pilchard shoals migrated to deeper waters (>100 m) and a third group migrated southward where pelagic fisheries off northern Chile benefited from recruitment (Alheit and Bernal 1993). Also, the silverside (*Odontesthes regia regia*), which sustains artisanal drift net fisheries, virtually disappeared from Peruvian water, possibly owing to migration into deeper waters or coastal areas off Chile. As a result of the different migrations, the pelagic food web collapsed and the withdrawal of pelagic fish shoals from areas almost devoid of food was reflected in reduced spawning and reproduction of anchovy and pilchard. Other important shoaling fish of the Humboldt upwelling, like the jack mackerel and Pacific mackerel, approached the shore between Ecuador and Chile (45°30'S) but were little affected by limited food resources.

Also, the shallow-water ichthyofauna reveals the impact of El Niño. More than 50 species (i.e., *Opisthonema* spp., *Cetengraulis mysticetus*, *Etrumeus teres*, *Scomberomorus sierra*, *Mycteroperca xenarcha*) that normally live further north or offshore occur in coastal waters off central Peru and northern Chile. Demersal fish catches off Peru decreased by about 60% because fish shoals dispersed and moved into deeper waters (i.e., hake below 1000 m) but nearly doubled off Chile (56,000–71,000 tons in 1982–1983), probably due to southward migration of part of the demersal stocks (Espino 1999). The limited food resources (i.e., anchovy populations) during strong El Niño events force fur seals, sea lions, and guano birds to migrate further south. The extension of offshore feeding excursions and dives to greater depths has led to high mortality of sea lions (50% of newborn) and fur seals (almost 100% of newborn) along the southern coast of Peru.

16.5 Need for Cautious Management

The peculiar structure and dynamics of the Peruvian coastal upwelling ecosystem make it one of the most productive marine systems worldwide. Owing to the immense natural nutrient input via upwelling and nutrient enrichment of waters around islands and capes by guano bird colonies, the assimilation of additional human-derived eutrophication is likely to be limited because areas of well-oxygenated shallow coastal water are extremely narrow. Despite both the extraordinary productive capacity (landings occasionally over 12 million tons year^{-1}) and resilience (recovery from anchovy overfishing and other resources; Arntz 1986; Alheit and Bernal 1993), coastal fishery resources have limited capacity for exploitation.

Warm and cold El Niño and La Niña events add to the high variability of the Peruvian upwelling ecosystem and further changes may be expected due to global warming. Therefore, an interdisciplinary approach for the comprehension of ecosystem functioning is necessary and monitoring programs and simulation models should be established to serve as a base for regulation of human activities and sustained exploitation. Despite some success in the 1980s (Pauly and Tsukayama 1987; Pauly et al. 1989; Jarre-Teichmann 1998) and recent progress in real time registration of El Niño and La Niña events by satellites and extended drift buoy systems, a holistic management approach has yet to be implemented for this upwelling ecosystem.

Acknowledgements. This work was supported by a grant from the Alexander von Humboldt Foundation to the first author during his sabbatical year at the Alfred Wegener Institute in Bremerhaven, Germany. We are grateful to the DePSEA Research Group, in particular Sonia Valle and Nora Malca. AWI contribution no. 1777.

References

Alheit J, Bernal P (1993) Effects of physical and biological changes on the biomass yield of the Humboldt Current Ecosystem. In: Sherman K, Alexander LM, Gold BD (eds) Large marine ecosystems: stress, mitigation, and sustainability. AAAS Press, Washington, DC, pp 53-68

Arntz WE (1986) The two faces of El Niño 1982-83. Meeresforschung (Rep Mar Res) 31:1-46

Arntz WE, Fahrbach E (1991) El Niño Klimaexperiment der Natur: die physikalischen Ursachen und biologischen Folgen. Birkhäuser, Basel

Arntz WE, Tarazona J (1990) Effects of El Niño on benthos, fish and fisheries off the South American Pacific coast. In: Glynn PW (ed) Global ecological consequences of the 1982-83 El Niño-Southern Oscillation. Elsevier Oceanography Series, Amsterdam, pp 323-360

Arntz WE, Tarazona J, Gallardo V, Flores L, Salzwedel J (1991) Benthos communities in oxygen deficient shelf and upper slope areas of the Peruvian and Chilean Pacific coast, and changes caused by El Niño. In: Tyson RV, Pearson TH (eds) Modern and ancient continental shelf anoxia. Geol Soc Spec Publ Lond 58:131-154

Bakun A (1990) Global climate change and intensification of coastal ocean upwelling. Science 247:198-201

Barber RT, Smith RL (1981) Coastal upwelling ecosystem. In: Longhurst AR (ed) Analysis of marine ecosystems. Academic Press, London, pp 31-68

Barber RT, Kogelschatz JE, Chavez FP (1985) Origin of productivity anomalies during the 1983-83 El Niño. Calcofi Rep XXVI:65-71

Chávez FP, Barber RT, Sanderson MP (1989) The potential primary production of the Peruvian upwelling ecosystem, 1953-1984. In: Pauly D, Muck P, Mendo J, Tsukayama I (eds) The Peruvian upwelling ecosystem: 1953-1984. ICLARM Conf Proc 18:50-63

DeVries TJ, Pearcy WG (1982) Fish debris in sediments of the upwelling zone off central Peru: a late Quaternary record. Deep Sea Res 28:87-109

Espino MA (1999) El Niño 1997-1998: su efecto sobre el ambiente y los recursos pesqueros en el Perú. Rev Peru Biol [Suppl] 6:266-278

Fossing H, Gallardo VA, Jörgensen BB, Hüttel M, Nielsen LP, Schulz H, Canfield DE, Forster S, Glud RN, Gundersen JK, Küver J, Ramsing NB, Teske A, Thamdrup B, Ulloa O (1995) Concentration and transport of nitrate by the mat-forming sulphur bacterium *Thioploca*. Nature 374:713-715

Gallardo VA (1977) Large benthic microbial communities in sulphide biota under Peru-Chile subsurface countercurrent. Nature 268:331-332

Guillén O (1983) Condiciones oceanográficas y sus fluctuaciones en el Pacífico sur oriental. FAO Fish Rep 3:607-658

Hill EA, Hickey BM, Shillington FA, Strub TP, Brink KH, Barton DE, Thomas AC (1998) Eastern ocean boundaries coastal segment. In: Robinson AR, Brink KH (eds) The sea, vol 11. Interscience, New York, pp 29-67

Jaimes E (1999) Condiciones meteorológicas a nivel global y local, cambio climático y El Niño 1997-98. Rev Peru Biol [Suppl] 6:291-299

Jarre-Teichmann A (1998) The potential role of mass balance models for the management of upwelling ecosystems. Ecol Appl 8:93-103

Muck P (1989) Major trends in the pelagic ecosystem off Peru and their implications for management. In: Pauly D, Muck P, Mendo J, Tsukayama I (eds) The Peruvian upwelling ecosystem: dynamics and interactions. IMARPE/GTZ/ICLARM, Manila, pp 386-403

Ñiquen M, Bouchon M, Cahuin S, Valdez J (1999) Efectos del fenómeno El Niño 1997-1998 sobre los principales recursos pelágicos en la costa peruana. Rev Peru Biol [Suppl] 6:304-315

Paredes C, Tarazona J, Canahuire E, Romero L, Cornejo O, Cardoso F (1998) Presencia de moluscos tropicales de la provincia panameña en la costa central del Perú y su relación con los eventos "El Niño". Rev Peru Biol 5:23-28

Pauly D, Tsukuyama I (eds) (1987) The Peruvian anchoveta and its upwelling ecosystem: three decades of change. ICLARM Stud Rev 15. IMARPE, Callao, Perú; GTZ, Eschborn, Federal Republic of Germany; and International Center for Living Aquatic Resources Management, Manila, Philippines

Pauly D, Muck P, Mendo J, Tsukayama I (eds) (1989) The Peruvian upwelling ecosystem: dynamics and interactions. IMARPE/GTZ/ICLARM, ICLARM, Manila, Philippines

Rosenberg R, Arntz WE, Chumán de Flores E, Flores LA, Carbajal G, Finger I, Tarazona J (1983) Benthos biomass and oxygen deficiency in the upwelling system off Peru. J Mar Res 41:263-279

Rowe GT (1985) Benthic production and processes off Baja California, northwest Africa and Peru: a classification of benthic subsystems in upwelling ecosystems. Int Symp Upw W Afr, Inst Inv Pesq, Barcelona 2:589-612

Strub PT, Mesías JM, Montecino V, Rutllant J (1998) Coastal ocean circulation off western South America. In: Robinson AR, Brink KH (eds) The global coastal ocean. The sea, vol 11. Interscience, New York, pp 273-313

Tarazona J, Valle S (1999) La diversidad biológica en el mar peruano. In: Halffter G (ed) Diversidad biológica en Iberoamérica, Programa CITED, vol III. Veracruz, México, pp 97-109

Tarazona J, Paredes C, Romero L, Blascovich V, Guzmán S, Sánchez S (1985) Características de la vida planctónica y colonización de los organismos epilíticos durante el fenómeno "El Niño". In: Arntz WE, Landa A, Tarazona J (eds) El Fenómeno "El Niño" y su impacto en la fauna marina. Special Issue Boln Inst Mar Perú-Callao, pp 41-49

Tarazona J, Salzwedel H, Arntz WE (1988) Oscillations of macrobenthos in shallow water of the Peruvian central coast induced by El Niño 1982-83. J Mar Res 46:593-611

Tarazona J, Canahuire E, Salzwedel H, Jeri T, Arntz WE, Cid L (1991) Macrozoobenthos in two shallows areas of the Peruvian upwelling ecosystem. In: Elliott, Ducrotoy JP (eds) Estuaries and coasts: spatial and temporal intercomparisons. ECSA 19 symposium, Amsterdam, Holland, pp 251-258

Vinogradov MY, Shushkina EA (1984) Succession of marine epipelagic communities. Mar Ecol Prog Ser 16:229-239

Wolff M (1988) Spawning and recruitment in the Peruvian scallop *Argopecten purpuratus*. Mar Ecol Prog Ser 42:213-217

Wyrtki K (1982) The southern oscillation, ocean-atmosphere interaction and El Niño. Mar Technol Soc J 16:3-10

Zuta S, Guillén O (1970) Oceanografía de las aguas costeras del Perú. Boln Inst Mar Perú Callao 2:157-324

17 The Gulf of Guayaquil and the Guayas River Estuary, Ecuador

R.R. Twilley, W. Cárdenas, V.H. Rivera-Monroy, J. Espinoza, R. Suescum, M. M. Armijos, and L. Solórzano

17.1 Introduction

The Gulf of Guayaquil (3°S, 80°W) of the coastal province of Guayas, Ecuador, is the largest (12,000 km^2) estuarine ecosystem on the Pacific coast of South America (Cucalón 1983). The Gulf has historically been defined by an outer and an inner estuary (Stevenson 1981). The outer estuary, referred to as the Gulf of Guayaquil, has a shelf boundary along the 81°W meridian with a distance of about 204 km at the mouth of the gulf. The boundary extends about 130 km inland to Puna Island (80°15'W) at the mouths of Morro and Jambeli channels. The inner estuary, referred to as the Guayas River estuary, can be classified as a tectonic estuary (Stevenson 1981) which extends about 74 km from its mouth at the northern shore of Puna Island to the tidal signature of the Guayas River (Fig. 17.1). To the northwest, a relatively large subestuary, known as the Salado, has freshwater input restricted to wastewater from the city of Guayaquil (population of 2.5 million) and to the southeast, the Churute subestuary is influenced by the Churute and Taura Rivers. More than 20 rivers, with a total watershed of 51,230 km^2, discharge into the Gulf of Guayaquil (Fig. 17.1). The Guayas River, which forms at the confluence of the Babahoyo and Daule Rivers, has a drainage basin area of about 32,800 km^2 (64% of total drainage basin) and is the major source of freshwater (20 km^3 year^{-1}; Stevenson 1981). The Taura and Churute River drain a watershed of 1,600 and 300 km^2, respectively. Along the southeast coast, the Jubones and Arenillas Rivers contribute freshwater to the Gulf of Guayaquil and represent 4,830 km^2 of watershed (9.4% of total). The depth of the Gulf of Guayaquil ranges from 183 m along the shelf to 18 m along the inner shoals. The depth of the Morro and Jambeli channels at the inner boundary of the Gulf is 56 and 22 m, respectively, and depth in the Guayas River estuary averages about 9 m.

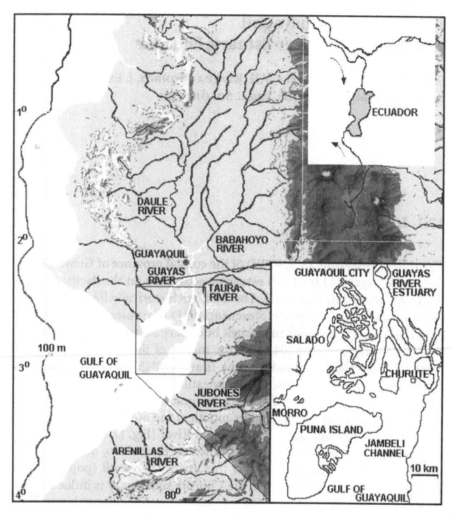

Fig. 17.1. Map of the coastal zone of Ecuador with details of the Gulf of Guayaquil and the Guayas River estuary

17.2 Environmental Setting

The oceanic region off the coast of Ecuador is formed by unique coastal ocean currents that influence the ecology of the coast (Wyrtki 1966). The region represents a transition zone of tropical and subtropical water masses that form an equatorial front. The equatorial front moves seasonally along the coast, mixing less saline (<34), tropical (>25 °C) water of the

Panama Current (2–3°S) with more saline (>34), subtropical cold (<22 °C) water of the Humboldt Current (Cucalón 1983, 1986). A seasonal flux of warmer waters from the north to the Gulf of Guayaquil, associated with higher precipitation and river discharge between January and April, sets up maximum temperature and salinity gradients from the inner to the outer boundaries and stratification of deeper Gulf waters. During the summer, increased meridional winds strengthen the force of the Humboldt Current and colder waters (known as Superficial Subtropical Water) move from the south into the Gulf of Guayaquil, causing outcrops of high-nutrient waters with increased biological productivity. Anomalous ocean currents in the Pacific, associated with El Niño phenomena, cause a shift in the dominance of the Panama current, which results in abnormally warm and less saline coastal waters, higher precipitation in coastal provinces, greater river discharge, and lower salinity in the estuaries. Therefore, both intra- and inter-annual variations in the estuary can be coupled to coastal oceanographic processes (Stevenson 1981).

Winds in the Gulf of Guayaquil are variable and influenced by the Inter-Tropical Convergence Zone (ITCZ). Most winds are from the south (1.5 m s^{-1}) while more northerly winds prevail during El Niño events. During the wet season northeast winds dominate (1.8 m s^{-1}), compared to southwest winds during the dry season (3.5 m s^{-1}). Depending on the location of dominant oceanographic currents off the coast, the precipitation in the Guayas River drainage system (mean 885 mm year^{-1}) ranges from 400 to 3,800 mm year^{-1} (Cucalón 1983, 1986). More than 95 % of the rainfall occurs during the summer (December to May). During an average year of precipitation, seasonal river discharge ranges from 200 m^3 s^{-1} during the dry season to 1,400 m^3 s^{-1} in the wet season (Fig. 17.2), but rates as low as 100 m^3 s^{-1} and as high as 2,400 m^3 s^{-1} may occur (Stevenson 1981). Annual mean temperatures vary between 24 and 27 °C, resulting in a potential evapotranspiration rate of about 1,300 mm year^{-1}.

Tides are 1.8 m near the upper boundary of the Gulf and increase to 3–5 m in the Guayas River estuary near the city of Guayaquil. The tide is semi-diurnal (M_2=12.42 h) with currents in excess of 3.5 m s^{-1} in the river and channels of the estuary. There is also evidence of a 3.5-h seiche along Puna Island (Stevenson 1981). The Guayas River estuary is partially mixed with tidal current speeds of up to 100 cm s^{-1}. The dominant upstream flux of mass and salt is associated with the tidal prism and water turnover time is 11 days, leading to large fluxes of water and nutrients across the boundaries that determine system behavior (Murray et al. 1975). Although vertical layers and water column structure develop in the Gulf of Guayaquil, these hydrographic features do not occur within the estuary (Stevenson 1981).

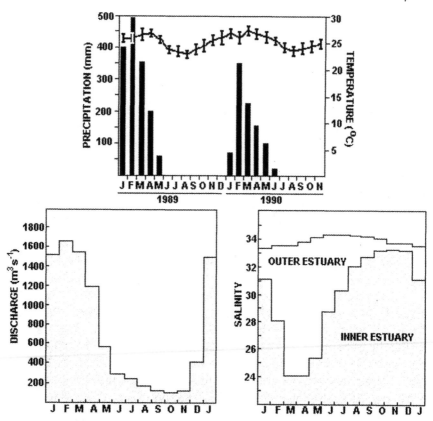

Fig. 17.2. Precipitation, air temperature, river discharge, and salinity in waters of the Gulf of Guayaquil

A thermocline (ΔT 2 °C within 20 m) develops during wet and dry seasons out on the shelf. Surface water temperatures augment from 21.5 °C at the mouth of the Gulf of Guayaquil to 25 °C at the inner boundary during the dry season (Stevenson 1981) and increase by 3 °C during the wet season. Water temperatures are generally higher in the Guayas River estuary and little stratified. The seasonal temperatures in the Guayas (22–30 °C, mean 27 °C), Salado (24–31 °C, mean 27 °C), and Churute (23–30 °C, mean 25 °C) subestuaries reflect oceanographic conditions off the coast. Salinity decreases from 34 in outer regions of the Gulf to 30 (20 during the wet season) near Puna Island where pronounced salinity stratification (0.3 m^{-1} depth) occurs along the eastside of the channel. Although surface isohalines in the Gulf change seasonally with river flow, salinities at >20 m depth remain constant (about 35 at 30 m depth). The mixed layer depth ranges from 15 to 25 m during the dry season and is <10 m during the wet season. Seasonal salinity patterns in the Guayas (0.08–33, mean 18), Churute (2–29, mean 16), and Salado (19–36, mean 28) subestuaries reflect freshwater discharge (Cárdenas 1995; Fig. 17.2). Average pH values in the subestuaries range from 7.5 to 7.7 at high and low river flow, respectively. However, pH increases significantly between the upper (7.3±0.1) and lower (8.01±0.2) Guayas River estuary (Cárdenas 1995). During the dry season, pH values tend to increase (0.2) from the inner to the outer Gulf. Changes of up to 0.5 during the wet season are attributed to higher productivity and detritus concentration in the inner region.

17.3 Biogeochemistry

Except during initial stages of high river flow, mean total suspended solid (TSS) concentrations in Guayas (362 mg l^{-1}), Churute (175 mg l^{-1}), and Salado (127 mg l^{-1}) subestuaries are higher at low flow. Concentrations tend to reflect the relative importance of freshwater input as a source of suspended particulate matter (Cárdenas 1995), although increased TSS may be associated with resuspension of bottom sediments during flood tides, particularly during low river flow (Meade 1972). Episodic or seasonal flow of the Guayas River may temporarily cause high concentrations of TSS, with monthly loads from 2.62×10^6 to 0.12×10^6 tons during high and low river flow, respectively. Water turbidity (Secchi depth) relates inversely with TSS (0.51–0.57 m in the Guayas and Churute, 1.20 m in the Salado) and varies between about 1 m in the inner and 13 m in the outer Gulf (Stevenson 1981).

During the wet season, dissolved oxygen (DO) concentrations in the Gulf are lower than 4.6 ml l^{-1} and are rarely saturated. During the dry season, DO (4.8–5.0 ml l^{-1}) is saturated or supersaturated and pronounced DO stratification (5.0 ml l^{-1} at surface, <2.0 ml l^{-1} at 70 m depth) occurs in the outer Gulf region (Stevenson 1981). Surface DO concentrations in the inner Gulf are 55–60 % saturated. Average DO concentrations in the subestuaries are similar (3.2–4.2 ml l^{-1}) and display little seasonal variation (Cárdenas 1995). Despite high total oxygen demand, owing to urban wastewater input and shrimp pond effluents (Solórzano 1989; Twilley 1989), tidal exchange and water turnover appear to maintain DO above biological stress levels and prevent development of hypoxia in the Guayas River estuary.

Nitrate, dissolved organic nitrogen (DON), and particulate nitrogen represent about 30 % each of the total nitrogen (TN) pool and nitrate accounts for about 70% of dissolved inorganic nitrogen (DIN; Cárdenas 1995). Along the inner boundary of the Gulf in the Jambeli Channel, nitrate concentrations increase from the surface (1–6 µM) to a depth of more than 10 m (>20 µM; Pesantes and Perez 1982). Similarly, nitrite concentrations are higher (0.6–0.9 µM) at 5 m depth than at surface and tend to increase during the wet season near the mouth of estuaries, particularly in the Jubones River (Stevenson 1981). The subestuaries have low nutrient loading and lack clear seasonal changes of NO_3^-, NH_4^+, and DON concentrations with river discharge. Average NO_3^-, NH_4^+, and DON concentrations in the Guayas, Churute, and Salado subestuary are 12, 3.0, 11 µM, 11, 2.4, 12 µM, and 7, 3.3, 7 µM, respectively (Cárdenas 1995).

Soluble reactive phosphate (SRP) concentrations in surface waters of the Gulf range from 0.6 to 1.0 µM and reach about 2.0 µM near Puna Island. Concentrations increase with depth (20 m: <1.0 µM; 70 m: >1.9 µM) and higher bottom concentrations are particularly evident near shoal areas along the southeast boundary of the Gulf proximate to the Jubones River. Soluble reactive phosphate (SRP) represents about 55% of the total phosphorus (TP) pool in the Guayas River estuary (Cárdenas 1995) and, as DIN, lacks a clear seasonal pattern (Fig. 17.3). The similarity of average SRP concentrations in Guayas (2.4 µM), Churute (2.6 µM), and Salado (2.2 µM) subestuaries may be the result of buffer mechanisms that maintain SRP concentrations within the 1–3 µM range. With a few exceptions, dissolved organic phosphorus (DOP) concentrations in the Gulf are approximately 2 µM (Cárdenas 1995) and mean concentrations in subestuaries vary between 0.7 and 0.9 µM.

Dissolved silicate distributions in surface waters of the Gulf are related to freshwater discharge. Concentrations in the outer Gulf are 5–10 µM, compared to 46 µM in the inner Gulf during the dry and 120 µM during the wet season (Stevenson 1981). Similarly, mean Si concentrations are conspicuously higher in the subestuaries (Guayas 124 µM, Churute 146 µM, Salado 56 µM) during the wet season (Cárdenas 1995). Elevated silicate levels during both the dry (153 µM) and wet seasons (206 µM) in the upper Guayas subestuary indicate the importance of the Guayas river as a major source of silicate (Fig. 17.3).

The Gulf of Guayaquil and the Guayas River Estuary, Ecuador

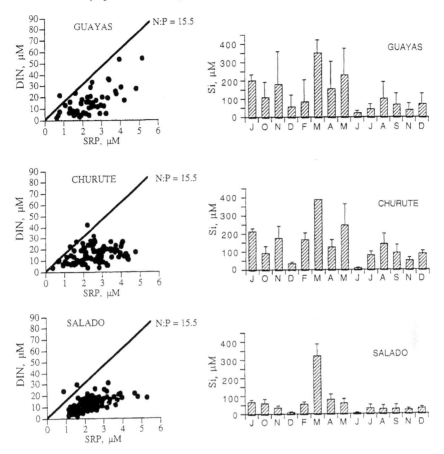

Fig. 17.3. Concentrations of soluble reactive phosphate (*SRP*), dissolved inorganic nitrogen (*DIN*), and silicate (*Si*) in the Guayas River estuary (after Cárdenas 1995)

17.4 Estuarine Habitats and Communities

17.4.1 Mangroves

Two species of *Rhizophora* (*R. harrisonii* = *R. brevistyla* and *R. mangle* = *R. samoensis*) have been confirmed for the coast of Ecuador (West 1963; Chapman 1976; Tomlinson 1986). *Rhizophora mangle* (L.) is apparently indistinguishable from *R. samoensis* which also occurs in the Pacific Islands of New Caledonia and Fiji (Tomlinson 1986). *Rhizophora racemosa* (GFM Mayer) is responsible for the hybrid *Rhizophora harrisonii* (Leachm) and has been listed for the Gulf of Guayaquil though other surveys (Tomlinson 1986) do not cite the species for the Pacific coast of South America. The dominant mangrove in the Gulf of Guayaquil is *Rhizophora harrisonii* followed by *R. mangle*. *Avicennia germinans* (L.), *Laguncularia racemosa* (L.) Gaertn., and *Conocarpus erectus* (L.) occur in less frequently inundated regions of the intertidal zone. Other vegetation in salinas includes *Salicornia fruticosa* (L.), *Maytenus octogona, Acrostichum aureum, Cryptocarpus pyriformis,* and *Batis maritima* (L.). Owing to the construction of shrimp ponds (147,978 ha in 1995) and urban expansion along the shore of Estero Salado, mangrove areas in the Gulf of Guayaquil and Guayas River estuary decreased from 159,247 ha in 1969 to 122,566 ha in 1995, representing a loss of 1,612–2,163 ha year^{-1} over the last 10 years (Fig. 17.4). Furthermore, defoliation particularly of *Rhizophora* sp. but also of *Avicennia* and *Laguncularia* by larvae of *Oiketicus kirbyi* has contributed to loss of mangroves (about 1,000 ha Churute Ecological Reserve), though impacts vary interannually (Sarango 1990).

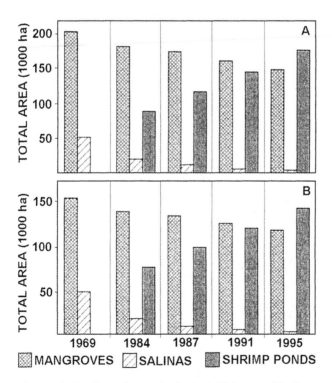

Fig. 17.4.**A** Land use changes in the intertidal zone of the four coastal provinces of Ecuador from 1969 to 1995, and **B** these land use changes in the combined provinces of Guayas and El Oro that represent the Gulf of Guayaquil

17.4.2 Plankton

A total of 159 species of phytoplankton have been identified in the Guayas River estuary (C. Cajas, unpubl. data), compared to earlier surveys of diatoms and silicoflagellates (189) and dinoflagellates (103; Jiménez 1980). Throughout the year, the phytoplankton community composition is dominated by diatoms (i.e., *Skeletonema costatum, Polymyxus coronalis, Coscinodiscus lineatus, Actinoptychus undulatus, Biddulphia sinensis, Thalassionema nitzschioides*; Cárdenas 1995). Mean values for the Churute, Guayas, and Salado subestuaries are 3.6, 3.3, and 2.0 × 10^5 cells ml^{-1}, respectively (Cárdenas 1995) and mean values for cyanobacteria and dinoflagellates vary between 5–19 and 9–80 cells ml^{-1}, respectively. Chlorophyll-*a* concentrations range from 1 to 5 mg m^{-3} (De Peribonio 1981; Stevenson 1981), with highest concentrations (5 mg m^{-3}) in the inner Gulf and lower Guayas River estuary and mean concentrations of 15, 14, and 6 mg m^{-3} in the Guayas, Churute, and Salado subestuaries, respectively (Cárdenas 1995). Toxic phytoplankton blooms (*Gonyaulax catenella, G. monilata, Gymnodium breve*) are a common occurrence in the Gulf of Guayaquil and the Guayas River estuary (Jiménez 1980; de Arcos 1982) and have caused fish kills and mortality of post-larvae in shrimp hatcheries (Jiménez 1989). Blooms also occur in the Guayas River (*Gyrodinium stratum, Prorocentrum maximum, Nitzschia* sp., and *Mesodinium rubrum*), Estero Salado (*Mesodinium rubrum*), and coastal waters of the Gulf of Guayaquil (*Thalassiosira* sp.; de Arcos 1982).

The zooplankton community in the Guayas River estuary is composed of 101 species. Arthropods, dominated by copepods (mean 300 ind. cm^{-3}), and protozoans are the major components (C. de Cajas, unpubl. data). High zooplankton biomass outside the Gulf of Guayaquil is largely comprised of copepods while fish eggs dominate zooplankton inside the Gulf (de Cajas 1982). Ichthyoplankton is represented by 29 families, the most abundant are Engraulidae (*Anchoa* sp., *Engraulis* sp., *E. ringens, Cetengraulis mysticetus, Anchovia*), Sciaenidae, Gobiidae, and Carangidae (Luzuriaga et al. 1998). Engraulidae and Gobiidae dominate in surface waters and Sciaenidae in deeper waters. The highest density of larvae occurs from May to July and is correlated with high concentrations of phytoplankton.

17.4.3 Benthos

Several macrobenthic species, like the oysters *Crassostrea columbiensis* (Hanley) and *C. iridiscens* (Hanley), the mangrove crab *Ucides occidentalis*, and *Mytella guyanensis* (Lamarck), *M. strigata* (Hanley), and *Ala-*

bina sp., live in the intertidal of mangrove habitats in the Guayas River estuary. Five species of penaeid shrimp, *Penaeus occidentalis* (Streets), *P. stylirostris* (Stimpson), *P. brevirostris* (Kingsley), *P. vannamei* (Boone), and *P. californiensis* (Holmes), occur in the Gulf, the last two being the most abundant (García-Sáenz et al. 1998). *Penaeus californiensis* is largely restricted to shallow marine waters while *Penaeus stylirostris* and *P. vannamei* occur near mangrove areas. The densities of post-larvae tend to be higher during the wet season and, although the abundance of juveniles lacks seasonal differences, densities appear to be inversely related to the distance of mangrove tidal channels from the coast (Zimmerman and Minello 1989). Higher density of post-larvae and juveniles occur in the El Morro Channel, which is close to spawning grounds (Loesch and Avila 1966; Zimmerman and Minello 1989). Additionally, site-specific DO concentrations correlate with the spatial distribution of shrimp in the Guayas River estuary. Low DO concentration (0.7–3 ml l^{-1}) retards growth of *P. vannamei* (Seidman and Lawrence 1985) and values of less than 2.1 ml l^{-1} lead to suspension of feeding (Villalón et al. 1989). The negative effect of DO may explain lower post-larvae densities in the Salado (2.1 ml l^{-1}) than in the Churute (2.3 ml l^{-1}), although both subestuaries have similar distances to spawning grounds. Artisanal fisheries capture post-larvae, juveniles, and gravid females of *Penaeus vannamei* and *P. stylirostris* to supply the shrimp aquaculture industry (McPadden 1985b), while offshore trawl fisheries capture pre-adults and adults. Trawl by-catch is composed of the dominant Mysidacea. Decapod crustaceans (family Caridea) are represented by Paleomonidae (50.6%), Alpheidae (25.3%), Hippolytidae (21.7%) and Pasiphalidae (2.4%) and blue crabs include *Callinectes* sp. (63.6%), *C. arcuatus* (36.1%), and *C. toxotes* (0.3%; García-Sáenz and Peláez 1998).

17.5 Productivity and Trophic Structure

Variations of net primary productivity (NPP) are more pronounced between inner and outer Gulf regions than between seasons (Stevenson 1981). While NPP in the outer Gulf is 200 mg C m^{-3} day^{-1} or less, values in the inner Gulf are 600–800 mg C m^{-3} day^{-1} during the wet season and 800–1,000 mg C m^{-3} day^{-1} during the dry season, differences being due to higher turbidity in the wet season. Based on Secchi disk depth, the volume estimates of NPP on an annual production per unit area range from 200 to 300 mg C m^{-2} day^{-1} in inner Gulf and from 300 to 1,000 mg C m^{-2} day^{-1} in outer Gulf regions. Integration of monthly spatial estimates of daily NPP results in an outer Gulf productivity of about 200 g C m^{-2} year^{-1} compared to 90 g C m^{-2} year^{-1} in the inner Gulf. Primary productivity in the subestuaries is potentially limited by DIN concentration (N:P ratio <5.8). DIN limitation appears to be a result of more rapid P than N regeneration from the organic matter pool (Cárdenas 1995). Uncoupling of DIN and SRP in the nutrient cycles leads to an increase in DON:DOP ratios (around 65) in the Guayas River estuary. An increase of SRP tends to increase the TP pool (SRP forms 50% of the TP pool) and reduce TN:TP ratios (about 12). Denitrification and nitrogen deposition has been implicated in low N:P ratios of sediment-water fluxes that could also affect the availability of DIN to primary producers (Smith et al. 1985).

High TSS concentration and tides (>3 m) cause elevated sediment accumulation rates (4,074–5,151 g m^{-2} year^{-1}) in mangrove forests of the Churute River subestuary. More than 75% of accumulated matter is associated with inorganic sediments. High nutrient concentrations (C:N ratios 24–30; N:P ratios 7.7 to >40; total phosphorus 0.23–0.71 mg g^{-1}), low salinity (<17), and the proximity of the Gulf to the equator result in tall mangrove trees (up to 30 m) and biomass between 300 and 400 tons ha^{-1} (Twilley et al. 1993). Mangrove litter fall rates vary from 6.47 to 10.64 tons ha^{-1} year^{-1}, with higher daily rates during the rainy season, while mean leaf litter standing crop (1.53–9.18 g m^{-2}) peaks during the dry season in September. Leaf litter dynamics in riverine mangroves of the Guayas River estuary depend on the relative effects of tides and the activity of the mangrove crab *Ucides occidentalis*. The leaf litter residence time on the forest floor is less than 1 day when crabs are active (December to July); however, when crabs aestivate (August to November), residence time averages 5.3 days. During the active season, crabs bury leaf litter in the forest soil at low tide, but at high tide leaf litter is exported to the estuary, regardless of whether crabs are active or not. Although the rate of mangrove leaf degradation is related to site, season (rainy > dry), and initial nitrogen content of senescent leaves, tidal processes are responsible for higher (10–20 times) litter turnover rates than expected based on leaf degradation alone. A model of leaf litter dynamics (Fig. 17.5) suggests that geophysical energy (tides, river discharge) controls the fate of mangrove leaf litter, though highest litter turnover rates are associated with crab activity (Twilley et al. 1986, 1997).

Particulate carbon to chlorophyll-*a* (PC:Chl*a*) ratios can distinguish between living (20 to >200) and detrital (>1,000) organic matter in suspended particulate material (SPM; Cifuentes et al. 1988). The lowest PC:Chl*a* ratio (104) occurs far from terrestrial and intertidal sources in the Morro Channel, while PC:Chl*a* ratios in the Guayas River near Guayaquil are 50 times higher. Average PC:Chl*a* (327), δ^{13}C (−26.7 to −23.1), and δ^{15}N

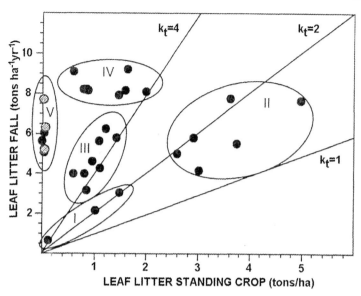

Fig. 17.5. Plots of leaf litter fall (*LF*) and leaf litter standing crop (*SC*) in mangrove wetlands around the world (after Twilley et al. 1998). *Diagonal lines* represent leaf turnover rates based on k_t ($k_t=LF/SC$) and groups are based on cluster analysis of data for each site. Results from sites in Ecuador are designated with *hatched circles*

(+4.4 to +9.1) values in the Guayas River estuary indicate significant degradation of phytoplankton and/or terrestrial inputs, with mangrove litter being a likely primary source of terrestrial organic detritus (Fig. 17.6; Twilley 1985; Twilley et al. 1996). However, (C:N)a (14.1±0.9) of the detritus is lower than that of senescent and decomposing mangrove leaves (Benner et al. 1990). It appears that bacterial colonization of SPM alters the character of terrestrial-derived organic matter and increases the nitrogen content of litter during decomposition (Twilley et al. 1986, 1996; Rivera-Monroy and Twilley 1996). Bacterial carbon production (11–120 µg C l^{-1} day^{-1}) and bacterial turn-over (20–600 µg C l^{-1} day^{-1}) increase in the Guayas River, suggesting that oligohaline regions of the estuary have the highest available carbon concentrations (Cifuentes et al. 1996). Despite the transformation of pristine mangrove stands into shrimp farms and massive river transport of water hyacinth rafts into the estuary, mangroves appear to supply most of the organic matter to the system while water hyacinths or organic matter from shrimp pond effluents only have local effects.

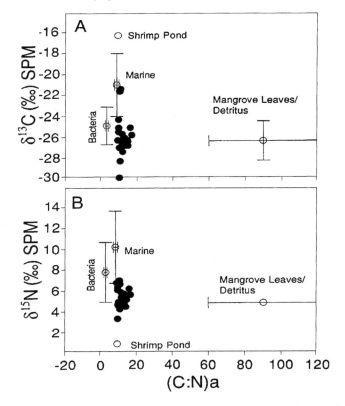

Fig. 17.6.A Stable carbon and B nitrogen isotope ratio of suspended particulate matter (*SPM*) vs. C:N atomic ratio. (After Cifuentes et al. 1996)

17.6 Coupling of Coastal and Estuarine Ecosystems

The migratory life cycle of shrimp represents an important linkage between estuarine and coastal offshore water habitats in the Gulf of Guayaquil. Shrimps reproduce in the coastal ocean and planktonic larvae are horizontally advected. The locomotion of post-larvae and transport by currents and tides provide the necessary mechanisms to recruit shrimp into nearshore and estuarine waters. Temporal and spatial variations (i.e., temperature and precipitation) in the Gulf and site-specific factors in the estuary influence recruitment processes. Higher densities of post-larvae (*P. vannamei, P. occidentalis, P. stylirostris*) occur during the wet season, especially during El Niño events (McPadden 1985a; Gaibor et al. 1992). Mangroves provide nursery habitats for the maturation of juvenile stages. However, while juveniles of *P. occidentalis* and *P. brevirostris* are rarely associated with mangroves, *P. californiensis*, *P. vannamei*, and *P. stylirostris* are abundant, though it is uncertain whether the habitats enhance their survival or their growth (Zimmerman and Minello 1986). After approximately 3 months, predominantly benthic adults (80–100 mm length) return to the coastal ocean to re-supply the estuary with a new generation of post-larvae (Garcia 1985).

17.7 Human Impacts

Changes in environmental quality of the Gulf of Guayaquil and the Guayas River estuary are associated with land use in upland and intertidal watersheds. Deforestation of natural vegetation, replacement by agro-ecosystems, and urban and industrial activities have changed the quality, quantity, and seasonality of river discharge, thus influencing distribution and turnover rates in the estuary and Gulf. The diversion of water for irrigation and industrial use after the construction of a dam at the Daule River resulted in a loss of freshwater to the Guayas River (15%) and the Guayas River estuary (13%; Arriaga 1989). Urban areas and industry are associated with point-source chemical inputs. About 86% of 54.83×10^6 m^3 of waste generated annually by 3.14 million people in the Guayas River basin is discharged into the aquatic ecosystems. Domestic and industrial wastes also introduce bacterial contamination and cause nutrient enrichment and eutrophication (Solórzano 1989; Pin et al. 1998). Despite BOD values of 0.65–2.88 mg l^{-1}, anoxic waters are rare and only occur near sewage outfalls of Guayaquil (<2.7 mg l^{-1}; Solórzano and Viteri 1981; Arriaga 1989; Solórzano 1989), because strong tides with large amplitude (3–5 m) effectively mix the water column in the Guayas River estuary. Irrigated agriculture (approx. 175,000 ha) in the Guayas River watershed is likely to introduce agrotoxins; however, only traces of pesticides have been detected at the beginning of the rainy season in the Daule River (Solórzano 1989). In contrast, mining activities in Guayas River basin add iron, copper (36.92–94.52 µg l^{-1}), and cadmium (0.1–14.5 µg l^{-1}) to the water column and sediments of Babahoyo, Daule, and Guayas rivers. Copper (>10 µg l^{-1}) and cadmium concentrations in the water and mercury concentrations in sediments of the Guayas River estuary are reason for concern (Solórzano 1989).

Shrimp pond construction has caused loss in area and function of mangrove habitats (Fig. 17.4; Twilley et al. 1997) and shrimp pond operation (pumping, fertilization, dredging, harvest of post-larvae) has impacted the environmental quality of the Gulf and Guayas River estuary (Csavas 1994).

Shrimp production increased from 5,000 tons in 1979 (US$ 56.9 million) to more than 100,000 tons in 1991 (US$ 482 million; Olsen and Arriaga 1989; Aiken 1990). An infectious disease (identified as a virus), known as Taura Syndrome (TS), affected juvenile shrimp (*Penaeus vannamei*) in densely stocked shrimp ponds, causing an estimated loss of US$ 100 million in the mid 1990s and threatening the Ecuadorian market (and several other markets in Latin America). The elevated biological productivity and economic value of the Gulf and Guayas River estuary and the impact shrimp pond construction and operation have on ecological processes and on environmental quality have led to a controversy on coastal resource management.

17.8 Preliminary Models and Ecosystem Management

A dynamic box model for the Guayas River estuary considers scenarios of mangrove to shrimp pond conversion in three regions of the estuary and different rates of river discharge (100, 50 and 10 % based on the 1989 flow) after construction of a dam (Twilley et al. 1998). Seasonally high river flow and tidal exchange with low water residence time (11 days) maintain good water quality. After a 90 % reduction of mangrove forests caused by shrimp pond construction, total nitrogen concentration would increase fivefold. However, if river discharge decreases to 10 % at the same time, nitrogen concentrations would increase 60-fold. Increases in nitrogen concentration are higher in the upper than in the lower estuary, thus the sensitivity of the environment to changes in intertidal and upland land use is site-specific and linked to the hydrography of the estuary. It thus appears that shrimp pond farming in the Guayas River estuary has managed to sustain high productivity and profits largely as a consequence of high river discharge (Twilley et al. 1998). However, the construction of the Daule-Peripa Dam threatens the industry by substantially diminishing river flow, with water quality problems leading to reduced, though spatially selective, profitability of the shrimp industry. If shrimp pond effluents were introduced into nearby mangrove forests, the natural potential of mangroves to remove excessive nutrient loads (Rivera-Monroy et al. 1999) could benefit shrimp pond management, enhance environmental quality in the estuary, and also possibly increase mangrove productivity.

The conceptual model argues that ecological and socioeconomic forcing functions, such as international markets, political processes, exchange rates, and monetary and trade policies, control coastal zone management decisions (Twilley et al. 1998). The goods and services of mangroves, such as habitat and water quality, receive no explicit value in the model (the same as occurs in society). Mangroves and tides provide the shrimp industry with clean water and productive habitats that enhance wild post-larvae supply and shrimp yield in ponds. With the loss of free services provided by natural resources, the costs of shrimp production (providing post-larvae from hatcheries, dredging to remove sediment, and pumping to control eutrophication) increase and profits of the shrimp industry decrease (Twilley 1989; Twilley et al. 1998). As a consequence, the combination of ecological models with economic analyses of the goods and services of mangroves may provide better techniques to evaluate the economic impacts of specific coastal zone management decisions.

References

Aiken D (1990) Shrimp farming in Ecuador, an aquaculture success story. World Aquacult 21:7–16

Arriaga L (1989) The Daule-Peripa dam project, urban development of Guayaquil and their impact on shrimp mariculture. In: Olsen S, Arriaga L (eds) Establishing a sustainable shrimp mariculture industry in Ecuador. University of Rhode Island, technical report series TR-E-6, pp 147–162

Benner R, Weliky K, Hedges JI (1990) Early diagenesis of mangrove leaves in a tropical estuary: molecular-level analyses of neutral sugars and lignin-derived phenols. Geochim Cosmochim Acta 54:1991–2001

Cárdenas WB (1995) Patterns of phytoplankton distribution related to physical and chemical characteristics of the Guayas river estuary, Ecuador. MSc Thesis, University of Southwestern Louisiana, Louisiana, USA

Chapman VJ (1976) Mangrove vegetation. Cramer, Vaduz

Cifuentes LA, Sharp JH, Fogel ML (1988) Stable carbon and nitrogen isotope biogeochemistry in the Delaware estuary. Limnol Oceanogr 33:1102–1115

Cifuentes LA, Coffin RB, Solorzano L, Cardenas W, Espinosa J, Twilley RR (1996) Isotopic and elemental variations of carbon and nitrogen in a mangrove estuary. Estuar Coast Shelf Sci 43:781–800

Csavas I (1994) Important factors in the success of shrimp farming. World Aquacult 25:34–56

Cucalón E (1983) Temperature, salinity, and water mass distribution off Ecuador during an El Niño event in 1976. Rev Cienc Mar Limnol 2:1–25

Cucalón E (1986) Oceanographic characteristics off the coast of Ecuador. In: Olsen S, Arriaga L (eds) Establishing a sustainable shrimp mariculture industry in Ecuador. University of Rhode Island, technical report series TR-E-6, pp 185–194

de Arcos TV (1982) Mareas rojas en aguas ecuatorianas. Rev Cienc Mar Limnol 1:115–125

de Cajas LC (1982) Estudios del zooplancton marino en aquas ecuatorianas Eastropac 1, 2 y 3. Rev Cienc Mar Limnol 1:147–163

De Peribonio RG (1981) Distribucion de clorofila a I feopigmento en el Golfo de Guayaquil. Revista de Ciencias del Mar y Limnologia 1:1–7

Gaibor N, Coello S, Garcia R, Luzuriaga de MC, Massay S, Ortega D, Villamar F, Mora E, Basantes A, Vicuña H (1992) Evaluacion de la pesqueria de postlarvas de camarón penaeido y su fauna acompañante. Informe Interno INP-PMRC (informal report at the National Fisheries Institute, Guayaguil, Ecuador)

Garcia S (1985) Reproduction, stock assessment models and population parameters in exploited penaeid shrimp populations. In: Rothlisberg BJ, Hill BJ, Staples DJ (eds) Second Australian National Prawn Seminar, NPS2, Cleveland, Australia, pp 139–158

García-Sáenz R, Peláez R (1998) Distribution y abundancia de callinectes (Crustacea: Portunidae) en el estuario interior del Golfo de Guayaquil, durante 1995. In: Comportamiento temporal u espacial de las caracteristicas fisicas, quimicas y biologicas del Gulfo de Guayaquil y sus afluentes Daule y Babahoyo entre 1994–96. Instituto Nacional de Pesca, Guayaquil, pp 369–383

García-Sáenz R, Peláez R, Lindao J, Calderon G, Morales G (1998) Distribucion y abundancia de larvas y post-larvas de camarones marionos y fauna acompañante en el estuario interior de Golfo de Guayaquil. In: Comportamiento temporal u espacial de las caracteristicas fisicas, quimicas y biologicas del Gulfo de Guayaquil y sus afluentes Daule y Babahoyo entre 1994–96. Instituto Nacional de Pesca, Guayaquil, pp 305–353

Jiménez R (1980) Marea roja en el Golfo de Guayaquil en abril de 1980. Bol Inform Inst Nac Pesca 11–13

Jiménez R (1989) Red tide and shrimp activity in Ecuador. In: Olsen S, Arriaga L (eds) Establishing a sustainable shrimp mariculture industry in Ecuador. University of Rhode Island, technical report series TR-E-6, pp 179–184

Loesch H, Avila Q (1966) Observaciones sobre la presencia de camarones juveniles en dos esteros de la costa del Ecuador. Bol Cient Tecn Inst Pesca Ecuador 1:1–30

Luzuriaga M, Ortega D, Elias E, Flores ME (1998) Relaciones de abundancia entre el fitoplancton e ictioplancton con enfasis en la familia Engraulidae en el Golfo de Guayaquil. In: Comportamiento temporal u espacial de las caracteristicas fisicas, quimicas y biologicas del Gulfo de Guayaquil y sus afluentes Daule y Babahoyo entre 1994–96. Instituto Nacional de Pesca, Guayaquil, pp 387–418

McPadden CA (1985a) A brief review of the Ecuadorian shrimp industry. Bol Cient Inst Pesca Ecuador 8:1–68

McPadden CA (1985b) The ecuadorian trawl fishery, 1974–1985. Bol Cient Inst Pesc Ecuador 9:1–25

Meade RH (1972) Transport and deposition of sediments in estuaries. Geol Soc Am Mem 133:91–120

Murray SD, Conlon Siripong A, Santoro J (1975) Circulation and salinity distribution in the Rio Guayas estuary, Ecuador. In: Cronin GG (ed) Estuarine research. Academic Press, New York, pp 345–363

Olsen S, Arriaga L (1989) Establishing a sustainable shrimp mariculture industry in Ecuador. University of Rhode Island, Technical Report Series TR-E-6

Pesantes F, Perez E (1982) Condiciones hidrográficas, físicas y quimicas en el estuario del Golfo de Guayaquil. Rev Cienc Mar Limnol 1:87–113

Pin G, García F, Castello M (1998) Microflora bacteriana de las aquas del estuario interior del Golfo de Guayaquil. In: Comportamiento temporal u espacial de las caracteristicas fisicas, quimicas y biologicas del Gulfo de Guayaquil y sus afluentes Daule y Babahoyo entre 1994–96. Instituto Nacional de Pesca, Guayaquil, pp 285–301

Rivera-Monroy VH, Twilley RR (1996) The relative role of denitrification and immobilization on the fate of inorganic nitrogen in mangrove sediments of Terminos Lagoon, Mexico. Limnol Oceanogr 41:284-296

Rivera-Monroy VH, Torres LA, Bahamon N, Newmark F, Twilley RR (1999) The potential use of mangrove forests as nitrogen sinks of shrimp aquaculture pond effluents: the role of denitrification. J World Maricult Soc 30:12-25

Sarango A (1990) El gusano de canasta (Leridoptero: Psychidiae) en los manglares Ecuatorianos de Churute. For Inform Bull 7:24-29

Seidman ER, Lawrence AL (1985) Growth, feed digestibility, and proximate body composition of juvenile *Penaeus vannamei* and *Penaeus monodon* growth a different dissolved oxygen levels. J World Maricult Soc 16:333-346

Smith CJ, DeLaune RD, Patrick WHJ (1985) Fate of riverine nitrate entering an estuary: I. Denitrification and nitrogen burial. Estuaries 8:5-21

Solórzano L (1989) Status of coastal water quality in Ecuador. In: Olsen S, Arriaga L (eds) Establishing a sustainable shrimp mariculture industry in Ecuador. University of Rhode Island, technical report series TR-E-6, pp 163-177

Solórzano L, Viteri G (1981) Investigacion quimica de una seccion del Estero Salado. Rev Ciencias Mar Limnol 3:41-48

Stevenson MR (1981) Seasonal variations in the Gulf of Guayaquil, a tropical estuary. Bol Cient Téc INP 4:1-133

Tomlinson PB (1986) The botany of mangroves. Cambridge University Press, Cambridge

Twilley RR (1985) The exchange of organic carbon in basin mangrove forests in a southwest Florida estuary. Estuar Coast Shelf Sci 20:543-557

Twilley RR (1989) Impacts of shrimp mariculture practices on the ecology of coastal ecosystems in Ecuador. In: Olsen S, Arriaga L (eds) Establishing a sustainable shrimp mariculture industry in Ecuador. University of Rhode Island, technical report series TR-E-6, pp 91-120

Twilley RR, Lugo AE, Patterson-Zucca C (1986) Litter production and turnover in basin mangrove forests in southwest Florida. Ecology 67:670-683

Twilley RR, Bodero A, Robadue D (1993) Mangrove ecosystem biodiversity and conservation: case study of mangrove resources in Ecuador. In: Potter CS, Cohen JI, Janczewski D (eds) Perspectives on biodiversity: case studies of genetic resource conservation and development. AAAS Press, Washington, DC, pp 105-127

Twilley RR, Snedaker SC, Yañez-Arancibia A, Medina E (1996) Biodiversity and ecosystem processes in tropical estuaries: perspectives from mangrove ecosystems. In: Mooney H, Cushman H, Medina E (eds) Biodiversity and ecosystem functions: a global perspective. Wiley, New York, pp 327-370

Twilley RR, Pozo M, Garcia VH, Rivera-Monroy VH, Zambrano R, Bodero A (1997) Litter dynamics in riverine mangrove forests in the Guayas River estuary, Ecuador. Oecologia 111:109-122

Twilley RR, Gottfried RR, Rivera-Monroy VH, Armijos MM, Bodero A (1998) An approach and preliminary model of integrating ecological and economic constraints of environmental quality in the Guayas River estuary, Ecuador. Environ Sci Policy 1:271-288

Villalón JR, Maugle PD, Laniado R (1989) Present status and future options for improving the efficiency of shrimp mariculture. In: Olsen S, Arriaga L (eds) Establishing a sustainable shrimp mariculture industry in Ecuador. University of Rhode Island, technical report series TR-E-6, pp 249-262

West RC (1963) Mangrove swamps of the Pacific Coast of Colombia. Annual Association of American Geography 46, pp 98-121

Wyrtki K (1966) Oceanography of the eastern equatorial Pacific Ocean. Oceanogr Mar Biol Rev 4:33–68

Zimmerman R, Minello TJ (1986) Recruitment and distribution of postlarval and early juvenile penaeid shrimp in a large mangrove estuary in the Gulf of Guayaquil during 1985. In: Olsen S, Arriaga L (eds) Establishing a sustainable shrimp mariculture industry in Ecuador. University of Rhode Island, Technical Report Series TR-E-6, pp 233–245

18 The Estuary Ecosystem of Buenaventura Bay, Colombia

J.R. CANTERA and J.F. BLANCO

18.1 Introduction

Buenaventura Bay at the Central Pacific Coast of Colombia is classified as a drowned valley (Fig. 18.1). The bay is located between two faults (NE–SW and NW–SE) on Tertiary consolidated rocky and sedimentary cliffs and on Quaternary mobile sediment platforms (Cantera 1991; Martínez 1993). The NE-SW fault is responsible for uplifting the northern shore of the bay, which is dominated by rocky cliffs, occasional rocky shores and intrusions of sand and silt (Gálvis and Mojíca 1993). Many depositional beaches of northern shore embayments are formed by sediments which originated from cliff erosion and decomposing mangrove litter. Depositional fans around the mouth of the Dagua and Anchicayá rivers characterize the southern shore of Buenaventura Bay. The composition of the fans varies from silt to sand, depending on the origin of sediments and the balance between river discharge and tidal flow (Lobo-Guerrero 1993). The addition of sediments and their transport by rivers and tides also causes their continuous deposition in the navigation channel and leads to a prograding southern coast (CAE 1995; Univalle 1997).

18.2 Environmental Settings

Buenaventura Bay is located within the Inter-Tropical Convergence Zone (ITCZ). As the low pressure belt of the zone moves north and south between 10°N and 3°S, it passes over Buenaventura Bay and causes two rainy periods from September to November and April to June with mean monthly precipitation exceeding 567 mm, interrupted by periods with lower precipitation (mean 374 mm; Eslava 1993). Furthermore, the high (3,500 m)

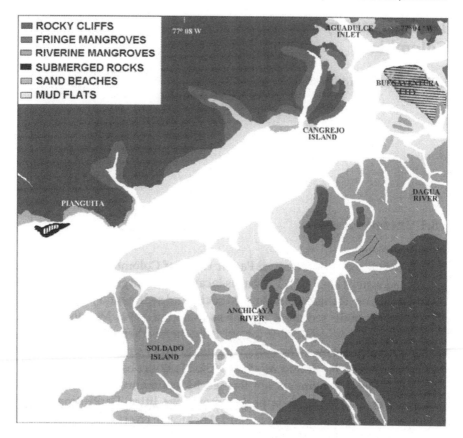

Fig. 18.1. Buenaventura Bay ecosystem with principal habitats

mountains of the "Cordillera Occidental", which run parallel and proximate (15–20 km) to the coast, cool the warm, wet air of onshore winds and cause intense precipitation. The effect of the ITCZ and orographic rain make Buenaventura one of the most humid places in the world, with a mean annual air temperature of 25.9 °C, 228–298 days of precipitation year^{-1}, a mean annual precipitation of 6,508 mm and a relative humidity of 80–95 % (Lobo-Guerrero 1993). However, high evapotranspiration rates (approx. 20–30 %) prevent excessive runoff from the land and groundwater accumulation. Total river discharge into Buenaventura Bay is about 427 m^3 s^{-1} (Dagua River 126 m^3 s^{-1}; Anchicaya River 112 m^3 s^{-1}) with peaks between September and November when temperatures decrease and runoff increases (Lobo-Guerrero 1993). The freshwater discharge into the bay accounts for 8–16 % of the tidal flow (1,254 m^3 s^{-1}; CAE 1995). Semi-diurnal tides (3.7–5 m) generate high surface currents in the navigation channel (1–2 m s⁻¹) and lower velocities in the inner bay and creeks (0.2–1.0 m s⁻¹).

At high slack water, salinity ranges from 18–27 at the mouth of Buenaventura Bay (Punta Soldado) to 4.8 at the Dagua River inlet in the inner bay. During low slack water, the freshwater discharge of Anchicaya River produces a low salinity zone (5) at the southern coast of the bay and traps seawater (19.8–21) near the northern shore. Salinities are predominantly stratified at the mouth of the rivers and also along the southern coast, especially at high tide (surface salinity 9.4 and at 13 m depth 20.5). Stratification is weak and occasionally homogeneous in the navigation channel and at the mouth of the bay, thus Buenaventura Bay can be considered a partially mixed estuary (circulation 3.08; stratification 0.62; Kjerfve 1990). Surface water temperatures experience annual variations between April (28 °C) and October (30.2 °C), with higher temperatures occurring during flood tides (+4 °C) though vertical temperature fluctuation (<2 °C) is not significant (CAE 1995). Anomalous temperatures (+3.5 °C) in the inner bay (October 1997) appear to be related to the occurrence of El Niño events (Univalle 1997). Water turbidity increases between the mouth (27–30 mg l⁻¹) and the inner bay (40 mg l⁻¹), owing to high total suspended sediment (TSS) input by rivers (i.e., Dagua 116 mg l⁻¹) and trapping of freshwater during ebb flow. The high suspension load accounts for mean water transparencies of 91.75±20.11 cm. Phytoplankton biomass peaks may further decrease transparency (50 cm) and cause high attenuation (up to 60 %) of light.

The dissolved oxygen concentrations in the bay (5.5–10 mg l⁻¹) tend to increase towards the mouth and during high tide and rainy seasons. The pH of the waters is alkaline (pH 7.2–8.5) but decomposition processes near mangrove habitats (Cantera 1993; Univalle 1997) and sewage discharge of Buenaventura City into the inner bay may lower the pH to 5.0–6.2 (CAE 1995). Nutrient concentrations in Buenaventura Bay (peak NO_3 0.539 mg l⁻¹; mean NO_2 0.03 mg l⁻¹; mean NH_4 0.221 mg l⁻¹; peak PO_4 0.160 mg l⁻¹; SiO_4 0.75–1.64 mg l⁻¹) increase with land-runoff. In trapping zones at the mouth of creeks (NO_3 1.5–8.9 mg l⁻¹, PO_4 0.1–0.5 mg l⁻¹, SiO_4 0.9–1.8 mg l⁻¹) and falls of rivers (NO_3 0.5–3 mg l⁻¹, PO_4 0.1–0.5 mg l⁻¹, SiO_4 0.7–6.4 mg l⁻¹) concentrations are five to ten times higher than in the outer bay, with a clear tendency to augment upstream (Cantera 1993).

18.3 Coastal Habitats and Communities

A combination of recent geological activity, wide tidal range, high precipitation, and elevated sediment input by rivers has bestowed Buenaventura Bay with a variety of habitats, like depositional intertidal sand beaches and mud flats, intertidal cliffs and rocky shores, mangrove swamps, and the pelagic zone of the estuary (Fig. 18.1).

18.3.1 Sandy Beaches

The discharge of sand by rivers and sediments from coastal erosion have formed sandy beaches at the mouth of Buenaventura Bay and shoals and sandy bars parallel to the coast (Fig. 18.1). The sandbars prevent shore erosion and some have developed into barrier islands (i.e. Soldado Island) occupied by beach vegetation and mangroves (Martínez et al. 1995). The morphodynamics of shoals and sand beaches are largely controlled by tidal currents and wave overwash during spring tides and ENSO events (Gonzalez et al. 1998) when large amounts of sand are deposited (Cantera 1991). Apart from the obvious action of tides and waves, the nature of sediments, soil humidity gradients, dissolved oxygen content and salinity of interstitial water, sand surface temperatures, and light penetration into sediments (2–10 cm) also determine the composition and distribution of species on beaches (Cantera et al. 1992). The abundance and diversity of species on sandy beaches is low and most organisms are infaunal, though shore and marine birds are constant visitors.

The upper littoral zone of sandy beaches is occupied by plants (*Cenchrus pauciflorus, Homolepis aturensis, Ipomoea pes-caprae, I. stolonifera, Canavalia maritima, Pectis arenaria, Stenotaphrum secundatum*) which have a stabilizing effect on the substrate. Insects (diptera, coleoptera, orthoptera) are abundant. Although the hermit crab *Coenobita compressus* accounts for 59 % of the marine invertebrates in this zone, ghost crabs (*Ocypode gaudichaudii, O. occidentalis*), that move over the beach in large groups at low tide, are the most conspicuous organisms. Insects (collembola, diptera) and crustaceans (amphipods, isopods) are found under tidal litter accumulations, with high densities (>1,000 ind. m^{-2}) of the amphipod *Talitrus* sp. The intertidal zone is occupied by the suspension-feeding clam *Donax assimilis* (17.6 ind. m^{-2}), olives (*Olivella* sp.; 34.3 ind. m^{-2}), and sand crustaceans (*Emerita rathbunae*; 6 ind. m^{-2}), which follow the advance and retreat of tides. Marine species, like annelid errant worms (Nereidae), tube worms, bivalves (*Iphigenia, Tellina, Sanguinolaria*), and seasonally abundant echinoderms (i.e., the sand dollars *Encope* and *Mellita*), are common in the lower intertidal (Cantera 1991).

18.3.2 Cliffs and Rocky Shores

The tectonic processes along the Pacific coast of Colombia have given origin to abundant steep cliffs (>45° slope) and to rocky shores with more gently sloping platforms, composed of boulders, pebbles, and gravel from cliff erosion. The slope, tidal- and wave-dependent dampening, substrate characteristics, wave impact, pore water temperature and salinity, and competition for space and food determine the species composition and structure of cliff and rocky shore communities along the northern coast and in the interior of Buenaventura Bay. As opposed to sand beaches, these substrates display clear vertical zonation patterns with a diverse flora and fauna and especially rocky shores are among habitats with particularly high species diversity (Shannon diversity index 3.48–3.63 bits/ind.; Cantera 1991). Desiccation-tolerant blue-green and green algae and a lichen inhabit the upper littoral (spray) zone of cliffs and rocky shores. The periwinkle *Austrolittorina aspera* is very abundant (115 ind. m^{-2}) and co-occurs with *Littoraria zebra* (3.6 ind. m^{-2}), the rock crab *Grapsus grapsus* (2 ind. m^{-2}), and the isopod *Ligia baudiana* (28 ind. m^{-2}). About 20 species occupy the upper intertidal zone where the periwinkle *L. zebra* (6.8 ind. m^{-2}) becomes increasingly abundant. The crab *Pachygrapsus transversus* (14 ind. m^{-2}) is common on less wave-exposed cliffs, while barnacles (*Tetraclita*, *Chthamalus*), limpets (Fissurellidae, Acmaeidae, Siphonariidae), crabs (i.e. *Grapsus grapsus*), and some green algae occur on exposed cliffs. The mid-intertidal zone is inhabited by 41 species and is dominated by bivalves of the families Mytilidae (*Brachidontes*), Isognomonidae, and Ostreidae (oysters), which occur together with crabs of the families Xanthidae (*Eriphia squamata* <1 ind. m^{-2}) and Grapsidae (*Pachygrapsus transversus* 12.0 ind. m^{-2}), as well as the red calcareous alga *Lithothamnion*. Apart from barnacles, anemones, and sponges, the gastropods *Acanthina brevidentata* and *Thais kiosquiformis* and some crabs (i.e. *Pachygrapsus transversus*) occur in the lower intertidal zone.

The outstanding feature of the lower intertidal is the increasing number of perforations (up to 54.8 % inside the bay and 60.2 % outside the bay) caused by abundant boring organisms. The dominant borers on protected, soft limestone are bivalves of the families Pholadidae (*Cyrtopleura crucigera* and *Dyplothyra curta*; up to 216.8 perforations m^{-2}) and Petricolidae (*Petricola denticulata*; 132 perforations m^{-2}). Species of the family Mytilidae (*Lithophaga aristata*, *L. plumula*; 62.9 ind. m^{-2}) perforate hard conglomerate rocks and sandstone substrates. Among crustaceans, only the ghost shrimp *Upogebia tenuipollex* (126 ind. m^{-2}) contributes to bioerosion of the rocky cliffs. By removing sediments from the cliff base, boring organisms form large cavities that weaken the base and cause the upper part to slide. Depending on the impact of storm waves, tides, and ENSO events, erosion rates in exposed areas may vary between 300 cm^3 m^{-2} month^{-1} for igneous rocky cliffs and 450 cm^3 m^{-2} month^{-1} for sedimentary rocky cliffs (Cantera et al. 1998). Several of the rocky islands and peninsulas originated from intense bioerosion of the northern shore of Buenaventura Bay.

18.3.3 Mangrove Swamps

Mangrove habitats in Buenaventura Bay are comprised of the red mangroves *Rhizophora mangle* and *R. racemosa* (Rhizophoraceae), the black mangroves *Avicennia germinans* and *A. tonduzii* (Avicenniaceae), the white mangroves ("comedero") *Laguncularia racemosa* and *Conocarpus erecta* (Combretaceae), the "Piñuelo" *Pellicera rhizophorae* (Theaceae), and the "Nato" *Mora oleifera* (Caesalpinaceae). These species are distributed along environmental gradients, like tidal range, wave exposure, sediment deposition, soil characteristics, microtopography, salinity and nutrients.

According to the classification by Cintron and Schaeffer-Novelli (1983), the mangrove habitats can be divided into three different physiographic types. Riverine mangroves are well developed (<35 m height, <20 cm dbh, ~100 trees per 0.1 ha) along tidal creeks and in the estuarine zone of rivers at the southern shore of the bay. Red and black mangroves colonize the unstable substrate near the sea front, whereas *Pellicera rhizophorae* and *Mora oleifera* occupy consolidated substrates (Fig. 18.2a). Bar mangroves grow behind sand ridges at Soldado Island near the mouth of the bay (<15 m height, <6 cm dbh, 185–766 trees per 0.1 ha). Red mangroves occupy unstable muddy substrates, while black and white mangroves colonize more stabilized sandy ridges (Fig. 18.2b). The fringe mangroves of protected areas along the northern shore of the bay occupy platforms resulting from bioerosion of sedimentary cliffs. Usually fringe mangroves develop slowly or are dwarfed, but exceptionally they exhibit development similar to riverine mangroves (<6–9 m height, 4–11 cm dbh, 10–250 trees per 0.1 ha). The unstable substrate of the frontal zone of platforms is colonized by red mangroves and *Pellicera rhizophorae* is present on stable internal platforms (Fig. 18.2c; Blanco et al. 1998).

Mangrove habitats offer niches for benthic macroalgae and animals, though few species are restricted to these habitats. Several green algae grow on prop roots (*Chaetomorpha californica*), pneumatophores (*Bood-*

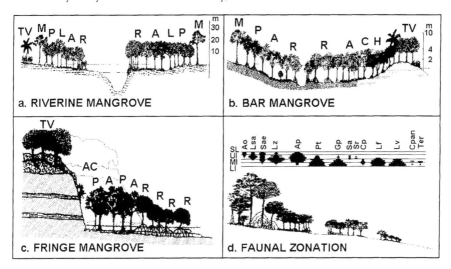

Fig. 18.2. Zonation of vegetation in different mangrove types (a-c) and faunal zonation (d) in Buenaventura Bay (*TV* terrestrial vegetation, *AC* ancient cliff, *P* Pellicera, *R* Rhizophora, *A* Avicennia, *M* Mora, *L* Laguncularia, *C* Conocarpus, *H* Hibiscus, *SL* supralittoral, *UI* upper intertidal, *MI* mid-intertidal, *LI* lower intertidal, *Ao* Armases occidentale, *Lsa* Littorinopsis scabra aberrans, *Sae* Sesarma aequatoriale, *Lz* Littoraria zebra, *Ap* Aratus pisonii, *Pt* Pachygrapsus transversus, *Gp* Goniopsis pulchra, *Sr* Sesarma rhizophorae, *Sa* Sesarma angustum, *Cp* Chthamalus panamensis, *Lf* Littorinopsis fasciata, *Lv* Littorinopsis varia, *Cpan* Clibanarius panamensis, *Ter* Teredo sp.)

leopsis verticillata, Cladophora graminea) and on the muddy substrate (*Cladophora albida*), which is also colonized by benthic diatoms (Naviculaceae, Nitzschiaceae) and blue-green algae (i.e. *Lyngbia aestuarii*). Red algae (*Bostrychia calliptera, B. tenella, Catenella caespitosa, C. impudica, Calloglossa stipitata*) grow on roots or form beds.

Similarly, faunal assemblages are associated with different mangrove biotopes (Fig. 18.2d). The tidal creeks are dominated by benthic fishes (Ariidae, Clupeidae, Centropomidae, Eleotridae, Mugilidae, Hemirhamphidae, Carangidae, Bothidae, Tetraodontidae), crustaceans (i.e., *Callinectes toxotes, C. arcuatus*), and marine (Penaeidae) and freshwater (Palaemonidae) shrimps, which reproduce or feed in the estuary. The actual mangrove trees are inhabited by both terrestrial and marine species. In the understory, epiphytic (orchids, bromeliads) and parasitic plants (Lorantaceae) provide protection and food for insects and birds. The tree trunks and roots are associated with crabs (*Aratus pisonii, Goniopsis pulchra, Pachygrapsus transversus, Sesarma* spp., *Armases* sp.), barnacles (*Chthamalus panamensis*), periwinkles (*Littoraria zebra, Littorinopsis scabra aberrans, L. fasciata, L. varia*), other snails (i.e., *Thais kiosquiformis*), the oyster *Crassostrea columbiensis*, and shipworms (Teredinidae). Although organisms move vertically with tidal oscillations (Blanco and Cantera 1999), clearly defined vertical zones are characteristic (Cantera 1991; Blanco 1995). Muddy bottoms among mangroves down to 30 cm depth are occupied by endofaunal organisms like bivalves (*Anadara, Mytella, Protothaca, Corbula*) and deposit-feeding polychaetes (i.e., *Nereis*). Most species live on the mud surface where the mud whelk *Cerithidea* and other gastropods (*Anachis, Theodoxus, Melampus, Marinula*) feed on detritus. Crustaceans (i.e., *Uca, Alpheus, Synalpheus*) are represented by 24 species and fishes, such as wizards or gobies (*Gobionellus, Gobiosoma, Gobiomorus*), flatfishes (Cynoglossidae, Bothidae, Soleidae), and toadfishes (*Batrachoides pacifici, Daector dowi*), are common (Cantera et al. 1983).

18.3.4 Mud Flats

Extensive mud flats occur around creeks in the inner bay. The sediments are generally less than 125 μm but are occasionally mixed with river-borne sand or gravel from cliff erosion. The structure of the community is typically related to characteristics of soft bottoms (i.e. grain size, organic matter, REDOX potential, temperature, etc.). Despite the apparent absence of organisms, mud flats are rich in species (157), notably deposit-feeders and detritivores, most of which occupy the aerobic upper subsurface layers of sediments, while some bivalves and polychaetes burrow into deeper layers. In general, tidal zones are not well defined. Owing to high temperature and desiccation stress, the upper littoral is poor in species, with few dominants like *Marinula, Cerithidea, Eurypanopeus*, and bivalves (*Protothaca, Tagelus, Chione, Polymesoda*) which remain buried most of time. *Theodoxus, Clibanarius*, and *Uca* dominate the upper intertidal zone but, where sediments are mixed with gravel, other species (*Petrolisthes, Anachis, Panopeus, Eurypanopeus*) are present. Species richness (32) increases in the mid-intertidal zone (423 ind. m^{-2}), with polychaetes (*Capitella, Glycera, Thelepus, Streblosoma, Neanthes*) representing the principal component (Cantera 1991). The lower intertidal zone (up to 67 species) diplays high diversity and evenness and is inhabited by gastropods (*Natica, Nassarius, Theodoxus, Anachis, Cerithium*), bivalves (*Tagelus, Anadara, Chione, Polymesoda*), stomatopods (*Squilla*), polychaete worms (Amphinomidae, Capitellidae, Glyceridae, Nereidae), crabs (*Panopeus, Eurypanopeus, Alpheus, Callinectes*), and benthic fishes (*Gobionellus, Gobiesox, Citharichthys, Symphurus*).

18.3.5 Pelagic Estuarine Environment

Organisms which live in the water column of Buenaventura Bay have to tolerate brackish and turbid conditions. Phytoplankton is dominated by high numbers of diatoms (*Thalassiosira* sp., *Biddulphia* spp., *Rhizosolenia* spp., *Chaetoceros* spp., *Nitzchia* spp., *Thalassiotrix mediterranea*, *T. frauenfeldii*, *Thalassionema* spp., *Coscinodiscus gigas*, *C. kurzii*). Dinoflagellates are scarce and restricted to the mouth of the bay, while volvocaceans frequently occur in polluted areas. Zooplankton diversity is high, owing to the large number of larvae (meroplankton) of species which spend their early life stages in estuarine waters, as well as the presence of holoplankton which is largely composed of copepods, arrowworms (Chaetognatha), medusae, and tintinids.

About 185 fish species of 50 families compose the nekton community (Rubio 1984), though marine fishes and shellfishes are generally present as larvae and juveniles. Scianidae is the most diverse family (25 species); however, families like Carangidae (15), Engraulidae (11), Clupeidae (9), Pomadasydae (8), Gobiidae (7), Ariidae (6), and Centropomidae (5) are also important. The herring *Lile stolifera* is the most abundant species, other representative species are pufferfish (*Sphoeroides annulatus*), anchovies (*Anchoa panamensis*, *A. spinifer*), mullet (*Mugil cephalus*), soapfish (*Dormitator latifrons*), mackerel (*Chloroscombrus orqueta*), herring (*Opisthonema libertate*), halfbeaks (*Hyporhamphus unifasciatus*), and silversides (i.e. *Melaniris pachylepis*). Invertebrates are poorly represented in the waters of the bay, prawns (Penaeidae and Palaemonidae) being most important, and crabs (Portunidae and Calappidae) and squids (*Loligunculla*) are frequent at the mouth of the bay. Typical estuarine fishes which reproduce and feed in the bay do not migrate upstream into freshwater areas or out of the bay into the ocean. Freshwater organisms do not migrate beyond the oligohaline zones of the river mouths and therefore are generally absent from Buenaventura Bay waters.

18.4 Trophic Relations

18.4.1 Primary Production

The different biological communities and habitats and the complex estuarine conditions in Buenaventura Bay provide for a diverse and productive ecosystem. The high nutrient concentrations at the mouth of creeks and the fall of rivers fuel primary production of phytoplankton, macroalgae, and mangroves. The tidal, high-turbidity zone of rivers has higher chlorophyll concentrations than non-tidal upstream zones (i.e. chlorophyll-a 0.16–0.10 µg l^{-1}; chlorophyll-b 0.11–0.06 µg l^{-1}; chlorophyll-c 0.23–0.1 µg l^{-1}). Owing to increased nutrient levels and lower water turbidity, concentrations increase in February-March and May-June, suggesting high phytoplankton biomass and photosynthetic oxygen production (Bejarano 1996). Mangroves appear to be highly productive, with red (*Rhizophora mangle* 960–4,800 kg ha^{-1} year^{-1}; *R. racemosa* 1,800–7,200 kg ha^{-1} year^{-1}) and black (*Avicennia germinans* 240–1,440 kg ha^{-1} year^{-1}) mangrove species contributing the most to total litter production (Prahl et al. 1990). The average annual litter production ranges from 9.6 tons ha^{-1} in bar mangroves to 11.4 tons ha^{-1} in riverine mangroves (Lasso and Cantera 1995), but is much lower in disturbed riverine mangroves (7.5 tons ha^{-1} year^{-1}; Garcia and Garces 1984). Owing to high photosynthetic rates (24–38 mg C per g dry wt. and day), the mangrove macroalgae community contributes about 23–26 % to the total annual mangrove production (Peña 1998).

18.4.2 Food Webs

The organic carbon of primary production processes enters the food web of Buenaventura Bay via herbivory or decomposition pathways as particulate (POM) or dissolved organic matter (DOM). Decomposition of the large (85.5 g m^{-2}) macroalgal biomass (Bejarano 1996) takes about 120 days, with rates of 0.58 g day^{-1} during the first 20 days which increase proximate to muddy bottom sediments. Herbivory on mangroves is largely restricted to leaves (84.9 %), though only a small part of the total leaf area (*A. germinans* 4.9 %, *L. racemosa* 4.7 %, *R. mangle* 2.3 %, *P. rhizophorae* 2.06 %) is removed. As a consequence, the grazing impact on well-developed mangrove stands in Buenaventura Bay ranges from 428.8 kg ha^{-1} year^{-1} in *A. germinans* and 413 kg ha^{-1} year^{-1} in *L. racemosa* to about 247.2 kg ha^{-1} year^{-1} in *R. mangle* and 200.8 kg ha^{-1} year^{-1} in *Pellicera rhizophorae* (Romero 1998). The principal primary consumers of mangrove leaf tissue are insects (Scotylidae, Saturnidae, Sphingidae), crabs (*Goniopsis pulchra*, *Aratus pisonii*, *Pachygrapsus transversus*, *Sesarma* spp., *Armases* sp.), and some periwinkles (Littorinidae). The decomposition of the leaf litter presents the most important energy pathway. The dead leaves are initially colonized by microorganisms (i.e. the fungi *Pestalotiopsis*, *Penicillium*, *Aspergillus*, *Fusarium*, *Trichoderma*, *Ceriosporopsis*, *Pontogenia*). The decomposing litter and microorganisms are then ingested by gastropods (*Theodoxus*, *Melampus*, *Ellobium*, *Mari*-

nula), crabs (*Sesarma* spp., *Armases*), amphipods (Talitridae), and some collembolan insects (*Archisotoma, Cryptopigas, Axelsonia*).

Particulate and dissolved organic matter derived from mangrove litter is the main food source for deposit-feeders (77 % of all species) in both mangrove sediments and in the sediments of adjacent mud flats and sand beaches. Filter- and suspension-feeding bivalves, polychaetes, and barnacles on mangrove prop roots, or species like the ghost crab *Ocypode* spp. on muddy and sandy sediments, consume dissolved organic matter and phytoplankton. Small filter-feeding fish (<15 cm total length) present a diverse and abundant trophic group. The herring *Opisthonema libertate* feeds exclusively on zooplankton (23–99 % of gut content) while the anchovy *Cetengraulis mysticetus* feeds on diatoms like *Thalassionema* (29.13 %) and *Skeletonema* (23.65 %). Benthic fish (i.e. *Mugil curema*) feed on bottom diatoms (*Navicula* 57 % in number of individuals; *Coscinodiscus* 8.5 %; *Fragillaria* 15.4 %) while prawns (Penaeidae) ingest zooplankton, phytoplankton, and DOM and POM.

Gastropods (*Natica unifasciata, Muricanthus radix, Thais kiosquiformis, T. biserialis*), crustaceans (*Goniopsis pulchra, Aratus pisonii, Panopeus purpureus, P. chilensis, Eurypanopeus transversus, Callinectes toxotes, Callinectes arcuatus*) and small fishes prey on benthic fauna in the mangroves. The crabs are omnivorous, for example, the diet of *Callinectes arcuatus* consists of fishes (17.5 %), mollusks (15.9 %), shrimps (12.9 %), and vegetal matter (7.7 %). Several abundant fish species, which are buried during low tide or enter with high tide, like the drum (*Stellifer oscitans*), feed on benthic crustaceans (0.7 % of the body weight), and the catfish (*Arius seemanni*) and the threadfin (*Polydactylus approximatus*) forage on crustaceans and fishes. During high tide several, mostly carnivorous marine fish visit estuarine waters and mangrove areas in search of food. The diet of the flatfish (*Cyclopsetta querna*) is composed of penaeid prawns (80.6 %) and goby fishes (19.4 %). The tusky drum (*Cynoscion phoxocephalus*) ingests the prawn *Xiphopenaeus* (2.2 %) and small pelagic fishes (0.68 %). Pelagic species like the jack (*Caranx caninus*) feed on crabs and larvae (3.2 % of body weight), the rocky snapper (*Lutjanus guttatus*) mostly ingests snapping shrimps (Alpheidae) and stomatopods (23.2 % of gut content), and the diet of grunt (*Pomadasys panamensis*) is composed of prawns and crabs (73–98 % of gut content) and small pelagic fishes (Gobiidae, Engraulidae, Clupeidae; 2–27 % of gut content).

In intertidal habitats, the predation by seabirds (*Pelecanus, Phalacrocorax, Larus, Sterna*) and migratory shorebirds represents an important export of energy (137.5 tons year^{-1} and 1.05 tons year^{-1}, respectively) out of Buenaventura Bay (Morales 1998). In contrast, terrestrial species like *Egretta* (2.7 tons year^{-1}) and *Butorides* (0.5 tons year^{-1}), which feed on intertidal

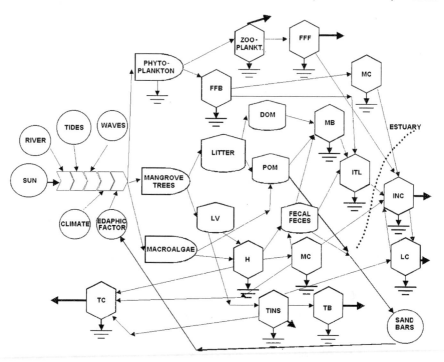

Fig. 18.3. Conceptual food web of Buenaventura Bay mangrove habitats (*FFB* filter-feeding benthos, *FFF* filter-feeding fishes, *MC* minor carnivores, *DOM* dissolved organic matter, *POM* particulate organic matter, *MB* microorganisms, *ITL* intermediate trophic level, *INC* intermediate carnivores, *LC* large carnivores, *TB* terrestrial birds, *TINS* terrestrial insects, *H* herbivores, *TC* terrestrial carnivores, *LV* live vegetation)

flats, and reptiles (*Iguana*), insects and birds of the tropical rain forest, which feed in mangrove habitats, export organic matter from Buenaventura Bay to adjacent terrestrial systems (Fig. 18.3).

18.5 Human Impact

Anthropic activities have significantly altered the Buenaventura Bay ecosystem. The sewage treatment and waste disposal facilities of Buenaventura City (population 220,600) are inadequate and elevated concentrations of heavy metals and hydrocarbons are being introduced by mining and port activities. Fisheries and logging-timbering activities have led to declining stocks of commercially important species in the bay and to a gradual decrease in biodiversity.

18.5.1 Pollution

The sewage contamination by Buenaventura City ($Q=820$ l s^{-1}, $BOD_5=14$ tons day^{-1}) has introduced localized but pronounced structural changes of the benthic soft-bottom community. In contrast to the impact of sewage on the soft-bottom community structure elsewhere, chronic contamination near most of the sewage outfalls in Buenaventura Bay has led to an intertidal benthic soft-bottom community with few (1–10) species, low total density (0.48–14.56 ind. m^{-2}), and high evenness (0.44–0.99; CAE 1995). Together, polychaetes (i.e., *Capitella* sp.), the fiddler crab *Uca* spp., the snapper shrimp *Alpheus columbiensis*, and the jackknife clam *Tagelus* sp. represent 82% of the total density with polychaetes accounting for 59%. However, the benthic soft-bottom community is dominated by *Capitella* sp. (90%) and nematodes around the main sewage outfall and near the petroleum dock, respectively (Cantera et al. 1992). In general, total community biomass of impacted areas (2–4.5 g m^{-2}) exceeds that of non-impacted areas (1.2–2 g m^{-2}).

Elevated heavy metal concentrations in waters and sediments at the mouth of the Anchicaya and Dagua Rivers and around Buenaventura City are largely a result of river discharge, land-runoff, and local sewage input (T. Fernández and A. Pion, unpubl. data). Additionally, dredging activities and natural resuspension influence the relationship between metal concentrations in the water column and sediments (Univalle 1997; Cortez 1998). Heavy metal surveys of biological tissue in mangrove cockles (*Anadara*), catfish (*Arius*), and swimming crabs (*Callinectes*) indicate concentrations close to permitted levels (Benítez 1995; Cortez 1998). Elevated mercury (5–10 µg l^{-1}) and lead (5–20 µg l^{-1}) concentrations in the blood of human populations living around Buenaventura Bay have been associated with seafood preference and frequent consumption (Benítez 1995).

18.5.2 Exploitation

For more than 400 years mangrove timber has been an important economic resource for the region (Prahl et al. 1990). Cutting of mangroves increased drastically during the 1960s, owing to demands by the leather industry for tanning dyes contained in the bark of the red mangrove, which led to intense exploitation of well-developed red mangrove forests (10–15 cm dbh). During the last 20 years, extensive mangrove areas were cleared for shrimp aquaculture and agricultural activities (i.e. banana, coconut, casaba). Today, the illegal cutting of mangrove trees represents up to 30% of their total density in Buenaventura Bay (Blanco and Cantera

1995) and clear-cut areas have been identified as the main disturbance for their development (Blanco et al. 1998). Although data on fishery landings in Buenaventura Bay are lacking, local fishermen catch significant quantities of commercially valuable fish (*Arius* spp., *Centropomus* spp., *Mugil* spp.), crustaceans (*Callinectes toxotes*, *C. arcuatus*, *Cardissoma crassum*, *Panulirus* sp., penaeid shrimps), and mollusks (*Anadara tuberculosa*, *A. grandis*, *Striostrea prismatica*, *Muricanthus radix*). Owing to highly efficient but hardly selective fishing gear, some of these species already show signs of overfishing.

18.6 Management Needs

Most environmental issues of Buenaventura Bay are related to land reclamation, activities in the watershed, dredging, and pollution. Despite the prohibition of commercial timber exploitation of any mangrove species until the year 2000 by the Regional Board for the Protection of the Environment (CVC), the growth of Buenaventura City and expanding aquaculture and agriculture have imposed increasing pressure, mainly on the mangroves in the bay. Dredging activities have altered natural hydrodynamic processes and sedimentation/erosion rates while contamination is seriously effecting benthic and pelagic communities and even local human populations. About 7% of the sewage (60 l s^{-1}) and 30% of solid wastes (85 tons) are discharged daily to Buenaventura Bay by the rivers. Furthermore, mining and deforestation in the watershed add high concentrations of TSS and cause increased sedimentation in the bay.

It is clear that sustainable management needs to integrate environmental issues of the bay and the watershed of rivers. The national and regional government should establish a permanent monitoring network of oceanographic conditions and of the water quality in the bay. They should promote joint research between environmental agencies and universities/research institutes, initiate a program of environmental education, and enhance the participation of the local population in problem identification, decision making, and protection measures. The plan for "Wetland Management and the Integrated Management of the Coastal Zone" by the Regional Bureau for the Protection of the Environment will significantly contribute to protection, research, education and management issues. Furthermore, zones of preservation, recovery, and multiple use (Sánchez-Paez and Alvarez-León 1998) recently established by the Colombian Ministry of the Environment have included Buenaventura Bay as a zone of preservation.

References

Bejarano AC (1996) Aporte de biomasa y detritus de las macroalgas bénticas del ecosistema manglar-estuario. Thesis, Universidad del Valle, Colombia

Benítez N (1995) Determinación de los niveles de metales pesados en agua, organismos marinos y sangre humana en la bahía de Buenaventura. Thesis, Universidad del Valle, Colombia

Blanco JF (1995) La malacofauna epibentónica como indicadora de condiciones naturales y de tensión en manglares del Pacífico Colombiano. Thesis, Universidad del Valle, Colombia

Blanco JF, Cantera JR (1995) Patrones estructurales de algunos manglares de la bahía de Buenaventura (Pacífico colombiano) y las condiciones hidrológicas y de intervención humana que los determinan. In: Restrepo JD, Cantera JR (eds) Delta del San Juan, Bahías de Málaga y Buenaventura, Pacífico Colombiano. Colciencias-Universidad Eafit-Universidad del Valle, Tomo II, pp 32-58

Blanco JF, Cantera JR (1999) The vertical distribution of mangrove gastropods and environmental factors relative to tide level on the Central Pacific Coast of Colombia. Bull Mar Sci 65(3):617-630

Blanco JF, Cantera JR, Bejarano AC, Lasso J (1998) Aplicaciones de los análisis de ordenación comunitaria para los estudios de estructura vegetal en manglares. Memorias XI Seminario Política, Ciencias y Tecnologías del Mar, Bogotá, Comisión Colombiana de Oceanografía, Universidad Jorge Tadeo Lozano

CAE (1995) Estudio de impacto ambiental por aguas residuales en la Bahía de Buenaventura. Comité de Acción Ecológica, Universidad del Valle, Colombia

Cantera JR (1991) Etude structurale des mangroves et des peuplements littoraux des deux baies du pacifique colombien (Málaga et Buenaventura). Rapport avec les conditions du milieu et les perturbations anthropiques. PhD Thesis, Université d'Aix-Marseille, France

Cantera JR (1993) Oceanografía. In: Leyva P (ed) Colombia Pacífico. Santafé de Bogotá, Fondo FEN-Proyecto Biopacífico, Tomo I, pp 13-23

Cantera JR, Arnaud PM, Thomassin B (1983) Biogeographic and ecological remarks on molluscan distribution in mangrove biotopes. I. Gastropods. J Molluscan Stud Suppl 12a:10-26

Cantera JR, Neira R, Tovar J (1992) Efectos de la polución domestica sobre la malacofauna bentónica de sustratos blandos en la costa Pacífica colombiana. Rev Cienc 7:21-39

Cantera JR, Neira R, Ricaurte C (1998) Bioerosión en acantilados del Pacífico colombiano. Santafè de Bogotá, Fondo FEN-Proyecto Biopacífico

Cintron G, Shaeffer-Novelli Y (1983) Introducción a la ecología del manglar. UNESCO, Montevideo

Cortez M (1998) Determinación de metales pesados, aguas, sedimentos y organismos marinos en la Ensenada de Tumaco, la Bahía de Buenaventura y Bahía Málaga. MSc Thesis, Universidad del Valle, Colombia

Eslava J (1993) Climatología. In: Leyva P (ed) Colombia Pacífico. Santafé de Bogotá, Fondo FEN-Proyecto Biopacífico, Tomo I, pp 137-147

Gálvis J, Mojíca J (1993) Geología. In: Leyva P (ed) Colombia Pacífico. Santafé de Bogotá, Fondo FEN-Proyecto Biopacífico, Tomo I, pp 80-96

García JH, Garces V (1984) Aporte de biomasa y notas ecológicas de un manglar intervenido, estero Limones, bahía de Buenaventura, costa Pacífica Colombiana. Thesis, Universidad del Valle, Colombia

Gonzalez JL, Morton RA, Correa ID, López GI (1998) Ocurrencia de "overwash" en islas barrera del Pacífico colombiano; caso de la Isla del Choncho, Delta del San Juan. Memorias XI Seminario Política, Ciencias y Tecnologías del Mar, Bogotá, Comisión Colombiana de Oceanografía, Universidad Jorge Tadeo Lozano

Kjerfve B (1990) Manual for investigation of hydrological processes in mangrove ecosystems. UNESCO/UNDP Regional Mangrove Project RAS/79/002 and RAS/86/120, occasional paper, New Delhi

Lasso J, Cantera J (1995) La caída de hojarasca como indicador de productividad: comparación entre un bosque riberino y uno de barra en la bahía de Buenaventura, Pacífico colombiano. In: Restrepo JD, Cantera JR (eds) Delta del San Juan, Bahías de Málaga y Buenaventura, Pacífico colombiano. Tomo I. Colciencias-Universidad Eafit-Universidad del Valle

Lobo-Guerrero A (1993) Hidrología e hidrogeología. In: Leyva P (ed) Colombia Pacífico. Santafé de Bogotá, Fondo FEN-Proyecto Biopacífico, Tomo I, pp 121–134

Martínez JO (1993) Geomorfología. In: Leyva P (ed) Colombia Pacífico. Santafé de Bogotá, Fondo FEN-Proyecto Biopacífico, Tomo I, pp 111–119

Martínez JO, Gonzalez JL, Pilkey OH, Neal WJ (1995) Tropical barrier islands of Colombia's Pacific coast. J Coast Res 11 (2):432–453

Morales G (1998) Flujo energético y disponibilidad de hábitats de forrajeo para las aves marinas y playeras del Pacífico colombiano. Thesis, Universidad del Valle, Colombia

Peña EJ (1998) Physiological ecology of mangrove-associated macroalgae in a tropical estuary. PhD Thesis, University of South Carolina, USA

Prahl H, Cantera JR, Contreras R (1990) Manglares y hombres del Pacífico colombiano. Santafé de Bogotá, Fondo FEN-Editorial Presencia

Romero IC (1998) La herbivoría foliar como proceso de transferencia de materia en el ecosistema de manglar de la bahía de Buenaventura. Thesis, Universidad del Valle, Colombia

Rubio E (1984) Estudios sobre la ictiofauna del Pacífico Colombiano. I. Composición taxonómica de la ictiofauna asociada al ecosistema manglar-estuario de la bahía de Buenaventura. Cespedesia 13(49/50):296–315

Sánchez-Paez H, Alvarez-León R (1998) Diagnóstico y zonificación preliminar de los manglares del Pacífico de Colombia. Ministerio de Ambiente, ITTO, PD 171-91 Rev 2 (F) Fase1

Univalle (1997) Dragado del canal de acceso al puerto de Buenaventura. Interventoría ambiental. Tomos: Metales Pesados, Bentos e Hidrología. Gobernación del Valle, Ministerio de Transporte, Buenaventura, Colombia

19 Eastern Pacific Coral Reef Ecosystems

P.W. Glynn

19.1 Introduction

Although not as common as coastal soft bottom, estuarine and mangrove ecosystems, reef-building coral communities and coral reefs occur in the eastern tropical Pacific region, ranging from northern Mexico to southern Ecuador, and are present on all offshore islands as well. Coral reefs are wave-resistant limestone structures built dominantly by the vertical accumulation of coral skeletons. In contrast, coral communities are loosely spaced to dense aggregations of coral colonies that veneer underlying substrates whose origin is other than the actively growing corals they support. This distinction is especially important in the eastern Pacific where coral communities can be easily confused with true structural coral reefs. Both coral reefs and coral communities consist of coral colonies that host zooxanthellae (endosymbiotic dinoflagellates), which are dependent on solar energy, and are thus confined to relatively shallow depths. Eastern Pacific coral reefs do not dominate coastal seascapes because they are relatively small and patchy in distribution, their presence depending upon a combination of requisite hydrographic and geomorphologic conditions. Also, eastern Pacific coral reefs are seldom visible to non-divers because they rarely build islands or possess emergent algal ridges, and are exposed for only brief periods during extreme low tides. However, where reefs are present they boost local biodiversity, provide habitats for a variety of fishes and shellfish, and offer attractive vistas for diver-oriented ecotourism. With the increasing attention and scientific interest directed toward eastern Pacific coral reefs since about the 1970s, our knowledge of this ecosystem has grown immeasurably (Cortés 1997) but the scope of new information is highly imbalanced. While much new understanding has been reached in the areas of systematics, population and community ecology, biogeography and reef growth history, no progress has been made at the ecosystem level focusing on trophic structure, nutrient cycling or productivity. Of necessity, therefore, the following essay will emphasize our level of understanding in the former disciplines.

This bias is not meant to diminish the significance of organic production and related topics, but hopefully will serve to alert workers of this deficiency.

19.2 Environmental Setting

Due to the great expanse of deep water (up to 8,200 km) separating central Pacific and western American shores, the far eastern Pacific is perhaps the most isolated coral reef region in the world (Dana 1975; Grigg and Hey 1992). However, evidence suggests that several Indo-Pacific shallow-water species are capable of crossing this barrier (Scheltema 1988; Jokiel 1990; Emerson and Chaney 1995; Lessios et al. 1998). Once central Pacific larvae or rafting organisms in eastward flowing currents accomplish this crossing they must settle and become established under sometimes harsh physico-chemical conditions and also survive various interactions (e.g. predation and competition) with resident eastern Pacific species. A significant factor influencing the distribution and development of eastern Pacific coral communities and coral reefs, recognized by such notable mid-19th century workers as James Dana and Charles Darwin, is the sea temperature climatology of the major regional current systems. Seasonal low temperatures of 18–20 °C are tolerated by most species of zooxanthellate corals, and brief exposures to 15–16 °C during upwelling episodes also are generally non-lethal. To a large degree, the asymmetrical distribution of eastern Pacific reef-building corals, located mostly north of the equator, reflects the flow of the relatively cool Peru Coastal and California Currents (Fig. 19.1). The northern-most coral reefs occur at the tip of Baja California (23.5°N) and are found as far south as about 3°S along the Ecuadorian coast. Coral communities are common along the eastern shores of Baja California (from about 26°N to the tip of the peninsula) and also occur abundantly throughout the eastern Pacific region slightly farther south than coral reefs, to at least Isla Santa Clara in the Gulf of Guayaquil, Ecuador (Fig. 19.1).

Fig. 19.1. The eastern Pacific coral reef region, spanning approximately 25° latitude and concentrated mostly north of the equator. Some of the known coastal coral reef areas are Cabo Pulmo-Los Frailes (*1*); Nayarit (*2*); Huatulco (*3*); Los Cóbanos (*4*); Gulf of Papagayo (*5*); Caño Island-Golfito (*6*); Gulf of Chiriquí (*7*); Gulf of Panamá (*8*); Utría (*9*); Gorgona Island (*10*); Cabo San Lorenzo-Salinas (*11*). *Arrows* denote synoptic surface flow of the major current systems. Costa Rica Coastal Current (*CRCC*). *Broken arrows* denote the Equatorial Under Current (*EUC*), which surfaces as it collides with the westernmost Galápagos Islands. All currents vary seasonally in their velocity and extent: the NECC and CRCC are best developed during the northern hemispheric summer

Eastern Pacific Coral Reef Ecosystem 283

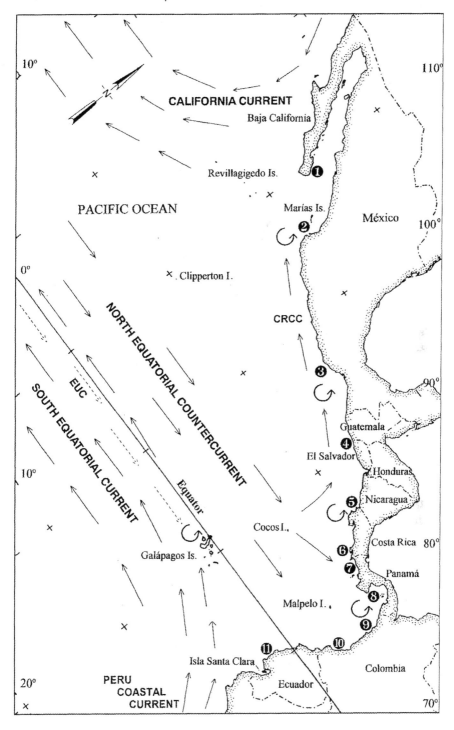

Within the eastern Pacific region, local upwelling systems also can influence the structure and distribution of coral communities. While coral reefs occur within the three major eastern Pacific upwelling centers in the Gulfs of Tehuantepec (site 3), Papagayo (site 5), and Panamá (site 8), they are usually more numerous, larger and older (with thicker framework accumulations) in non-upwelling areas, such as Caño Island/Golfito area (site 6), Gulf of Chiriquí (site 7), and Gorgona Island (site 10; Fig. 19.1) and their associated biota are somewhat different. For example, zooxanthellate hydrocorals (*Millepora*, two species) and the crown-of-thorns sea star [*Acanthaster planci* (Linnaeus)] are associated with coral reefs in the Gulf of Chiriquí, but not in the Gulf of Panamá (Glynn 1974). Another area subject to strong upwelling, devoid of coral communities and reef development, is the western shore of the Galápagos Islands where the Equatorial Undercurrent surfaces (Pak and Zanfeld 1973). Here, astride the equator, sea surface temperatures (SSTs) dip below 15°C, nutrients are usually in high concentration, and the littoral marine communities consist predominantly of macroalgae (Glynn and Wellington 1983). Anomalously high SSTs also can affect coral reef development. The magnitude and duration of positive sea temperature anomalies that accompanied the 1982–1983 and 1997–1998 El Niño-Southern Oscillation (ENSO) events correlated with local patterns of coral bleaching (loss of symbiotic zooxanthellae) and mortality (Glynn et al. 1988). Coral mortality and subsequent bioerosion have resulted in significant degradation of reefs throughout the equatorial eastern Pacific and the virtual elimination of the coral reef biotope from the Galápagos Islands.

Several other factors, like the availability of stable substrates, degree of exposure to wave assault, and water quality, can influence eastern Pacific reef development. Since much of the mainland coast supports mangrove ecosystems, freshwater lagoons, estuaries and sandy/muddy beaches, coral reefs occur along rocky coastal stretches that offer suitable conditions for their development (i.e. sites 1, 2, 3, 4, 6, 7; Fig. 19.1). Unlike coral reefs in most other biogeographic regions, eastern Pacific reefs are constructed mainly of branching corals which are not typically well consolidated by crustose coralline algae or underwater cementation. Therefore, reef frameworks are relatively fragile and tend to develop best on rocky shores sheltered from the full force of open seas, such as in bays and on the leeward sides of offshore islands. This dependency on sheltered locations is well illustrated by the distribution of coral reefs along the Huatulco coast (site 3; Figs. 19.1, 19.2C). Paradoxically, patch reefs even occur inside the Punta Entrada estuary (Gulf of Chiriquí), their presence due presumably to a saltwater wedge that sustains relatively high salinity conditions in this marginal setting. Coral reefs also occur in the Panamá Bight (sites 8, 9) where nearshore surface salinities are 30 psu (~‰) or lower at the height of the rainy season.

19.3 The Eastern Pacific Coral Reef Region

With the closure of the Central American seaway, about 3.5–3 Ma (Coates and Obando 1996), former connections between the tropical marine biota of the eastern Pacific, Indo-Pacific and western Caribbean Sea were severed for the last time. During earlier geologic time, between 18 and 14 Ma, the western portion of the circumtropical Tethys Sea became separated from the Atlantic Ocean by the emergence of a land corridor resulting from the collision of the African plate and southern India with Eurasia. Since this separation the species composition, ecological processes and structure of amphi-American coral reef communities have changed radically (Glynn 1982). The eastern Pacific region is now an impoverished outpost of the central and western Pacific coral reef domain and no longer shares any reef-building coral species with the Atlantic region. A total of 69 western Atlantic and 534 west/central Pacific coral species have been described. As many as 30 to 40 coral species are found on any given Caribbean reef and twice this number or more on most west/central Pacific reefs. By contrast, known eastern Pacific zooxanthellate corals (scleractinians and hydrocorals in the genus *Millepora*) number 41 species with about 10 species occupying any particular site where they contribute more or less significantly toward reef building. Branching coral colonies in the genus *Pocillopora*, specifically *Pocillopora damicornis* (Linnaeus) and *Pocillopora elegans* Dana, are preeminent in reef building, forming intermeshing compact frameworks not uncommonly attaining 2–3 m in relief. Species with massive, dome-shaped colony morphologies, such as *Porites lobata* Dana, *Pavona clavus* Dana and *Pavona gigantea* Verrill, can also sometimes build mono- to paucispecific reef frameworks or contribute to pocilloporid reef building. Remarkably, these examples demonstrate that only one or a few coral species are capable of building coral reefs.

Marginal coral framework development has been reported at Cabo Pulmo-Los Frailes off SE Baja California (Fig. 19.1, site 1), and some of the coral communities in this area accommodate a diverse alcyonarian fauna. Fringing coral reefs occur in the Nayarit area and coral communities are present at Bahía Banderas, which experiences local upwelling (Fig. 19.1, site 2; Fig. 19.2A; Reyes Bonilla 1993; Carriquiry and Reyes Bonilla 1997). Upwelling here is intensified topographically by a submarine canyon that approaches close to the coast. Possibly one of the largest coral reefs in the

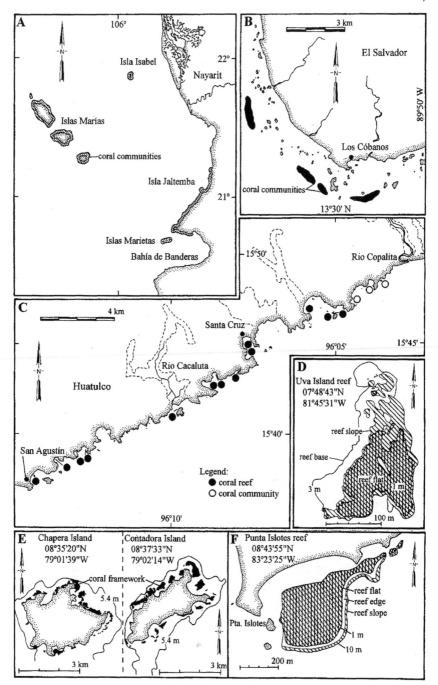

Fig. 19.2A-F. Selected areas of coral reef development along eastern Pacific continental and island shores. A Nayarit, México; B Los Cóbanos, El Salvador; C Huatulco, México; D Gulf of Chiriquí, Panamá; E Pearl Islands, Gulf of Panamá, Panamá; F Golfito, Costa Rica

entire eastern Pacific is located on the NE side of María Cleofas Island, Marías Islands (Fig. 19.2A). This reef is still undescribed, but a preliminary reconnaissance reveals that it is constructed dominantly of pocilloporid corals and forms a nearly closed ring several km in circumference (H.W. Chaney, pers. comm.). A coral reef tract extends along the southern coast of México at Huatulco (Fig. 19.1, site 3). Several fringing pocilloporid reefs, with 1–5 m vertical buildups, are present in this area in bays, inlets or behind islands, generally oriented away from zones of high turbulence (Fig. 19.2C). Although the extent of reef development off the coasts of Guatemala, El Salvador and Nicaragua is virtually unknown, a few small fringing reefs have been reported at Los Cóbanos, El Salvador (Fig. 19.1, site 4; Fig. 19.2B). Both Costa Rica and Panamá have seasonal upwelling (Fig. 19.1, sites 5, 8) and non-upwelling (Fig. 19.1, sites 6, 7) areas with relatively few weakly developed and several robust coral reefs in these respective environments. Coral reefs in the upwelling area of the Gulf of Papagayo (Fig. 19.1, site 5) are mostly dead, highly eroded and covered with filamentous algae. Evidence suggests that the entire dead coral reef tract in the Gulf of Papagayo possibly succumbed to extreme cooling during upwelling in the Little Ice Age, about 300 years ago (Glynn et al. 1983).

Fig. 19.3. Punta Islotes fringing reef (*light area in center*), oblique aerial view photographed at about 300 m elevation, 9 March 1989. A patchwork of exposed coral colonies is visible at the east shoulder (*right-hand side*) of the reef during a –0.3-m low tidal stand. (Courtesy J. Cortés)

Pocilloporid fringing reefs are present at Caño Island (Guzmán and Cortés 1989) and a few fringing reefs, with significant structural contributions from pocilloporid and poritid corals, occur in the innermost reaches of Golfito (Fig. 19.1, site 6; Figs. 19.2F, 19.3). The Golfito reefs have sparse live coral cover, a result of heavy siltation stress due to forest clearing and various sorts of other damage (e.g. mining and road construction) to the surrounding watershed. Numerous fringing and patch pocilloporid reefs occur in the Gulf of Chiriquí (Fig. 19.4A,B), particularly on the sheltered sides of islands. A coral reef on the south side of Bahía de las Damas, Coiba Island, covering about 160 ha, is the largest known reef adjacent to the eastern Pacific mainland (Fig. 19.1, site 7). In the upwelling area of the Pearl Islands, Gulf of Panamá (Fig. 19.1, site 8; Fig. 19.2E), most of the small pocilloporid reefs are located on the N or E sides of islands, away from the full effect of cool, upwelling currents. Farther to the south, along the Chocó coast of Colombia near Utría (Fig. 19.1, site 9), remarkably there are a few coral reefs present in this area in spite of exceptionally high annual rainfall (7–8 m) and voluminous river discharge with high sediment loading. Like Golfito, the fringing reefs of the Colombian Chocó reveal extensive dead coral cover, presumably also due to relatively recent stresses resulting from poor land use practices (Vargas Angel 1996). The fringing pocilloporid reefs on the lee (east) side of Gorgona Island (Fig. 19.1, site 10), with moderate to high coral cover, are the best developed in Colombia and reveal relatively few anthropogenic disturbances (von Prahl et al. 1979). At least one well-developed fringing reef occurs as far south as 1°28'S, on the coast at Machalilla, Ecuador (Fig. 19.1, site 11). Several large (10–20 m diameter) coral patches, some with 100- to ≥200-year-old colonies of *Pavona clavus*, occur along the leeward shore of Isla La Plata, about 30 km offshore of Machalilla. As in Costa Rica and Panamá, tracts of dead branching corals and dead and eroded massive colonies still evident on coral reefs in Colombia and Ecuador bear witness to the damaging effects of the 1982–1983 El Niño event.

The oceanic islands of the eastern Pacific, the Revillagigedo Islands, Clipperton Atoll, Cocos, Malpelo and the Galápagos Islands, also support abundant reef-building coral communities and coral reefs (Fig. 19.1). Clipperton Atoll, with dominantly massive poritid colonies covering about 370 ha, is probably the largest coral reef in the eastern Pacific. Even though Clipperton is a well-developed atoll with high coral cover, its reef-building coral fauna is notably depauperate, consisting of only seven species of scleractinian corals and one species of hydrocoral. Since the euphotic zone extends deeper at the oceanic island sites, the depth distribution of zooxanthellate corals also extends to 30–60 m or slightly deeper, in contrast to continental areas where corals are limited to 10–15 m (Fig. 19.5). Malpelo

Eastern Pacific Coral Reef Ecosystem

Fig. 19.4A,B. Uva Island patch reef during extreme, midday low water exposure (20 February 1992, courtesy C.M. Eakin). **A** Panoramic view of reef flat. **B** Close-up view of bleached *Pocillopora* colonies; the upper branches are pale, due to prolonged emersion and high temperatures during a series of extreme low tides

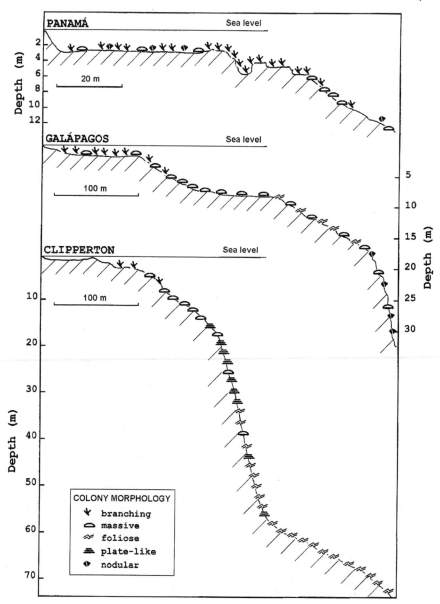

Fig. 19.5. Schematic near-shore profiles and coral zonation at three eastern Pacific sites: Secas Islands, Panamá (Glynn 1976); Bartolomé Island, Galápagos Islands (Glynn and Wellington 1983); NE sector, Clipperton Atoll (Glynn et al. 1996)

Island is a collection of small islands and rocks whose steep sides descend below the photic zone. While this site supports abundant coral populations, the absence of shallow shelves has prevented the accumulation of reef frameworks. Coral frameworks were also present at Cocos and the Galápagos Islands before the 1982–1983 ENSO, but most of these were severely damaged and subsequently have become greatly reduced in size, even disappearing in some instances, due to the grazing and boring activities of echinoids and other bioeroders.

19.4 Ecological Processes

Most studies on the ecology of interacting species and, where known, the influence of such processes on coral community structure (i.e., species composition, relative abundances, zonation and other spatial distributions) have been conducted on only one or a few coral reefs; therefore, the generality of these ecological processes on eastern Pacific reefs needs to be substantiated. The nutrient-rich eastern Pacific waters favor not only phytoplankton, which blocks downwelling light utilized by corals, but also internal bioeroders, and benthic filamentous and macroalgae, which compete directly with corals for space (Birkeland 1977). Under conditions of extremely high nutrient availability, e.g. during upwelling pulses, coral tissues are often invaded or overgrown by benthic algae, resulting in partial to total colony mortality. Dinoflagellate blooms have also been observed to cause coral mortality in Costa Rica and Panamá, presumably from large amounts of mucus produced by dinoflagellates, which adheres to and smothers the corals. Reef-associated mollusks, crustaceans and fishes also die in large numbers, perhaps from toxins and oxygen depletion.

In spite of generally high abundances of fish and sea urchin herbivores on many reefs, benthic algae often grow in profusion and compete with corals for space. This might be one of the reasons why pocilloporid reef frames are so common in the eastern Pacific region. Pocilloporid species are among the strongest coral competitors with relatively high rates of calcification. Colonies in close proximity typically grow vertically, forming rigid meter-sized blocks that are not easily invaded by other benthos. Diadematid sea urchins, notably *Diadema mexicanum* A. Agassiz, graze on algae and are often abundant on reefs, and a cidarid pencil sea urchin in the Galápagos Islands, *Eucidaris galapagensis* Döderlein, grazes on algae as well as live corals. Sea urchins can attain extraordinarily high abundances on coral reefs, with population densities commonly ranging between 50 and 100 m^{-2} in Panamá and the Galápagos Islands following the

1982–1983 ENSO disturbance. Other common herbivores are damselfish, which can effect coral zonation by protecting corals from grazers and corallivores. Damselfish typically cultivate turf algae, which they defend in patches roughly 25–50 cm in diameter. In the Pearl Islands, Panamá, *Stegastes acapulcoensis* (Fowler) establishes territories in shallow reef areas (1–3 m depth) where it finds shelter in stands of *Pocillopora* spp. (Wellington 1982). In such areas, *Pocillopora* survivorship is enhanced by the damselfish, which exclude potential grazers and corallivores. As the damselfish expand their territories they selectively bite and kill massive coral species, such as *Pavona gigantea*, thus promoting the survival and growth of *Pocillopora*. In deeper (4–8 m) reef areas of low topographic complexity, damselfish abundance declines and massive corals are favored. *Stegastes acapulcoensis*, therefore, plays a pivotal role in the maintenance of coral zonation, favoring the presence of branching species at shallow depths and massive species in deeper waters.

Numerous corallivores are found on eastern Pacific reefs. About 20 species of mollusks, crustaceans, a sea star, sea urchins and fishes are known to feed on live corals. *Jenneria pustulata* (Lightfoot), an ovulid gastropod, is the most important mollusk predator. It is commonly found grazing on the soft tissues of *Pocillopora* spp., leaving behind a bone-white, unblemished skeleton. Usually just a few individuals graze on coral colonies, resulting in only partial mortality. Occasionally large aggregations of *Jenneria*, up to 100 individuals, will attack and completely denude a large coral colony. Two species of hermit crabs, *Trizopagurus magnificus* (Bouvier) and *Aniculus elegans* (Stimpson), feed effectively on pocilloporid corals by scraping branches with their chelipeds, thereby removing both soft tissues and skeletal fragments. While *T. magnificus* can attain high mean population densities of 28 ind. m^{-2} on some reefs in Panamá, they have not been seen to cause the death of whole coral colonies. Earlier regarded as micropredators or parasites, xanthid crabs in the genus *Trapezia* (three species) and a shrimp in the genus *Alpheus* (one species) are now regarded as mutualistic symbionts of pocilloporid corals. These crustaceans are obligate symbionts, taking up shelter among the coral host's branches where they feed on mucus, entrapped detritus and plankton, and perhaps occasionally on coral tissues themselves. Recent studies have demonstrated that the movements of the crabs around the basal branches of corals have a beneficial effect in promoting increased circulation and a higher survivorship of polyps in potentially stagnant parts of the colony. Additionally, both crabs and shrimp symbionts vigorously defend their coral hosts from attack by the crown-of-thorns sea star *Acanthaster planci*. Stimson (1990) has shown that Hawaiian corals hosting crabs are stimulated by their guests to produce fat-bodies, lipid-filled globules that serve as a supplemental energy-rich food source.

Acanthaster planci has been observed preying on corals in several eastern Pacific mainland coral communities, ranging from the Gulf of California to Panamá, including most oceanic island localities. *Acanthaster* is potentially the most important corallivore in the eastern Pacific, due to its large size and high rate of consumption (about 0.5 m^2 of coral tissue per month). This corallivore is especially interesting ecologically because of its preferential feeding behavior; it tends to prey on massive, encrusting and nodular coral species, often avoiding branching corals such as *Pocillopora* spp. *Acanthaster* does, however, prey on small colonies of *Pocillopora* and their broken branches, which harbor few or no crustacean guards. The sea star generally avoids intact colonies or continuous stands of *Pocillopora*. The coral's nematocysts and crustacean guard defenses are usually sufficient to thwart *Acanthaster's* feeding. If a foraging sea star attempts to mount a *Pocillopora* colony, the crustacean guards will quickly snap and pinch (*Alpheus*) or pinch, jerk and clip tube feet and spines (*Trapezia*) in order to discourage the attack. Some examples of community-scale effects of *Acanthaster's* selective feeding behavior are: (1) the presence of smaller and younger colonies of preferred coral prey, (2) high relative abundances of non-preferred prey, and (3) the survival of preferred prey species surrounded by protective stands of pocilloporid corals. Unlike many Indo-Pacific reef areas that experience large outbreaks of *Acanthaster*, with tens of thousands of sea stars stripping reefs of most of their coral cover, eastern Pacific sea star populations have not been observed to exceed about 25–30 ind. ha^{-1}. The population sizes of adult sea stars seem to be limited by shrimp (*Hymenocera picta* Dana) and worm [*Pherecardia striata* (Kinberg)] predators. These invertebrates are abundant on Panamanian pocilloporid reefs, and numerous attacks on sea stars have been witnessed. In a typical attack scenario, a pair of shrimp will cut through the aboral dermis of *Acanthaster* and then begin to remove and eat such internal organs as the hepatic caeca and gonads. The shrimp ride along with the sea star, feeding on its soft parts for several days. Eventually, polychaete worm scavengers are attracted to the wounded sea star and will invade its internal body cavities through the incisions made by the shrimp. As many as 15 to 20, 5- to 10-cm-long worms will continue feeding on *Acanthaster's* soft parts until no internal organs remain and the sea star dies.

Other echinoderms feeding on live corals include two other sea stars and two species of sea urchins. The club-spined sea urchin, *Eucidaris*, is perhaps the most persistent corallivore, often observed grazing on pocilloporid corals in the Galápagos Islands. This sea urchin abrades both the soft tissues and skeletons of corals, and figures prominently in bioerosion. At least eight fish species, representative of five families, feed on live corals, with their feeding strategies ranging from removing mainly live tissue and causing little damage to the skeleton (Chaetodontidae, Pomacentridae) to abrading or breaking apart colonies in the feeding process (Balistidae, Scaridae, Tetraodontidae). Individual tetraodontids, *Arothron meleagris* (Bloch and Schneider), can typically ingest 10 g (dry weight) of pocilloporid branch tips daily. During the feeding forays the pufferfish also remove an approximately equal mass of branch tips, which are not ingested but fall to the bottom. The balistid, *Pseudobalistes naufragium* (Jordan and Starks), has the interesting habit of biting off 3- to 5-cm knobs from colonies of *Porites lobata*. This behavior exposes endolithic bivalves (*Lithophaga* spp.), which are then eaten by the balistid. If the discarded poritid fragments remain on a coarse sand bottom they often survive and continue to grow as disks or potato-shaped colonies, encircling the parent colony (Guzmán 1988). The often high level of disturbance of bottom sediments by fishes and invertebrates on rubble substrates also serves to promote the survival of coral fragments (coralliths), which frequently assume a spherical shape due to frequent colony rotation and nearly equal growth on all surfaces.

Clearly, large amounts of coral tissue and skeletal material are consumed and bioeroded on eastern Pacific reefs, in accordance with Highsmith's (1980) model relating elevated rates of bioerosion with increasing rates of productivity. Under usual conditions, coral reefs have continued to grow and accumulate calcium carbonate at relatively high rates. For example, before the 1982-1983 ENSO disturbance, reef-wide maximum calcification rates amounted to 10 kg m^{-2} yr^{-1} on a reef at Uva Island in the Gulf of Chiriquí, Panamá (Fig. 19.1, site7; Fig. 19.2D), and to 8-16 kg m^{-2} year^{-1} in the Galápagos Islands (Glynn et al. 1988). After the ENSO event, CaCO$_3$ loss due to bioerosion exceeded gains, resulting in net reef framework losses. In a model comparison of CaCO$_3$ budgets before and after the 1982-1983 ENSO, Eakin (1996) estimated that the 2.5-ha Uva Island reef was eroding at a rate of 4,800 kg year^{-1}. However, the rates of bioerosion were highly variable with some reef zones continuing to show positive net accretion. By ejecting sea urchins from their algal lawn territories, damselfish significantly retard bioerosion in shallow reef zones. Most of the erosion is due to sea urchins (*Diadema*), with boring sponges, bivalves and fishes also contributing significantly. On Galápagos reefs, damselfish also eject *Eucidaris*, but the high abundance of the sea urchins and low rates of coral recruitment are leading to a rapid disappearance of reef frameworks (Reaka-Kudla et al. 1996).

19.5 Nutrient Cycling, Carbon Production and Trophic Relationships

Perhaps the greatest gaps in our knowledge of eastern Pacific coral reef ecosystem function relate to nutrient cycling, community metabolism, organic production, and trophic relationships. These sorts of studies have not yet been undertaken, although data on inorganic carbon production, i.e. rates of calcium carbonate accumulation, have recently been forthcoming. Unlike the oft quoted simile of highly productive reef oases surrounded by desert-like oligotrophic environments, nearshore eastern Pacific reefs are generally exposed to waters of high nutrient and suspended sediment inputs. This holds true for reefs near drainage basins receiving high river discharge as well as for reefs in upwelling centers, such as the Gulf of Tehuantepec and the Panamá Bight. Whereas dissolved nitrate concentrations in waters surrounding oceanic reefs are typically $0.1–0.3\,\mu g$ atom l^{-1}, eastern Pacific reefs are commonly exposed to inorganic N levels an order of magnitude or higher during upwelling pulses. Since coral reefs are present under these conditions, it can be concluded that the requisite elements for reef growth are met.

A pivotal question begging attention is why vigorous algal growth in nutrient-rich environments does not exclude coral cover and prevent net reef accretion? Invertebrate and fish herbivore populations probably play an important role in preventing benthic algae from outcompeting corals. The high abundances of grazers on reefs in eastern Pacific upwelling areas may be sufficient to control algal growth and thus allow corals a competitive advantage, especially rapidly calcifying species such as *Pocillopora*. In a comparative study of the intensity of fish grazing on sponges in eastern Pacific and Caribbean nearshore areas, Birkeland (1987) demonstrated that grazing pressure was 25 times greater at Pacific than Caribbean sites. He attributed this interoceanic difference to high Pacific nutrient input that supports high phyto- and zooplankton production, which in turn favors large population densities of grazing and browsing fishes. Since some coral communities are occasionally overwhelmed by benthic algae, e.g. at Uva Island, Panamá (Glynn and Maté 1997) and Urvina Bay, Galápagos Islands (Macintyre et al. 1993), it is possible that such situations are simply temporary responses to nutrient pulses and not long-lived. A study designed to elucidate rates of algal production and consumption relative to coral recruitment and survival should shed light on the dynamics of reef growth in the eastern Pacific region.

Core-drilling and radiocarbon dating reveal that the best developed coral reefs in Panamá and Costa Rica range between 4 and 10 m thick and are

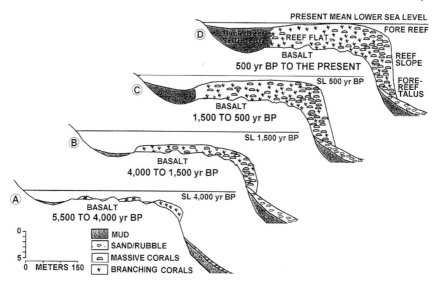

Fig. 19.6A–D. Proposed growth history of the Punta Islotes fringing reef, Golfito, Costa Rica. **A** Initiation of framework development on basalt foundation, 5,500–4,000 years B.P. **B,C** Establishment and growth of reef framework during sea level rise, 4,000–500 years B.P. **D** Vertical growth of reef flat to sea level and recent decline, 500 years B.P. to Present (*SL* sea level; modified after Cortés et al. 1994)

4,000 to 5,000 years of age (Glynn and Macintyre 1977; Macintyre et al. 1993; Cortés et al. 1994). One pocilloporid reef in the Gulf of Chiriquí, Panamá, has a maximum thickness of 13.4 m and began growing about 5,600 years ago. The mean accumulation rates of Gulf of Chiriquí reefs range from 1.6 to 7.5 m per 1,000 years. Coral reefs in the upwelling environment of the Pearl Islands, Gulf of Panamá, are thinner (2–6 m), more youthful (approx. 4,500 years), and have a lower mean accumulation rate of 1.3 m per 1,000 years. Maximum reef development in Costa Rica is exemplified by the Punta Islotes reef in Golfito, a mixed pocilloporid-poritid framework, which has attained 9 m in thickness and was established about 5,500 years ago (Figs. 19.3, 19.6). Pocilloporid corals formed the initial framework on a basalt substrate with poritid corals contributing significantly to vertical growth with rising sea level from about 1,500 years B.P. to the present. This reef, achieving accumulation rates of up to 8.3 m per 1,000 years, was one of the fastest growing reefs in the eastern Pacific region. Coral reefs present on offshore islands, e.g. at Caño and Cocos Islands (Costa Rica) and the Galápagos Islands (Ecuador), are thinner (2–3 m) and younger (500–2,000 years) than mainland reefs, perhaps as a result of ENSO disturbances, interrupted reef growth and subsequent severe bioerosion.

19.6 Functional Interfaces with Adjacent Biotopes

Since sea grass communities are non-existent in the eastern Pacific and coral reefs are usually distant from mangrove shores, there is likely little nutrient and organic matter exchange between reefs and benthic, plant-dominated ecosystems. Coral reefs and mangrove communities are often in close proximity or even juxtaposed in the Caribbean. Possibly the predominant coarse-grained calcareous sediments in the Caribbean, compared with mostly fine-grained siliciclastic sediments in the eastern Pacific, provide for more efficient circulation and oxygen exchange, thus allowing for the frequent commingling of mangrove and reef-associated taxa. Perhaps the most important inter-community interactions involve mangroves serving as: (1) sediment and nutrient sinks, improving the quality of water from terrestrial sources, (2) nursery habitats for invertebrates and fishes that migrate to reefs as they mature, and (3) coral reefs providing protective barriers, diminishing the impact of wave assault and substrate erosion (Ogden 1997). The extent of such co-dependencies in the eastern Pacific has not been investigated.

Planktonic productivity is often high in the eastern Pacific, and abundant suspension-feeding populations (e.g. oysters, endolithic sponges and bivalves, sedentary polychaetes, and barnacles) and fish planktivores associated with reefs attest to this rich food source.

Coral-dominated communities are usually surrounded by rubble/sand biotopes, and these commonly grade into sand/mud biotopes in deeper waters. Rhodolith communities are sometimes present adjacent to reefs, usually in deeper water (10–100 m) on sandy substrates with strong circulation. These communities consist dominantly of free-living calcareous red algae, but also support a high diversity of invertebrate and fish species. Several reef-associated fishes forage on animals living among and beneath the rhodoliths. Most of the associated corals are unattached and have colonized the rhodolith communities from nearby coral communities either as larvae or as partial or whole adult colonies transported down-slope by physical disturbances (e.g. storm-generated turbulence) or bioturbation (organism-induced movements).

19.7 Natural and Anthropogenic Impacts

For the past few decades, western American coral reefs have sustained significant damage from both anthropogenic and natural disturbances. The four categories of natural coral reef disturbances noted in UNEP/IUCN (1988) are coral bleaching, hurricane/storm damage, coral diseases and mass *Diadema* mortalities. In eastern Pacific coral reef ecosystems, the importance of these disturbances varies greatly.

Coral bleaching and mortality that accompanied the 1982-1983 El Niño-Southern Oscillation (ENSO) event were the most devastating, widespread and sustained thus far documented in the eastern Pacific region. Coral mortality resulting from this ENSO was geographically variable. Where SST anomalies reached 1-2 °C (Caño Island, Costa Rica, and Gulf of Chiriquí, Panamá), mean reef coral mortalities ranged between 50 and 75 %; at localities with 3 and 4 °C anomalies (Gulf of Panamá, Panamá, and the Galápagos Islands, Ecuador) mean coral mortality was 85 and 97 %, respectively (Glynn 1990). Galápagos coral communities and coral reefs were devastated, and low rates of coral recruitment and intense bioerosion in subsequent years have virtually eliminated coral reef biotopes from this region. Controlled microcosm tolerance experiments have substantiated the field results observed during the ENSO disturbance, demonstrating that prolonged high sea temperatures were the major causal factor (Glynn and D'Croz 1990). A comparison of coral mortality during three recent ENSO events also demonstrates a high degree of variability. In 1982-1983 the equatorial eastern Pacific (EEP), i.e., mainland Costa Rica, Panamá, Colombia and Ecuador, including offshore Cocos Island and the Galápagos Islands, was severely affected. However, only moderate coral bleaching and no detectable mortality were observed in parts (Galápagos Islands) of the EEP during the 1987 ENSO, but bleaching and mortality (about 10 %) occurred at the southern tip of Baja California, México, an area apparently not affected in 1982-1983 (Reyes Bonilla 1993; Podestá and Glynn 1997). While still under study, the 1997-1998 ENSO had stronger effects in México than in 1987, and caused coral mortality throughout the EEP, but apparently not so strong as in 1982-1983. Several secondary disturbances following the ENSO, e.g. strong upwelling pulses, phytoplankton blooms, extreme midday low water exposures, intense bioerosion and high predator concentrations, have continued to cause coral mortality and the erosion of coral reef frameworks.

Disturbances by storm damage are infrequent (hurricanes normally form off the Pacific coast and then move toward the west), and coral diseases and mass sea urchin mortalities are unknown. Some disturbances not included in the UNEP/IUCN listing can have severe local effects. On a decadal time scale, tectonic events and volcanism affect only small areas. One such event in 1954, resulting from a volcanic eruption and the uplift of a shallow shelf, exposed several km^2 of coral communities in the Galápagos Islands (Colgan 1990). While several species of corallivores occur in the eastern Pacific (Glynn 1982), their effects are usually low level and localized. Preferential feeding by the sea star *Acanthaster planci* can influence the species composition of coral communities, but no outbreaks resulting in mass coral mortality are known.

Unfortunately, as awareness of the occurrence of coral reefs in the eastern Pacific increases they become more exposed to various hazards from humankind (Glynn 1997). In a recent analysis, estimated anthropogenic threats to coral reefs were considered to be high off the Pacific coasts of Costa Rica, Panamá and Colombia based on: (1) coastal development, (2) overexploitation and destructive fishing practices, (3) inland pollution and erosion, and (4) marine pollution (Bryant et al. 1998). High levels of siltation caused by accelerated coastal erosion have degraded coral reefs in Costa Rica, Colombia and Ecuador. Although all other coral reef areas in the eastern Pacific were classified as low risk, it is highly likely that significant damage is also occurring at many other sites, but is simply not reported. This will become evident as studies expand in the region. For example, construction projects related to harbor and hotel development are believed to have killed two coral reefs at Huatulco, México. In addition, the harvesting of corals for the curio trade continues to be a serious problem (Guzmán and Cortés 1993). This activity has virtually eliminated pocilloporid corals from Acapulco (México), Bahía Culebra (Costa Rica), Taboga Island (Panamá), and parts of the coast of Ecuador. At Isla Pelado, Ayangue, and Salango Island, Ecuador, prodigious numbers of corals have been removed in recent years. According to a local fisherman, truckloads of live corals have been removed from Salango Island, which falls within the jurisdiction of the Parque Nacional de Machalilla. Over-fishing may well be responsible for some ecological imbalances on coral reefs that could prolong recovery from other disturbances, but such effects have not been studied in the eastern Pacific. For example, a marked reduction in (1) fish herbivores could favor benthic algal growth over coral recruitment, and (2) fish invertivores could favor increased abundance of bioeroding sea urchins.

19.8 Management Needs

At least 22 existing and 6 proposed marine protected areas (MPAs) harboring coral communities and/or coral reefs are presently recognized in the tropical eastern Pacific (Table 19.1). This is a major increase from 7 years ago when only seven existing and one proposed MPA were listed regionally (UNEP/IUCN 1988). Most existing MPAs are in México, Costa Rica, Panamá, Colombia and Ecuador. Important coral reef areas are located

Table 19.1. Existing (E) and proposed (P) marine protected areas with coral communities and/or coral reefs in the tropical eastern Pacific region

Country	Marine-protected area[a]	Coral type[b]	Anthropogenic threats[c]
México[d]	Gulf of California Islands, Biosphere Reserve (E)	Coral communities	Harvesting of corals; anchor damage; diving
	Cabo Pulmo, Biosphere Reserve (E)	Not true structural reef	Harvesting of black corals, mollusks and fishes; anchor damage; diving
	Los Cabos region (23°N), Parque Marina (E)	Coral communities	Diving; sewage pollution
	Los Arcos (20°N), Jalisco, Parque Marina (E)	Coral communities	Harvesting of corals; diving
	Revillagigedo Archipelago, Biosphere Reserve (E)	Coral reefs, coral communities	Siltation caused by feral sheep
	Bahías de Huatulco, Oaxaca, Parque Nacional Marítima Terrestre (E)[e]	Coral reefs, coral communities	Anchor damage; diving
	Colola-Maruata, Michoacán (P)	Sparse coral communities on rocky coast	Overexploitation of marine invertebrates; anchor damage
	Morro Ayutla-Chacagua, Oaxaca (P)	Sparse coral communities on rocky coast	Incidental damage by fishers
Guatemala	None	Minimal coral presence, unsuitable habitat	Unknown
El Salvador	Los Cóbanos (P)	Structural (?) coral reefs, coral communities	Unknown
Honduras	None	Minimal coral presence, unsuitable habitat	Unknown
Nicaragua	None	Unknown	Unknown
Costa Rica	Area de Conservación Guanacaste (E)	Coral reefs, coral communities	Overfishing
	Parque Nacional Manuel Antonio (E)	Coral communities	Pollution; overfishing
	Parque Nacional Isla del Caño (E)	Coral communities, coral reefs	None
	Parque Nacional Corcovado (E)	Coral communities, coral reefs	None
	Isla del Coco, World Heritage Biosphere, UNESCO (E)	Bioeroded and recovering coral reefs, coral communities	Feral pigs causing erosion

Table 19.1 (*continued*)

Country	Marine-protected area[a]	Coral type[b]	Anthropogenic threats[c]
Costa Rica	Reserva Absoluta Cabo Blanco (E)	Coral communities	Overfishing; sedimentation
	Parque Marino Ballena (E)	Coral communities	Overfishing (?); sedimentation
Panamá	Parque Nacional de Coiba (E)	Coral reefs (largest known in E Pacific)	Boat groundings; anchor damage; refuse disposal; shellfish exploitation
	Refugio de Vida Silvestre, Isla Iguana (E)	Coral reef	Unknown
	Islas de las Perlas (P)	Coral reefs, coral communities	Destructive shellfish exploitation (dynamiting, dredging)
	Islas Pacheca and Pachequilla (P)	Coral communities	Unknown
	Bahía de los Muertos, Gulf of Chiriquí (P)	Coral reefs	Unknown
	Refugio de Vida Silvestre de Isla Caña (E)	Unknown	Unknown
	Refugio de Vida Silvestre Isla Taboga (E)	Coral communities	Anchor damage; refuse disposal
Colombia	Parque Nacional Natural Utría (E)	Coral reefs	Sedimentation; anchor damage; boat groundings
	Parque Nacional Isla Gorgona (E)	Coral reefs	Anchor damage; oil pollution; fishing activities
	Isla del Malpelo, Santuario de Flora y Fauna (E)	Coral communities	Fishing activities
Ecuador	Parque Nacional de Machalilla (E)	Southern-most coral reefs in E Pacific	Unknown
	Reserva de Recursos Marinos Galápagos (E)	Bioeroded coral reefs, coral communities	Anchor damage; fishing activities; harvesting of corals

[a] Largely from Kelleher et al. (1995) with supplemental information
[b] From a variety of sources and personal observations
[c] From Kelleher et al. (1995), other sources and personal observations
[d] Reyes Bonilla (2000)
[e] Decreed in January 1999

within some of the MPAs, but also numerous coral reefs in good-to-pristine condition are wholly without protective sanction. In México, for example, the large reef system in the Marías Islands (Fig. 19.1, site 2) is not presently protected by any governmental agency although a prison in the Marías Islands has restricted public access to the area. The Mexican federal government has just decreed Bahías de Huatulco as a national park (Parque Nacional Marítimo Terrestre) which includes an 11,000-ha coastal zone and the coral reefs therein contained. There is a good chance that this coral reef tract will now receive adequate protection under government control and the vigilance of local enterprises cognizant of the value of this resource for ecotourism. The Galápagos marine reserve, which includes a 40-nautical-mile (74 km) perimeter around the archipelago, is the largest protected area in the eastern Pacific. Even though Galápagos coral reefs and coral communities have been severely degraded by recent ENSO events, the cooperative management and research efforts of the Galápagos National Park Service and the Charles Darwin Research Station should help minimize anthropogenic disturbances and thus allow natural recovery processes to unfold.

Notwithstanding the recent increase in MPAs, no consistent or sustained management practices are in effect specifically protecting coral reefs. Unfortunately, virtually all of the protected coral areas in the eastern Pacific are "paper parks", established with good intentions but with little support or chance for success. Persistent problems are the lack of enforcement capability and a generally poor public appreciation for the value of coral reefs. To a large extent, the exploitation of reefs is a 'free-for-all', following no system of tenure or rules of resource utilization. Even though coral reefs are among the most productive of all marine ecosystems, they typically exhibit a low yield. Export fisheries simply cannot be sustained by coral reefs (Birkeland 1997). Perhaps the greatest hope for the sustainable exploitation of coral reefs lies with local communities, since people living nearby benefit most, offering a strong incentive for their protection and management (White et al. 1994).

A potentially disastrous enterprise that is resurrected every few years is the plan to construct a sea-level canal across Central America. In light of the numerous known perturbations caused by introduced species, including the extirpation of native species, there would exist a real threat to coral reefs and other marine ecosystems with an unchecked mixing of Pacific and Atlantic waters. If such a project is initiated, every effort must be made to avert the interchange of Pacific and Atlantic biotas. Some workers have suggested that the 1983 mass mortality of *Diadema* in the Caribbean was caused by an exotic pathogen transported across the Panama Canal in a ship's balast water. While it is nearly impossible to predict the impacts of invading exotic species, a few foreseeable threats underline the gravity of this problem: (1) *Acanthaster planci* could cause massive coral mortality throughout the Caribbean region; (2) coral diseases, unknown in the eastern Pacific, could spread quickly among Pacific corals; (3) the endemism of Caribbean/Pacific corals, and numerous other taxa, could become compromised.

Although eastern Pacific coral reefs are probably among the most depauperate known, and are small and scattered in distribution, they nonetheless represent unique ecosystems that have innumerable benefits to many coastal communities of tropical western America now and in the future. The sustained use of this valuable resource depends upon a program of protection and appropriate management practices that requires urgent implementation and coordination at both national and local levels.

Acknowledgements. Thanks are due S.D. Cairns for information on regional differences in coral species diversity, and J. Cortés, H. Guzmán, G.E. Leyte Morales, J. Maté, H. Reyes Bonilla, and B. Vargas Angel for information on marine protected areas. R. Araujo and R. Carter kindly assisted in the preparation of illustrations. A.C. Baker, R.N. Ginsburg, I.G. Macintyre, A.M. Szmant, and G.M. Wellington are thanked for their help in sharpening the focus of this essay. My knowledge of eastern Pacific coral reef ecology has been gained largely through research support from the Smithsonian Institution, the National Science Foundation (USA), and the National Geographic Society.

References

Birkeland C (1977) The importance of rate of biomass accumulation in early successional stages of benthic communities in the survival of coral recruits. Proc 3rd Int Coral Reef Symp 1:15–21

Birkeland C (1987) Nutrient availability as a major determinant of differences among coastal hard-substratum communities in different regions of the tropics. In: Birkeland C (ed) Comparison between Atlantic and Pacific tropical marine coastal ecosystems: community structure, ecological processes, and productivity. UNESCO reports in marine science, vol 46. UNESCO, Paris, pp 45–97

Birkeland C (1997) Implications for resource management. In: Birkeland C (ed) Life and death of coral reefs. Chapman and Hall, New York, pp 411–435

Bryant D, Burke L, McManus J, Spalding M (1998) Reefs at risk, a map-based indicator of threats to the world's coral reefs. World Resources Institute, Washington, DC, p 56

Carriquiry JD, Reyes Bonilla H (1997) Community structure and geographic distribution of the coral reefs of Nayarit, Mexican Pacific. Cienc Mar 23:227–248

Coates AG, Obando JA (1996) The geologic evolution of the Central American Isthmus. In: Jackson JBC, Budd AF, Coates AG (eds) Evolution and environment in tropical America. University of Chicago Press, Chicago, pp 21–56

Colgan MW (1990) El Niño and the history of eastern Pacific reef building. In: Glynn PW (ed) Global ecological consequences of the 1982–83 El Niño-Southern Oscillation. Elsevier Oceanogr Ser 52. Elsevier, Amsterdam, pp 183–232

Cortés J (1997) Biology and geology of eastern Pacific coral reefs. Proc 8th Int Coral Reef Symp 1:57–64

Cortés J, Macintyre IG, Glynn PW (1994) Holocene growth history of an eastern Pacific fringing reef, Punta Islotes, Costa Rica. Coral Reefs 13:65–73

Dana T (1975) Development of contemporary eastern Pacific coral reefs. Mar Biol 33:355–374

Eakin CM (1996) Where have all the carbonates gone? A model comparison of calcium carbonate budgets before and after the 1982–1983 El Niño at Uva Island in the eastern Pacific. Coral Reefs 15:109–119

Emerson WK, Chaney HW (1995) A zoogeographic review of the Cypraeidae (Mollusca: Gastropoda) occurring in the eastern Pacific Ocean. Veliger 38:8–21

Glynn PW (1974) The impact of *Acanthaster* on corals and coral reefs in the eastern Pacific. Environ Conserv 1:295–304

Glynn PW (1976) Some physical and biological determinants of coral community structure in the eastern Pacific. Ecol Monogr 40:431–456

Glynn PW (1982) Coral communities and their modifications relative to past and prospective Central American seaways. Adv Mar Biol 19:91–132

Glynn PW (1990) Coral mortality and disturbances to coral reefs in the tropical eastern Pacific. In: Glynn PW (ed) Global ecological consequences of the 1982–83 El Niño-Southern Oscillation. Elsevier, Amsterdam, pp 55–126

Glynn PW (1997) Assessment of the present health of coral reefs in the eastern Pacific. In: Grigg RW, Birkeland C (eds) Status of coral reefs in the Pacific. Sea Grant College Program, University of Hawaii, Honolulu, pp 33–40

Glynn PW, D'Croz L (1990) Experimental evidence for high temperature stress as the cause of El Niño-coincident coral mortality. Coral Reefs 8:181–191

Glynn PW, Macintyre IG (1977) Growth rate and age of coral reefs on the Pacific coast of Panamá. Proc 3rd Int Coral Reef Symp, Miami 2:251–259

Glynn PW, Maté JL (1997) Field guide to the Pacific coral reefs of Panamá. Proc 8th Int Coral Reef Symp 1:145–166

Glynn PW, Wellington GM (1983) Corals and coral reefs of the Galápagos Islands. University of California Press, Berkeley

Glynn PW, Druffel EM, Dunbar RB (1983) A dead Central American coral reef tract: possible link with the Little Ice Age. J Mar Res 41:605–637

Glynn PW, Cortés J, Guzmán HM, Richmond RH (1988) El Niño (1982–83) associated coral mortality and relationship to sea surface temperature deviations in the tropical eastern Pacific. Proc 6th Int Coral Reef Symp 3:237–243

Glynn PW, Veron JEN, Wellington GM (1996) Clipperton Atoll (eastern Pacific): oceanography, geomorphology, reef-building coral ecology and biogeography. Coral Reefs 15:71–99

Grigg RW, Hey R (1992) Paleoceanography of the tropical eastern Pacific Ocean. Science 255:172–178

Guzmán HM (1988) Distribución y abundancia de organismos coralívoros en los arrecifes coralinos de la Isla del Caño, Costa Rica. Rev Biol Trop 36(2A):191–207

Guzmán HM, Cortés J (1989) Coral reef community structure at Caño Island, Pacific Costa Rica. PSZNI Mar Ecol 10:23–41

Guzmán HM, Cortés J (1993) Arrecifes coralinos del Pacífico oriental tropical: revisión y perspectivas. Rev Biol Trop 41:535–557

Highsmith RC (1980) Geographic patterns of coral bioerosion: a productivity hypothesis. J Exp Mar Biol Ecol 46:177–196

Jokiel PL (1990) Long-distance dispersal by rafting: reemergence of an old hypothesis. Endeavour New Ser 14(2):66–73

Kelleher G, Bleakley C, Wells S (1995) A global representative system of marine protected areas, vol IV. South Pacific, northeast Pacific, northwest Pacific, southeast Pacific and Australia/New Zealand. The International Bank for Reconstruction and Development/The World Bank, Washington, DC

Lessios HA, Kessing BD, Robertson DR (1998) Massive gene flow across the world's most potent marine biogeographic barrier. Proc R Soc Lond B 265:583-588

Macintyre IG, Glynn PW, Cortés J (1993) Holocene reef history in the eastern Pacific: mainland Costa Rica, Caño Island, Cocos Island, and Galápagos Islands. Proc 7th Int Coral Reef Symp 2:1174-1184

Ogden JC (1997) Ecosystem interactions in the tropical coastal seascape. In: Birkeland C (ed) Life and death of coral reefs. Chapman and Hall, New York, pp 288-297

Pak H, Zaneveld JRV (1973) The Cromwell Current on the east side of the Galápagos Islands. J Geophys Res 78(33):7845-7859

Podestá GP, Glynn PW (1997) Sea surface temperature variability in Panamá and Galápagos: extreme temperatures causing coral bleaching. J Geophys Res 102(C7): 15749-15759

Reaka-Kudla ML, Feingold JS, Glynn PW (1996) Experimental studies of rapid bioerosion of coral reefs in the Galápagos Islands. Coral Reefs 15:101-107

Reyes Bonilla H (1993) Biogeografía y ecología de los corales hermatípicos (Anthozoa: Scleractinia) del Pacífico de México. In: Salazar Vallejo SI, González NE (eds) Biodiversidad marina y costera de México. Com Nal Biodiversidad y CIQRO, México, pp 207-222

Reyes Bonilla H (2000) Coral reefs of the Pacific coast of México. In: Cortés J (ed) Coral reefs of Latin America. Elsevier, Amsterdam (in press)

Scheltema RS (1988) Initial evidence for the transport of teleplanic larvae of benthic invertebrates across the east Pacific barrier. Biol Bull 174:145-152

Stimson J (1990) Stimulation of fat body production in the coral *Pocillopora* by mutualistic crabs of the genus *Trapezia*. Mar Biol 106:211-218

UNEP/IUCN (1988) Coral reefs of the world, vol 1: Atlantic and eastern Pacific. UNEP regional seas directories and bibliographies. IUCN, Gland, Switzerland and Cambridge, UK/UNEP, Nairobi, Kenya

Vargas Angel B (1996) Distribution and community structure of the Utría reef corals, Colombian Pacific. Rev Biol Trop 44(2):627-635

von Prahl H, Guhl F, Grögl M (1979) Gorgona. Futura Grupo Editorial, Bogotá, Colombia

Wellington GM (1982) Depth zonation of corals in the Gulf of Panamá: control and facilitation by resident reef fishes. Ecol Monogr 52:223-241

White AT, Hale LZ, Renard Y, Cortesi L (eds) (1994) Collaborative and community-based management of coral reefs: lessons from experience. Kumarian Press, West Hartford

20 The Tropical Pacific Coast of Mexico

F.J. Flores-Verdugo, G. de la Lanza-Espino,
F. Contreras Espinosa, and C.M. Agraz-Hernández

20.1 Introduction

The tropical Pacific coast of Mexico (Fig. 20.1) between the Piaxtla sandbars and the Guatemalan border extends for 2,530 km along the shores of southern Sinaloa, Nayarit, Jalisco, Colima, Michoacan, Guerrero, Oaxaca, and Chiapas State (Contreras-Espinosa 1993). The region is tectonically influenced by the Cocos Plate and classified as a collision coast. The dominant geological features are quaternary alluvial plains, metamorphic rocks as well as cretaceous igneous and calcareous sedimentary rocks. The continental platform is narrow (12–25 km) but widens to 50–75 km along the coast of Sinaloa, Oaxaca, and Chiapas. Much of the coast is composed of continuous sand beaches interrupted only by rivers and lagoon inlets. Between Jalisco and Oaxaca the coast is predominantly rocky with narrow beaches and small lagoons surrounded by mangroves. With the exception of the Santiago and Balsas rivers, most of the approximately 70 rivers have small drainage basins and seasonally intermittent flow (Arriaga-Cabrera et al. 1998b). Approximately 32 barred inner shelf coastal lagoons (2,545 km^2) as well as mangroves and freshwater wetlands comprise important ecosystems.

20.2 Climate

The climate ranges from semi-arid to humid (Chiapas), though most of the region is dry sub-humid (800–1,200 mm year^{-1}; Garcia 1973). Winds from the north-west (<6 m s^{-1}) prevail during the winter (November–April) along the Sinaloa and Nayarit coast while S-SW winds, which are related to tropical depressions and hurricanes, dominate as far south as Oaxaca. Strong northerly winds ("northers"), which originate in the Gulf of Mexi-

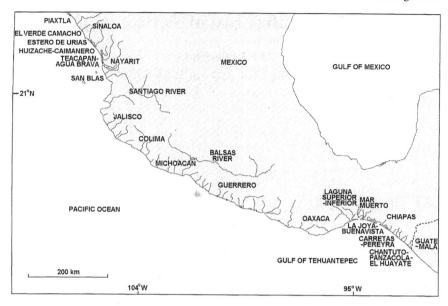

Fig. 20.1. The tropical Pacific coast of Mexico with coastal states and major coastal ecosystems

co, prevail in the Gulf of Tehuantepec (Lankford 1977). The mean annual temperature is >26 °C and heavy rains occur mainly in the summer (July–September) with an annual mean of 800 mm. Precipitation differences between years are largely a result of frequency and intensity of hurricanes and tropical storms.

20.3 Coastal Ecosystems and Biota

20.3.1 Lagoons

The coastal lagoons of the semi-arid Sinaloa coast are shallow (El Verde Camacho mean depth 1.00 m; Huizache-Caimanero mean depth 0.65 m). During the dry season, when river flow is absent, the El Verde Camacho lagoon (0.5 km^2) inlet is closed and estuarine conditions (salinity 4–35 psu) are presumably maintained by groundwater infiltration. During the dry season mangrove (*Laguncularia racemosa*) litter tends to accumulate in the lagoon and is only exported to the ocean when the inlet opens with river flow in summer (Flores-Verdugo et al. 1987). In contrast, more than

70% of the shallower Huizache-Caimanero lagoon (17.5 km^2) dries up during the dry season. The exposed bottom substrate cracks, increasing nutrient release during flooding in summer (Arenas and de la Lanza 1981). High concentrations of nutrients favor elevated biomass (>1 kg dry wt. m^{-2}) of the sea grass *Ruppia maritima* (de la Lanza and García-Calderon 1991). The Estero de Urías lagoon permanently exchanges water with the ocean. However, the lagoon receives urban wastewater, cooling water from a thermoelectric plant, effluents from the fishing industry, and elevated nutrient discharge from shrimp ponds (Paéz-Osuna et al. 1997a), leading to eutrophication which is often associated with excessive growth of the alga *Ulva lactuca*.

Coastal lagoon systems of the humid Chiapas coast are characterized by high water temperatures (29–35.5 °C) and salinity varies between 0–34.5 psu (Chantuto-Panzacola), 0–22.7 psu (Carretas-Pereyra, 3.7 km^2), and 12–39 psu (La Joya-Buenavista, 62 km^2) during the rainy (September) and dry season (April), respectively. Reduced water exchange with the ocean (La Joya-Buenavista) and elevated phosphorus input from rivers (Carretas-Pereyra and Chantuto-Panzacola) tend to cause high total phosphorus concentrations (up to 42 µg l^{-1}) compared to those of nitrite/nitrate (0.5–2.9 µg l^{-1}), thus favoring elevated chlorophyll-*a* levels (55–80 mg m^{-3}). High ammonia concentrations (Carretas-Pereyra, 34 µg l^{-1}) are likely to be due to large quantities of mangrove detritus (Flores-Verdugo et al. 1992; Contreras-Espinosa 1993). Hypersaline (salinity up to 55 psu) conditions are common in the Superior, Inferior, and Mar Muerto lagoons (about 1,400 km^2) in Oaxaca State.

Coastal lagoons assume an important function in the life cycles of marine fishes, and commercial species, such as mullet (*Mugil*), snapper (*Lutjanus*), and snook (*Centropomus*), are abundant in both lagoons and adjacent marine waters. In general, the abundance and distribution of the ichthyofauna is controlled by seasonal changes in precipitation (Yañez-Arancibia 1978).

20.3.2 Mangroves and freshwater wetlands

Although mangrove distribution is discontinuous, extensive forests are common in southern Sinaloa and Nayarit (Teacapán-Agua Brava and San Blas) and Chiapas (Chantuto-Panzacola-El Huayate). The structure of mangrove forests increases with higher rainfall and freshwater runoff towards the south (Flores-Verdugo et al. 1992). Mangrove forests are dominated by red (*Rhizophora mangle*), white (*Laguncularia racemosa*), black (*Avicennia germinans*), and button mangrove (*Conocarpus erectus*). *Rhizo-*

phora harrisonni has only been reported for the Carretas-Pereyra lagoon (Rico-Gray 1981). Common algal associates are five species of the red alga *Bostrychia* as well as *Chaetomorpha antennina*. Mangroves occasionally also occur mixed with forested freshwater wetland composed of *Pachira acuatica* ("zapotón").

The Teacapán – Agua Brava – Marismas Nacionales mangrove complex represents the largest mangrove area (1,500 km^2) at the Pacific coast of Mexico and Central America. While the black mangrove (*A. germinans*) dominates in Teacapán, the white mangrove (*L. racemosa*) prevails in Agua Brava. Mangrove density can be as high as 3,203 trees ha^{-1} and basal areas range from 14 to 29.6 m^2 ha^{-1}. High concentrations of humic substances (>150 mg l^{-1}), which add a reddish color to the water, are derived from mangroves detritus during the wet season (Flores-Verdugo et al. 1992).

Like San Blas (100 km^2), Chantuto-Panzacola-El Huayate (350 km^2) is a complex system of mangroves and freshwater wetlands with high faunal richness. Tall mangrove trees (up to 30 m), high stem density (1,722 ha^{-1}) and basal area (41.7 m^2 ha^{-1}) warrant elevated monthly litter production (362–509 g dry wt. m^{-2}; Flores-Verdugo et al. 1992; Contreras-Espinosa 1993). Macrophytes, such as *Nymphae blanda*, *Cabomba* sp., *Pistia stratiotes*, *Salvinia* sp., *Azolla* sp., *Eichornia crassipes*, and *Neptunia* sp. are common and cover extensive areas during the rainy season. The high productivity sustains important fisheries resources. Two species of river crocodiles (*Crocrodylus acutus*) and *Caiman crocrodylus*, the latter endemic to Chiapas, occur in the mangroves and endangered species, like the ocelot *Felis pardalis*, the margay *Felis wiedii*, and the jaguar *Felis onca* have been observed in several mangrove forests (Flores-Verdugo et al. 1992).

Large concentrations of aquatic bird species (165 in Huizache-Caimanero; Contreras-Espinosa 1993), like the great blue heron (*Ardea herodias*), the wood stork (*Mycteria americana*), the roseate spoonbill (*Ajaia ajaja*), the egret (*Egretta* spp.), the brown pelican (*Pelecanus occidentalis*), the white pelican (*Pelecanus erythrorhyncus*), and the osprey (*Pandion haliaetus*), are common components in mangroves and freshwater wetlands of Sinaloa, Nayarit, and Chiapas (Olmstead 1993). About 80% of Pacific migratory shore bird populations concentrate in the San Blas mangroves and freshwater wetlands during the winter.

20.4 Human Impacts and Management Needs

20.4.1 Fisheries

The region is characterized by diverse (more than 28) indigenous cultures, each exploiting natural resources in a sustainable way. Numerous shell middens (*Tivela* and *Crassostrea*) in Chantuto-Panzacola and Teacapán-Marismas Nacionales date back to 1,500 B.P. and pigment extraction from the purple sea snail (*Purpura panza*) since prehispanic times suggests important fishing activities by indigenous populations prior to the discovery of the Americas (Flores-Verdugo et al. 1997).

In rocky coastal areas lobsters (*Panulirus inflatus* and *Panulirus gracilis*) and octopus are important fisheries resources for fishermen and fishing intensity appears to be related to the local tourist industry (Briones-Fourzán 1995). At sandy beaches and coastal lagoons, catches of the shrimps *Penaeus vannamei* and *P. californiensis* in Sinaloa/Nayarit and *P. vanamei* and *P. occidentalis* in Chiapas may represent up to 90 % of total fisheries. Shrimp fishing (*Litopenaeus* and *Farfantepenaeus*) in the Gulf of Tehuantepec is largely restricted to upwelling areas, which are affected by strong winds and El Niño events.

Beaches are important nesting grounds for olive Ridley (*Lepidochelis olivacea*), black (*Chelonia agassizi*), hawksbill (*Eretmochelis imbricata*), and leatherback (*Dermochelis coriacea*) turtles (Marquéz-Millan 2000). After catches of marine turtles increased from 2 tons in 1963 to about 14.5 tons in 1968, fishing was finally banned in 1972/1973. However, poaching of eggs and adult turtles continued and on some beaches turtle nests decreased from 200,000 before 1960 to less than 2,000 in 1996, though olive Ridley turtle populations appear to be making a recovery.

20.4.2 Environmental Problems

Most coastal lagoon and mangrove ecosystems of the tropical Pacific coast of Mexico have suffered the impact of increased sedimentation (i.e., 1 cm year^{-1} in the Huizache-Caimanero lagoon; de la Lanza and García-Calderon 1991) as a result of agriculture and deforestation activities in the drainage basin of rivers. Furthermore, urban, industrial, and agricultural freshwater demands have reduced river flow, and groundwater extraction (up to four wells per km^2 in the coastal plain of Teacapán) has lowered the water table in wetlands. Reduced river flow has led to seasonal closure of inlets (i.e., El Verde Camacho lagoon), impeding migration of aquatic organisms, many of which are important fisheries resources (Flores-Verdugo et al. 1987). The extraction of freshwater has diminished to occurrence of brackish conditions in lagoons and increased the interstitial salinity in mangrove soils. As a consequence, highly productive ecosystems have been transformed into little productive hypersaline systems with salt pans (i.e., Huizache-Caimanero lagoon). Additionally, mangroves are sensitive to changes in flooding periods. The opening of an artificial channel in the Teacapán-Agua Brava complex altered the system from predominantly freshwater-brackish to marine, causing the death of 24% (280 km^2) of the white and black mangrove forests (Flores-Verdugo et al. 1997).

Pollution is caused by organochlorine pesticides which are added to lagoon sediments (i.e., Carretas-Pereyra 120 ng g^{-1}) from surrounding agricultural lands (Rueda et al. 1997). Also, gold, silver, iron, and asbestos mining as well as the introduction of polycyclic aromatic hydrocarbons from oil refineries add to the degradation of productive coastal systems (Botello et al. 1998). Along the entire coast valuable mangrove wetlands are being transformed into marinas and hotels and shrimp aquaculture is expanding, particularly in Sinaloa and Nayarit (94 km^2 in 1993 to about 200 km^2 in 1997; Páez-Osuna et al. 1997b). The discharge of shrimp pond effluents is likely to impact coastal waters (i.e., Estero de Urías) by adding elevated concentrations of nitrogen and phosphorus (Paez-Osuna et al. 1997a).

The detrimental impact of overfishing, wetland conversion, environmental degradation, disrespect for indigenous cultures, increasing development of tourism, etc. (Arriaga-Cabrera et al. 1998a) calls for an ecological management program which warrants sustainable use of natural resources and nature conservation (Bojorquez 1993). However, lack of information, strong economic pressure, and insufficient reinforcement pose serious limitations for sound management despite the establishment of a Ministry for the Environment, the requirement of environmental impact assessments, and the creation of several biological reserves.

References

Arenas V, de la Lanza G (1981) The effect of dried and cracked sediments on the availability of phosphorous in coastal lagoons. Estuaries 4(3):206–212

Arriaga-Cabrera L, Vazquez-Dominguez E, González-Cano J, Hernández S, Jimenez-Rosenberg R, Muñoz-Lopez E, Aguilar-Sierra V (1998a) Regiones prioritarias marinas de Mexico. Comision Nacional para el Conocimiento y Uso de la Biodiversidad (CONABIO/USAID/FMCN/WWF), Mexico, DF

Arriaga-Cabrera L, Vazquez-Dominguez E, González-Cano J, Hernández S, Jimenez-Rosenberg R, Muñoz-Lopez E, Aguilar-Sierra V (1998b) Regiones hidrologicas prioritarias: fichas técnicas y mapa (1:4,000,000). Comision Nacional para el Conocimiento y Uso de la Biodiversidad (CONABIO/USAID/FMCN/WWF), Mexico, DF

Bojorquez LA (1993) Programa de Ordenamiento Ecologico para el Desarrollo Acuícola de la Region costera de Sinaloa y Nayarit. Proyectos de Ordenamiento Ecologico de Regiones Geográficas con Actividades Productivas Prioritarias. Instituto Nacional de Ecología de la Secretaría de Desarrollo Social (Mexico) and Environmental and Regional Development Department of the Organization for American States (United Nations)

Botello AV, Villanueva SF, Diaz GG, Escobar-Briones E (1998) Polycyclic aromatic hydrocarbons in sediments from Salina Cruz, Oaxaca, México. Mar Pollut Bull 36(7): 554–558

Briones-Fourzán P (1995) Biología y pesca de las langostas de México. In: González-Farías F, de la Rosa-Vélez T (eds) Temas de oceanografía biológica en México. Universidad Autónoma de Baja California, Ensenada, México, pp 207–236

Contreras-Espinosa F (1993) Ecosistemas Costeros Mexicanos. CONABIO & Universidad Autónoma Metropolitana, Iztapalapa Mexico DF

De la Lanza G, García-Calderón JL (1991) Sistema Lagunar Huizacha y Caimanero, Sin: Un estudio Socio-ambiental, Pesquero y Acuícola. Hidrobiológica 1:1–35

Flores-Verdugo FJ, Day J Jr, Briseño-Dueñas R (1987) Structure, litterfall, decomposition and detritus dynamics of mangroves in a Mexican coastal lagoon with an ephemeral inlet. Mar Ecol Prog Ser 35:83–90

Flores-Verdugo FJ, González-Farías F, Zamorano DS, Ramirez-García P (1992) Mangrove ecosystems of the Pacific coast of Mexico: distribution, structure, litterfall and detritus dynamics. In: Seeliger U (ed) Coastal plant communities of Latin America. Academic Press, San Diego, pp 269–288

Flores-Verdugo FJ, Gonzáles-Farías F, Blanco-Correa M, Nuñez-Pastén A (1997) The Teacapan-Agua Brava-Marismas Nacionales mangrove ecosystem on the Pacific coast of Mexico. In: Kjerfve B, Drude de Lacerda L, Diop EHS (eds) Mangrove ecosystem studies in Latin America and Africa. UNESCO, ISME and US Forest Service, Paris, pp 35–46

Garcia E (1973) Modificaciones al Sistema de Clasificación Climatico de Koppen (Adaptación a condiciones de la Republica Mexicana). Instituto de Geografía, UNAM, Mexico

Lankford RR (1977) Coastal lagoons of Mexico. Their origin and classification. In: Wiley M (ed) Estuarine processes, vol 2: circulation, sediments and transfer of material in the estuary. Academic Press, New York, pp 182–215

Marquéz-Millan R (2000) The Ridley sea turtle populations in Mexico. In: Abreu-Grobois FA, Briseño-Dueñas R, Marquez R, Sarti L (eds) Proceedings of the 18th International Sea Turtle Symposium. US Dep Commer NOAA Tech Memo, NMFS-SEFSC-436, 3–7 March 1998, Mazatlan, Sinaloa, Mexico, p 19

Olmstead I (1993) Wetlands of Mexico. In: Whigham DF (ed) Wetlands of the world. Kluwer, Dordrecht, pp 637–677

Paéz-Osuna F, Guerrero Galván SR, Cruz-Fernández AC, Espinoza-Angulo R (1997a) Fluxes and mass balances of nutrients in a semi-intensive shrimp farm in the North-Western Mexico. Mar Pollut Bull 34(5):290–297

Paéz-Osuna F, Guerrero Galván SR, Cruz-Fernández AC (1997b) The environmental impact of shrimp aquaculture and coastal pollution in Mexico. Mar Pollut Bull 35(1):65–76

Rico-Gray V (1981) *Rhizophora harrisonnii* (Rhizophoraceae), un nuevo registro de las costas de México. Bol Soc Bot México 41:163–165

Rueda L, Botello AV, Díaz G (1997) Presencia de plaguicidas organoclorados en dos sistemas lagunares del estado de Chiapas, Mexico. Rev Int Contam Ambient 13(2):55–61

Yañez-Arancibia A (1978) Taxonomía, ecología y estructura de las comunidades de peces en lagunas costeras con bocas efímeras del Pacífico de México. Special Publication, Cen Cienc del Mar y Limnol Nat Univ Autono Mex 2:1–306

21 Upwelling and Lagoonal Ecosystems of the Dry Pacific Coast of Baja California

S.E. Ibarra-Obando, V.F. Camacho-Ibar, J.D. Carriquiry, and S.V. Smith

21.1 Introduction

Baja California, between 32°30'N, 117°W and 23°N, 110°W, is a 1300-km-long and 30- to 240-km-wide peninsula, which is flanked by the Pacific Ocean in the west and the Gulf of California in the east. The continental shelf along the Pacific coast of Alta and Baja California (42–22.9°N) is typically narrow (usually <20 km) but widens to 50–70 km in the central and southern embayments. The coastal geomorphology is composed of fault-line, dunes, elevated wave-cut benches, barrier beaches, islands, spits, and bay barrier coasts. Numerous and extensive, mainly sandy plains and valleys occur along the coast, and muddy marsh flats can be found at the mouth of the streams. Coastal lagoons and estuaries represent a small part of the coastline and are partially protected by bay-mouth sand spits (Wiggins 1980).

In general, the Baja Californian coast can be divided in three latitudinal regions based on coastal geomorphology and offshore and inshore regions delimited by the edge of the continental shelf (200 m depth). The northern third comprises the region between Bahía de Todos Santos (31°30'N) and Bahía Sebastian Vizcaíno (27°30'N), and represents an extension from the Southern California Bight. The region in the middle third includes two important upwelling areas, Punta Eugenia and Punta Abreojos, which represent the typical upwelling ecosystem of Baja California. The region in the southern third, centered in Bahía Magdalena, has been described as the region with the strongest latitudinal gradient in physical and chemical variables, and is considered as a subtropical-tropical interface.

21.2 Oceanography and Climate

The climate and oceanography of Alta and Baja California are largely controlled by the California Current (CC) System, which represents the eastern limb of the large-scale, anticyclonic North Pacific gyre, and has a pronounced influence on the structure and function of the coastal upwelling and lagoonal ecosystems. The major alongshore currents of the CC System are an offshore, southward flow (California Current) and a nearshore, northward flow (California Countercurrent). The western boundary of the CC lies about 850–900 km off the coast. The California Current is a surface current (0–300 m) with an average speed of typically <25 cm s^{-1} (peak velocity 50 cm s^{-1}) which transports cool, low-salinity, oxygen- and nutrient-rich, subarctic water towards the North Equatorial Current. The California countercurrent is an important nearshore surface feature along the Baja Californian coast throughout the year. During the summer, the Countercurrent is covered by southward flowing surface waters, owing to persistent northwesterly winds, and remains as an undercurrent at depths of 200–500 m. In fall and winter, the countercurrent is strong and reaches the surface inshore of the California coast (Davidson Current). The area influenced by the countercurrent may extend up to 500 km offshore at depths below 200 m. Owing to its broad seaward extent under the California Current, it has been described as the California Undercurrent domain (McLain and Thomas 1983). The coastal ocean off Alta and Baja California is a classic wind-driven coastal upwelling system. The strong atmospheric pressure gradient between a thermal low-pressure cell over the heated land mass and higher barometric pressure over the cooler ocean is largely responsible for creating vigorous alongshore winds that drive the coastal upwelling system. During the warmer seasons, strong northerly and northwesterly winds induce offshore transport of surface waters. Upwelling of cool, nutrient-enriched water from the depth balances the resulting loss of surface water near the coast.

Cooling and stabilization of the onshore airflow by contact with upwelled waters leads to a cool summer climate at the adjacent coast. During upwelling seasons, the coastal inland under direct influence of coastal stratus and fog is characterized by clear atmospheric conditions which lead to strong daytime heating by short-wave solar radiation, particularly in interior valleys, and rapid nighttime, long-wave radiate cooling (Bakun 1990). As a whole, the Mediterranean-type climate of the region is hot and dry with prevalence of desert conditions, though considerable local variations may occur. The northern portion of the peninsula is characterized by a Temperate Zone temperature regime (12–18 °C), the middle part of the peninsula has a semi-warm regime (18–22 °C) which becomes a warm regime (22–26 °C) at the southern tip of the peninsula. In general, the northern half of Baja California receives most of the annual rainfall (mean 225 mm) from November to March. In contrast, most of the annual precipitation (mean 160 mm) at the southern peninsula occurs during the summer months. However, both precipitation regimes experience extreme inter-annual variability (i.e. 32.5°N, range 63–433 mm and 24.4°N, range 46–608 mm).

21.3 The Upwelling Ecosystem of Baja California

Although most intense in spring and early summer (March to June), upwelling off western Baja California persists during the entire year (Bakun and Nelson 1977), and advected upwelled waters bathe the entire coastline of the peninsula. However, the influence of upwelling is characteristically stronger between Punta Baja (30°N) and Punta Abreojos (26°N), owing to negative wind stress curl (maximum Ekman convergence) which favors the formation of fronts and convergent patches of upwelled water. North and south of this region divergent Ekman conditions prevail and upwelling (although reduced) is stronger offshore than inshore. The coastal ecosystem off Baja California (~250 km) is the southern extension of the cold California Current ecosystem. However, owing to latitudinal temperature gradients between regions off Northern California (mean annual SST 12 °C, seasonal SST 3.5 °C) and Southern Baja California (mean annual SST 23 °C, seasonal SST >10 °C), the abundance of several species changes. Furthermore, limited communication between semi-permanent cyclonic eddies (Fig. 21.1) north and south of Punta Eugenia (28°N) appears to induce discontinuities among a wide variety of taxa. The distribution of oceanic euphausiids and copepods, which are indicators of subarctic equatorial and central waters, terminates at Punta Eugenia. Several coastal pelagic fish species, like the northern anchovy (*Engraulis mordax*), the Pacific sardine (*Sardinops caeruleus*) and the Pacific hake (*Merluccius productus*), display distinct discontinuities (i.e. morphometric and meristic measurements), suggesting a restricted flow of genes between populations north and south of 28°N. A provincial boundary in the vicinity of Punta Eugenia appears to separate northern warm-temperate faunal elements from biota of uncertain origin with reduced tropical elements, high endemism, and insular characteristics of a province tentatively called Panamic between Punta Eugenia and Cabo St. Lucas (Hewitt 1981).

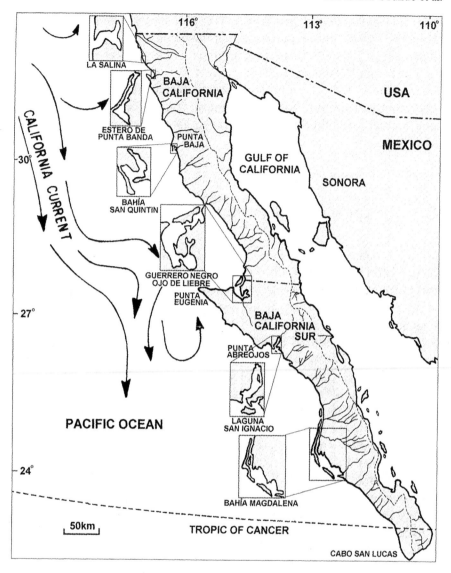

Fig. 21.1. The Pacific coast of Baja California with main coastal lagoons

21.3.1 Biotic Components and Trophic Relations

Kelp forests (i.e., *Macrocystis pyrifera*, *Laminaria farlowii*, *Pterygophora californica*, *Pelagophycus porra*) are among the most prominent coastal marine communities associated with cold California Current waters, though the number of species gradually decreases towards the south. The phytoplankton community of the California Current ecosystem is characterized by an assemblage of diatoms (*Chaetoceros decipiens, Rhizosolenia robusta, R. styliformis, Skeletonema costatum, Asterionella japonica, Thalassionema nitzschiodes*), dinoflagellates (*Ceratium extensus, Peridinium* sp., *Gonyaulax catenella, G. polyhedra*), and coccolithophores (*Umbilicosphaera sibogae*). *Chaetoceros* and *Rhizosolenia* (up to 40%) tend to dominate nutrient-rich waters where they compete with *Asterionella* while *Skeletonema* appears to prefer moderate nutrient concentration, and thus blooms at different times than *Chaetoceros* and *Asterionella*. Dinoflagellates, with *Ceratium* as the dominant genus, bloom profusely when surface waters are warm in the summer or when warm water masses are mixed with California Current waters. Protozoans are largely represented by planktonic (*Globigerina bulloides, Globigerinoides ruber*) and benthic foraminifera (*Buliminella elegantissima, Elphidium translucens, Quinqueloculina costata*), radiolaria (*Stylatractus* spp., *Acanthastaurus* spp.), and tintinnidae (*Tintinnus fraknoii*).

The zooplankton in shallow waters (<100 m) off Baja California is dominated by copepods (*Euchaeta media, Temora discaudata, Rhincalanus nasutus, Pleurommamma gracilis*). Deeper waters (>200 m) are typical of *Calanus pacificus* and *C. californicus*, euphausids (*Nyctiphanes simplex, Euphasia eximia, E. brevis, E. hemigibba*), caetognaths (*Sagitta bierii, S. euneritica, S. minima*), the pteropod *Limacina inflata*, the decapod *Pleuroncodes planipes*, the salp *Thalia democratica*, and squid and fish larvae (*Vinciguerria lucetia*). Most of these species are herbivores, such as the euphasiids (which rank second to copepods in abundance), while others, such as chaetognaths (also commonly known as arrowworms), are predators and feed mostly on copepods and other small zooplankton. In general, tropical oceanic populations (e.g., *Euchaeta* sp.) dominate the epipelagic layer of the water column while deeper water populations are dominated by temperate species (e.g., *Calanus* sp.).

Micronekton in the California Current is composed of cephalopods (*Abraliopsis felis, Loligo opalescens*), myctophid (lantern) fishes (*Triphoturus mexicanus, Stenobrachius leucopsarus*), Pacific saury (*Cololabis saira*), rockfishes (*Sebastes macdonaldi*), cnidarians (*Liriope tetraphyla, Rhopalonema velatum*), and pelagic shrimps (*Sergestes similis*). An inverse relation between larvae of medusae and anchovies suggests that medusae prey upon anchovies. Also, adults of the shrimp *Sergestes similis* most often prey on larvae of fish but are a favorite prey for the squid *Loligo opalescens*. Copepods and euphasids are ingested by lantern fishes, which follow their vertical diurnal migrations, and by the Pacific saury; thus, these species represent an important link between zooplankton and high-level carnivores such as tuna, squid, billfishes, seabirds, sharks, seals, and whales.

Sea urchins (*Strongylocentrotus purpuratus, S. franciscanus*), abalone (*Haliotis sorenseni, H. rufesens, H. fulgens, H. cracherodii*) and lobsters (*Panulirus interuptus, P. inflatus, P. gracilis*) find a habitat in the kelp forests. A large number of fish, like the white seabass (*Atractoscion nobilis*), kelp bass (*Paralabrax clathratus*), topsmelt (*Atherinops affinis*), jack mackerel (*Trachurus symmetricus*), kelp perch (*Brachyistius frenatus*), black perch (*Embiotoca jacksoni*), rubberlip seaperch (*Rhacochilus toxotes*), blacksmith (*Chromis punctipinnis*), senorita (*Oxyjulis californica*), kelp rockfish (*Sebastes atrovirens*), yellowtail (*Seriola lalandi*), and California halibut (*Paralichthys californicus*), are associated with kelps. The bat ray (*Myliobatus californica*) and torpedo ray (*Torpedo californica*) as well as the horn shark (*Heterodontus francisci*) and leopard shark (*Triakis semifasciata*) are common in kelp habitats.

Top carnivores are represented by species with wide latitudinal distribution, such as the shortbeak common dolphin (*Delphinus capensis*), bottlenose dolphin (*Turniops truncatus*), Pacific white-sided dolphin (*Lagenorynchus obliquidens*), killer whale (*Orcinus orca*), humpback whale (*Megaptera novaengliae*), blue whale (*Ballaenoptera musculus*), California sea lion (*Zalophus californianus*), and the harbor seal (*Phoca vitulina*). Sea turtles (*Eretmochelys imbricata, Lepidochelys olivacea, Chelonia mydas*) are also common and the gray whale (*Eschrichtius robustus*) is a regular visitor with breeding grounds in the Laguna Ojo de Liebre and Laguna San Ignacio during the winter.

The trophic dynamics of upwelling off Punta San Hipólito (27°N) during two successive seasons (Walsh et al. 1977) indicate that the initial stage of upwelling is accompanied by a transition from an oceanic (dinoflagellate–herbivorous copepod–carnivorous red crab–striped dolphin) food chain to upwelling food chains (Walsh et al. 1974). Major components of phyto- and zooplankton are *G. polyedra* and the copepod *Acartia* sp., while the red crab *P. planipes* and the Pacific white-sided dolphin *Lagenorynchus obliquidens* appear to be the major carnivores, though the role of *L. obliquidens* as top predator is still uncertain. Despite similar upwelling intensity, nutrient content and phytoplankton growth rates, the production and biomass of phytoplankton vary between years (i.e. 1972 ~8 g C m^{-2} day^{-1} and 1973 ~4 g C m^{-2} day^{-1}), possibly in response to grazing pressure by the red crab *P. planipes*. Red crabs, which during the onset of upwelling feed at higher trophic levels, tend to move to lower trophic levels during advanced stages of upwelling, probably following changes in phytoplankton composition from smaller dinoflagellates (*G. polyedra*) to larger diatoms (*Coscinodiscus excentricus*). After continued consumption and lower phytoplankton density, red crabs use detritus as an alternate food source and could thus replace phytophagous clupeids (*E. mordax* and *S. caeruleus*) in the nekton niche off Baja California (Walsh et al. 1977).

Furthermore, changes in frequency and intensity of upwelling, which must be strong enough to infuse nutrients and enrich the trophic pyramid but avoid turbulent mixing of the water column to maintain concentrations of food organisms (Bakun 1990), are likely to influence the reproductive success of pelagic fish and the replacement of species. For instance, the Pacific sardine (*Sardinops caeruleus*) was the dominant species in the California Current ecosystem but was replaced by the northern anchovy (*Engraulis mordax*) in the 1950s. Both anchovy and sardine populations follow 60-year cycles in the Santa Barbara Basin off Southern California and over the last 1,700 years, nine major recoveries and subsequent collapses of the sardine have occurred (Baumgartner et al. 1992).

21.3.2 El Niño Effect on the California Current Ecosystem

In the North Pacific basin, Southern Oscillation (ENSO) episodes are associated with a cyclonic circulation and a strong (40–90 cm s^{-1}) and broad inshore countercurrent that is responsible for the displacement of southern, typically warm-water species towards higher latitudes along the west coast of North America. Synchronous with the development of El Niño in the tropics, large-scale anomalies of a winter barometric pressure field over the North Pacific cause anomalous atmospheric circulation in the region (Mysak 1986). Their impact on the California Current system off Baja California includes enhanced countercurrent intensity, reduced upwelling or increased downwelling, widespread and intensive warming of upper mixed layer, depressed salinity and thermo- and nutriclines, higher dissolved oxygen, lower inshore nutrients, and higher sea levels (Lynn et al. 1995).

Biological changes during ENSO events are related to reduced primary productivity (Torres-Moye and Alvarez-Borrego 1987), delayed phytoplankton blooms (Lenarz et al. 1995), changes in kelp forest carrying capacity (Dayton et al. 1998), arrival of early life stages of some warm water species, and low reproductive success of some species of rockfish. Zooplankton abundance is primarily influenced by large-scale variations of the California Current flow. While increased southward transport leads to biomass enhancement, reduced transport causes abnormally low zooplankton biomass, probably reflecting primary productivity oscillations in response to nutrients. Since large-scale environmental forcing of the California Current ecosystem has disruptive effects on the ecology of the region, processes like competition and predation might not play a dominant role in determining the composition of species assemblages or in modulating the differences in abundance of the species (Chelton et al. 1982).

21.4 Lagoonal Ecosystems

The better known coastal lagoons along the Pacific coast of Baja California are La Salina, Estero de Punta Banda, Bahia San Quintín, the lagoonal complex of Guerrero Negro–Ojo de Liebre, Laguna San Ignacio, and Bahia Magdalena (Fig. 21.1). Although the lagoons differ in geological origin, the sediments are mostly sandy and alluvial. Since land-to-ocean stream flux is limited, the advected nutrient-rich waters of the ocean directly (i.e. Bahía San Quintín; Millán-Núñez et al. 1982) or indirectly (Estero de Punta Banda and Laguna Ojo de Liebre; Amador-Buenrostro et al. 1995) affect most biogeochemical processes in the lagoons (Fig. 21.1). Owing to a dominance of evaporation over precipitation and lack of inflowing surface or groundwater, moderately hypersaline conditions are common and salinity tends to increase from the mouth (approx. 33.5) towards the inner part (approx. 40). However, average salinity and residence times vary between 35 and 14 days (Bahía San Quintín; Camacho-Ibar et al. 1997), 36 and 16 days (Laguna Guerrero Negro), and 45 and 100 days (Laguna Ojo de Liebre; Postma 1965), respectively. Pacific tides are predominantly semidiurnal with Lower Low Water (LLW) following Higher High Water (HHW). These conditions cause high velocities during ebb tide which flush suspended sediments and inhibit or retard the sedimentary filling processes typical in these environments.

Most of the coastal lagoons are pristine or are barely modified and they represent highly productive and diverse environments. Only some seasonally ephemeral lagoons (La Salina, 32°N, and Laguna Figueroa, 30°N) have low diversity because few species can tolerate the prolonged dry periods and extreme salinity variations. Owing to their high productivity (i.e. Laguna Ojo de Liebre; Phleger and Ewing 1962; Bahía Magdalena; Millán-Núñez et al. 1987) and protection, the lagoons and estuaries are feeding and resting sites and millions of migrant shorebirds and waterfowl overwinter in these areas. The tidal vegetation, sea grasses and algae between California (34°N) and the northern half of Baja California are similar and thus seem to belong to the same biogeographic province. However, between 27–24°N (Bahía Magdalena) mangroves (*Rhizophora mangle*, *Laguncularia racemosa*) replace salt marshes, and tropical and subtropical red algae (*Corallina vancouverensis*, *Gelidium robustum*, *Gracilaria robus-*

ta, *Laurencia pacifica, Pterocladia capillaceae, Eisenia arborea, Sargassum sinicola, Caulerpa sertularioides, Colpomenia tuberculata*) become abundant.

21.4.1 Bahía San Quintín

21.4.1.1 Environmental Setting

Bahía San Quintín (30°27'N, 116°00'W) has an area of 42 km² and an average depth of ~2 m. An important part of the watershed (~2,000 km²) consists of agricultural lands. Despite a high ratio of watershed:bay area, the only significant surface flow into the bay occurs during extreme rainfall events though groundwater from the San Simón basin may contribute about 1×10^3 m³ day^{-1} in the winter. Since total evaporation in summer (164×10^3 m³ day^{-1}) and winter (91×10^3 m³ day^{-1}) exceeds total rainfall (4×10^3 m³ d^{-1} and 67×10^3 m³ day^{-1}, respectively), Bahía San Quintín is a net evaporative system. The bay can be considered a negative estuary with summer and winter salinities in the interior (34.7–33.8) always higher than in the adjacent ocean (33.8–33.6; Millán-Núñez et al. 1982).

The lack of continuous freshwater input to Bahía San Quintín emphasizes the exchange of dissolved and particulate materials with the adjacent ocean. The horizontal advection of upwelled waters, which surface immediately off the mouth, into the bay occurs mostly by tide-induced currents. The distribution of salinity, temperature and dissolved inorganic nutrients in Bahía San Quintín therefore depends, apart from water residence time and internal biogeochemical processes, upon the alternation of upwelling events (mostly in spring and summer) and spring-neap tidal cycles (mostly in autumn and winter). An "erratic" behavior of nutrient concentrations may be related to turbulence and irregular mixing, owing to bathymetry, winds and solar radiation input (Millán-Núñez et al. 1982). Salinity and phosphate concentrations usually increase from the mouth to the inner bay (Fig. 21.2). The spatial distribution of dissolved inorganic nitrogen (DIN) follows no clear overall trend, owing partially to complex biogeochemical cycling and because nitrogen acts as a limiting nutrient throughout the year (Fig. 21.2). The N/P ratios at the mouth of the bay (~16 in winter) decrease to extremely low values (<0.5 in summer) at the innermost part of the bay, probably as a result of prolonged water residence and high N requirement by sea grasses, which are dominant primary producers (Fig. 21.2). An inverse relationship between silicate and salinity in different years (Fig. 21.2) may reflect consumption by diatoms vs. regeneration processes.

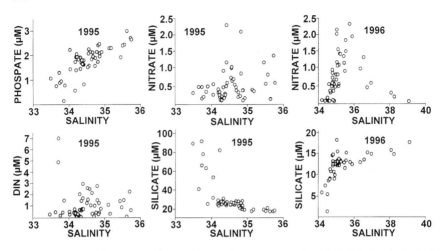

Fig. 21.2. Mixing diagrams of inorganic nutrient concentrations in Bahía San Quintín in August 1995 and August 1996

21.4.1.2 Biotic Components

Terrestrial vegetation around Bahía San Quintín is composed of at least 227 species of vascular plants, some of which are endemic to the region (i.e. *Cammissonia cherantifolia, Dudleya attenuata, Marah macrocarpys, Rhus integrifolia, Salicornia subterminalis, Atriplex julacea, Galvezia juncea, Harfordia macroptera, Hazardia berberidis, Mammillaria sf. Louisae, Ferocactus*). Coastal sand dunes between Ensenada and San Quintín are among the most diverse (25 species) of the peninsula and some species (*Cakile maritima, Ambrosia chamissonis, Carpobrotus chilensi*) reach their southern distribution limit in San Quintín (M. Salazar, pers. com.). Common salt marsh species include *Spartina foliosa, Salicornia virginica, S. bigelovii, Batis maritima, Jaumea carnosa, Suaeda esteroa, Triglochin maritima, Frankenia grandifolia, Distichlis spicata,* and *Cordylanthus maritimus* ssp. *maritimus*. The eelgrass *Zostera marina* and widgeon grass (*Ruppia maritima*) are typical sea grasses in the lagoon while two species of surfgrass (*Phyllospadix torreyi* and *P. scouleri*) are found on rocky shores.

Diatoms (i.e. *Nitzchia, Rizosolenia, Cocconeis*) are the principal component of the phytoplankton community at the mouth of the bay while dinoflagellates and microflagellates have higher abundance in the inner bay (Lara-Lara et al. 1980). Despite the shallow euphotic zone, phytoplankton primary productivity rates at the mouth of Bahía San Quintín in summer (mean 27 mg C m^{-3} h^{-1}, max. 44 mg C m^{-3} h^{-1}) are among the highest reported for upwelling systems off Western Baja California (Lara-Lara et al. 1980). Although chlorophyll concentrations and surface primary productivity decrease from the mouth to the interior, mean assimilation ratios (25 mg C mg^{-1} Chl-*a* h^{-1}) do not vary significantly (Millán-Núñez et al. 1982). About 60% of the gross primary production in the Bahía San Quintín system (~38,380 tons C year^{-1}) depends on the phytoplankton contribution while benthic primary producers contribute approximately 40%.

High primary productivity appears to support relatively large populations of consumers at higher trophic levels (Phleger and Ewing 1962) and the coastal lagoons provide protection and feeding, nursery and acclimation areas for juveniles of several fish species. About 120 species of fishes, the California halibut (*Paralichthys californicus*) and the spotfin croaker (*Roncador stearnsii*) being the most abundant, occur in Bahía San Quintín. Most fish (81%) have northern affinity, about 6% are southern species and only four species have wide distribution (Rosales-Casian 1997), while the signal blenny (*Emblemaria walkerii*) is endemic to the bay (J. Rosales-Casian, pers. comm.). More than 25,000 migratory shore birds spend the winter in Bahía San Quintín, including 30–50% of Mexico's black brant (*Branta bernicla*) population, as well as birds of prey like the American peregrine falcon (*Falco peregrinus anatum*), the burrowing owl (*Athene cunicularia*), and the short-eared owl (*Asio flammeus*). The area hosts the largest reproductive populations of threatened or endangered bird species and subspecies, like the yellow-footed clapper rail (*Rallus longirostris levipes*), Belding's savannah sparrow (*Passerculus sandwichensis beldingi*), California least tern (*Sterna antillarum*), snowy plover (*Charadrius alexandrinus*), black rail (*Laterallus jamaicensis*), and California gnat catcher (*Polioptila californica atwoodi*; Massey and Palacios 1994). The bay is the only nesting site for Foster's tern (*Sterna forsteri*) along the Mexican Pacific coast.

21.4.1.3 Ecosystem Metabolism

The San Quintín coastal ecosystem is geographically isolated and comprised of a terrestrial compartment (San Quintín Valley) and a marine compartment (Bahía San Quintín). A mass balance budget model (Gordon et al. 1996) was used to estimate the net metabolism in Bahía San Quintín (Camacho-Ibar et al. 1997). As a result of evaporation, the residual flow of seawater and dissolved properties from the ocean into the bay exceeds freshwater input from precipitation, runoff and groundwater flow. However, residual flow is an order of magnitude lower than the exchange of water volume (approx. 6×10^6 m^3 day^{-1} in summer) between the bay and ocean. Higher respiration of organic detritus than primary production leads to excess dissolved inorganic phosphorus (DIP) inside the bay; thus, especially in summer the bay is a net heterotrophic system. In contrast, an overall net deficit of dissolved inorganic nitrogen (DIN) in the bay suggests a net denitrifying system. Bahía San Quintin is therefore a net source of N to the atmosphere and DIP to the adjacent ocean (Table 21.1). However, the fate of excess dissolved inorganic carbon is not clear. Preliminary data show an increase in alkalinity, suggesting that sulfate reduction is an important pathway for net heterotrophy. If this is the case, the system will act as a source of alkalinity to the ocean rather than a source of CO_2 to the atmosphere.

The exchange of water, C, N and P among the marine compartment, the ocean, and to some degree the atmosphere, is largely due to natural forcing. In contrast, the effects of natural forcing (i.e., freshwater input) and material fluxes between the terrestrial and marine compartment are almost insignificant (Fig. 21.3). The terrestrial compartment exchanges these materials principally through anthropogenic forcing as a result of agriculture. Although aquaculture activities (mainly oysters) in Bahía San Quintín are among the most extensive on the peninsula, their effect on C, N and P fluxes in the bay is small as compared to natural fluxes. The culture of oysters and clams represents less than 10% of the bay area. It is totally sustained by the naturally high primary productivity of phytoplankton and sea grasses, and annual extraction of about 2,000 metric tons wet product (approx. 33 tons C) is at a sustainable level. In contrast, agriculture has a major effect on material fluxes. Since natural primary productivity in the arid San Quintín Valley is low (Fig. 21.3), high crop yields require artificial supply of water and application of fertilizer. Much of the fertilizer (approx. 1000 tons N and 230 tons P) remains in the San

Table 21.1. Results of the stoichiometrically linked water-salt-nutrient budgets for samples collected on three dates at Bahia San Quintin

	Steps 1 and 2			Step 3		Step 4	
	Residual volume (10^3 m^3 day^{-1})	Exchange volume	Residence time (days)	ΔP	ΔN	Nfix-denit	p-r
				(mmol m^{-2} day^{-1})		(mmol m^{-2} day^{-1})	
Aug-95	160	6,222	14	0.17	−0.14	−2.8	−17
Feb-96	23	3,230	28	0.01	−0.52	−0.7	−1
Aug-96	160	5,536	16	0.19	0.3	−2.7	−20

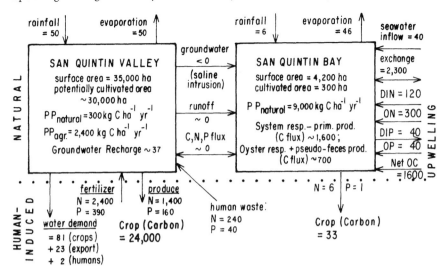

Fig. 21.3. Natural and human-induced C, N, P (tons year^{-1}) and water (10^6 m^3 year^{-1}) mass fluxes in the San Quintín coastal system (estimates for 1997)

Quintín Valley, though nitrogen may be partially lost to the atmosphere through denitrification and P will eventually be transported to the aquifer. However, the most limiting natural resource for agriculture is water, and overexploitation of groundwater has caused saline intrusion in the coastal aquifer since the 1970s. Although at present little water is lost to evaporation or infiltration, owing to drip-irrigation, this practice has lead to an increase in cultivated area, and the corresponding increase in water use was from 32 to 81 × 10^6 m^3 year^{-1}. If water extraction continues at current rates, the groundwater resource in San Quintín could be depleted or contaminated within one decade.

21.5 Natural Resources and Human Impact

The population of the state of Baja California has increased from about 48,000 inhabitants in 1930 to about 1,700,000 in 1990. In the state of Baja California Sur it has increased during the same time from about 47,000 to 320,000 (INEGI 1991a); thus present environmental problems associated with human population reflect a static view of a rapidly changing situation, unless environmental regulations can help counteract the effects of change.

Historically, subsistence plant agriculture was based on "dry land farming" or made limited use of water from "arroyos". Today, such practice remains widespread and appears to have little impact on the environment, although native vegetation may temporarily be displaced, as some of the native grasses and other desert scrub vegetation are slow to return. Modern horticulture based on irrigation, with most products being exported to the U.S.A., is an example of widespread natural resource utilization in Baja California (4,000 ha in 1960 to 240,000 ha in 1996; SPP 1981; INEGI 1991b, 1997). Apart from over-exploitation of groundwater, the use of agrochemical nutrients, herbicides, and pesticides potentially contaminates the groundwater and soil. However, one of the most detrimental effects of modern horticulture has been on the desert landscape. The coastal scrub between 31°30'N and 30°19'N has almost disappeared. As a consequence, some endemic plants and animals, like the Baja California legless lizard (*Aniella geronomensis*), the California meadow wole (*Microtus californicus equivocus*), and the kangaroo rat (*Dipodomys gravipes*) have also disappeared. Loss of habitat has reduced the distribution range of some species and population size of others, including the wren cactus (*Campylorhynchus brunneicapillus bryanti*), the California gnatcatcher (*Polioptila californica atwoodi*), and the ornate shrew (*Sore ornatus*; E. Mellink, pers. comm.).

Fisheries of a variety of species have been present in Baja California for a long time. However, only coastal ocean fisheries, mainly of sardines, tuna, sharks, chub mackerel, Gulf coney, skipjack tuna, Pacific bonito, spotted sand bass and squids, are economically important. A few species (i.e. abalone, lobster, clams and crabs) are both "coastal" and commercially significant and especially abalone and lobster appear to be overfished. The formation of natural salt deposits in evaporation pans of some lagoons (i.e., La Salina, Laguna Guerrero Negro, Laguna San Ignacio) has been exploited for commercial salt production without harm to the environment (Fig. 21.1). Laguna Guerrero Negro has the largest salt pan in the world and thousands of migratory birds are present in ponds and whales enter the lagoon. However, the expansion of salt flats in Laguna San Ignacio Lagoon has raised international concern about a possible loss of gray whales nursery grounds.

References

Amador-Buenrostro A, Argote-Espinoza ML, Mancilla-Peraza M, Figueroa-Rodríguez M (1995) Short term variations of the anticyclonic circulation in Bahía Sebastián Vizcaíno, Baja California. Cienc Mar 21:201–223

Bakun A (1990) Global climate change and intensification of coastal ocean upwelling. Science 247:198–201

Bakun A, Nelson CS (1977) Climatology of upwelling related processes off Baja California. CalCOFI Rep 19:107–127

Baumgartner TR, Soutar A, Ferreira-Bartrina V (1992) Reconstruction of the history of Pacific sardine and northern anchovy populations over the past two millennia from sediments of the Santa Barbara Basin, California. CalCOFI Rep 33:24–40

Camacho-Ibar VF, Carriquiry JD, Smith SV (1997) Bahía San Quintín, Baja California (a teaching example). In: Smith SV, Ibarra-Obando SE, Boudreau PR, Camacho-Ibar VF (eds) Comparison of carbon, nitrogen and phosphorus fluxes in Mexican coastal lagoons. LOICZ Rep Stud no 10, LOICZ, Texel, The Netherlands

Chelton DB, Bernal PA, McGowan JA (1982) Large-scale interanual physical and biological interaction in the California Current. J Mar Res 40:1095–1125

Dayton PK, Tegner MJ, Edwards PB, Riser KL (1998) Sliding baselines, ghosts, and reduced expectations in kelp forest communities. Ecol Appl 8:309–322

Gordon DC Jr, Boudreau PR, Mann KH, Ong J-E, Silvert WL, Smith SV, Wattayakorn G, Wulff F, Yanagi T (1996) LOICZ biogeochemical modelling guidelines. LOICZ Rep Stud no. 5, LOICZ, Texel, The Netherlands

Hewitt R (1981) Eddies and speciation in the California Current. CalCOFI Rep 22:96–98

INEGI (1991a) XI censo general de población y vivienda, 1990. Perfil Sociodemográfico. Baja California y Baja California Sur. Instituto Nacional de Estadística, Geografía e Informática, México

INEGI (1991b) Anuario estadístico del Estado de Baja California Sur. Instituto Nacional de Estadística, Geografía e Informática, México

INEGI (1997) Anuario estadístico del Estado de Baja California y Baja California Sur. Instituto Nacional de Estadística, Geografía e Informática, México

Lara-Lara R, Alvarez-Borrego S, Small LF (1980) Variability and tidal exchange of ecological properties in a coastal lagoon. Estuar Coast Shelf Sci 11:613–637

Lenarz WH, Ventresca DA, Graham WM, Schwing FB, Chavez F (1995) Explorations of El Niño events and associated biological population dynamics off Central California. CalCOFI Rep 36:106–119

Lynn R, Schwing FB, Hayward TL (1995) The effect of the 1991–1993 ENSO on the California Current system. CalCOFI Rep 36:57–71

Massey B, Palacios E (1994) Avifauna of the wetlands of Baja California. Stud Avian Biol 15:45–57

McLain DR, Thomas DH (1983) Year to year fluctuations of the California countercurrent and effects on marine organisms CalCOFI Rep 24:165–181

Millán-Núñez R, Alvarez-Borrego S, Nelson DM (1982) Effects of physical phenomena on the distribution of nutrients and phytoplankton productivity in a coastal lagoon. Estuar Coast Shelf Sci 15:317–335

Millán-Núñez R, Ripa-Soleno E, Aguirre-Buenfil LA (1987) Preliminary study of the composition and abundance of the phytoplankton and chlorophytes in Laguna Ojo de Liebre, BCS. Cienc Mar 13:30–38

Mysak LA (1986) El Niño, interannual variability and fisheries in the Northeast Pacific Ocean. Can J Aquat Fish Sci 43:464–497

Phleger FB, Ewing GC (1962) Sedimentology and oceanography of coastal lagoons in Baja California, Mexico. Geol Soc Am Bull 73:145–182

Postma H (1965) Water circulation and suspended matter in Baja California lagoons. Neth J Sea Res 2:566–604

Rosales-Casian J (1997) Inshore fishes of two coastal lagoons from the northern Pacific coast of Baja California. CalCOFI Rep 38:180–192

SPP (1981) Manual de estadísticas básicas del Estado de Baja California. Secretaría de Programación y Presupuesto, México

Torres-Moye G, Alvarez-Borrego S (1987) Effects of the 1984 El Niño on the summer phytoplankton of a Baja California upwelling zone. J Geophys Res 92(14):383–386

Walsh JJ, Kelley JG, Whitledge TE, MacIsaac JJ, Huntsman SA (1974) Spin-up of the Baja California upwelling ecosystem. Limnol Ocean 19:553–572

Walsh JJ, Whitledge TE, Kelley JG, Huntsman SA, Pillsbury RD (1977) Further transition states of the Baja California Upwelling ecosystem. Limnol Ocean 22:264–280

Wiggins IL (1980) Flora of Baja California. Stanford University Press, Stanford, California

22 The Colorado River Estuary and Upper Gulf of California, Baja, Mexico

S. Alvarez-Borrego

22.1 Introduction

Since the times of early explorers the Gulf of California has been described as an area of high fertility (León-Portilla 1972), owing mainly to tidal mixing and upwelling processes (Alvarez-Borrego et al. 1978). The shallow northernmost part of the Gulf, the Upper Gulf of California, and the Colorado River delta are an area of reproduction and nursery for many fish species (Guevara-Escamilla et al. 1973). They are also habitat for endangered endemic species, like the totoaba (*Totoaba macdonaldi*) and the small dolphin "vaquita" (*Phocoena sinus*). In 1993, the Mexican Federal Government decreed the area as a Reserve of the Biosphere (Reserva de la Biósfera del Alto Golfo de California y Delta del Río Colorado; Fig. 22.1).

22.2 Meteorology

The moderating effect of the Pacific Ocean upon the climate of the Gulf of California is greatly reduced by the almost uninterrupted, up to 3,000-m-high mountain chain of Baja California. The climate of the Gulf is therefore more continental than oceanic, with maximum air temperatures of >40 °C in summer and large diurnal and annual (about 18 °C) temperature ranges (Roden 1964). In winter, air temperature differences between the Gulf and the Pacific coast of Baja California are small and temperatures decrease towards the interior of the Gulf, while in summer differences may exceed 10 °C and temperatures increase towards the interior of the Gulf. The northern Gulf has two seasons (Mosiño and García 1974). During the mid-latitude winter (November–May), 3- to 6-day events of northwesterly winds (8–12 m s^{-1}) transport cool, dry desert air over the Gulf and during the sub-

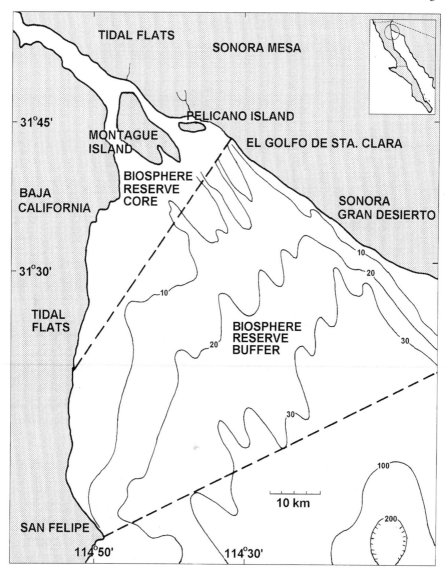

Fig. 22.1. The Upper Gulf of California and the Colorado River delta

tropical summer (June–October), large-scale pressure fields drive weak southeasterly winds (2–5 m s^{-1}) primarily along the Gulf axis. Cross-Gulf winds are particularly intense in the northwestern region during upper level trough passages over Baja California (Badan-Dangon et al. 1991). The northern half of the Gulf is dry and desert-like, with a mean annual precipitation of almost zero and mean evaporation of 1.1 m year^{-1}. Despite heat loss from the water between October and January, the mean annual net surface heat flux is into the sea (69 W m^{-2} near the coast and 83 W m^{-2} offshore). These conditions require an export of heat and salt out of the northern Gulf to achieve a balance, thus the mean annual thermohaline circulation must have an inflowing component at depth (Lavín and Organista 1988). Since inflowing deep water has higher inorganic nutrient concentrations than outflowing surface water, the thermohaline circulation in the northern Gulf is a natural fertilization mechanism.

22.3 Environmental Settings

The Colorado River forms a 2- to 8-km-wide and 50-km-long estuarine basin, which opens at the mouth into a 16-km-wide delta. Numerous submarine channels and shoals extend for about 50 km in a northwest-southeast direction. The semidiurnal tide regime has a range of 6.95 m at San Felipe, which is even larger at the entrance of the delta (Gutiérrez and González 1989). Extensive low lands on the western and eastern shores and the low and flat Montague and Pelicano islands are periodically flooded by tides (Fig. 22.1). Tides produce current velocities of 1.5 m s^{-1} over the shallow platform adjacent to the delta and up to 3 m s^{-1} inside the estuarine basin (Meckel 1975; Cupul 1994). The evolution of the Colorado River delta has been controlled by the interaction between constructive and destructive processes, like sediment supply (about 160×10^6 tons year^{-1}) from the Colorado River and one of the largest tidal regimes in the world, respectively (Carriquiry and Sánchez 1999). Human intervention, in particular the construction of the Hoover and Glen Canyon Dams in the Colorado River basin, has almost eliminated the discharge of sediments into the delta (Meckel 1975) that has become subjected to destructive processes, such as strong tidal currents and wind-induced waves. The delta is passing through an erosional phase and exports sediments to the northern Gulf of California at rates similar to those of unaltered river flow prior to 1935 (van Andel and Shor 1964; Carriquiry et al. 1992; Cupul 1994).

The construction of the dams and the diversion of freshwater for irrigation, industrial and municipal uses also had a profound effect on the amount and timing of freshwater into the Upper Gulf of California. Before the filling of Lake Mead (1935), freshwater discharge had a seasonal modulation with a peak in June (Fig. 22.2). In the absence of freshwater flow from the Colorado River, surface salinity generally increases northwestward, from about 36 outside the estuary to about 38.5 at the northern end of Montague Island in winter, and from 38 to about 40 in summer, thus

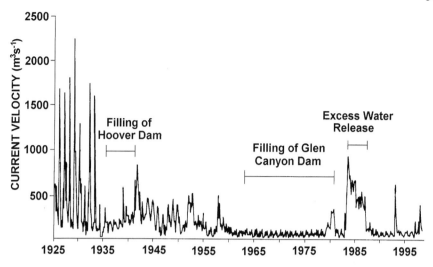

Fig. 22.2. Freshwater flow from the Colorado River across the US-Mexican border. (Modified after Lavín and Sánchez 1999)

establishing inverse-estuary conditions. Lower salinity values, both in winter (34) and in summer (38.5), at the internal extreme of the estuary indicate the importance of fresh groundwater flow from the river (Alvarez-Borrego et al. 1973; Hernández-Ayón et al. 1993). However, during a short (March/April 1993) release of freshwater (Fig. 22.2), surface salinity decreased towards the head of the estuary and increased from 32 approximately 10 km south of Montague Island to about 35.4 in the Upper Gulf near San Felipe. In the estuary, a salt wedge-type stratification was established during neap tides (Lavín and Sánchez 1999).

The depth of the Upper Gulf of California is less than 30 m, with shallower waters at the Baja California than at the Sonora side. During winter, the mean residual circulation (>5 cm s^{-1}) in the Gulf of California is influenced by wind, rather than by tidal forcing (Argote et al. 1995, 1998). Winds induce cyclonic circulation during winter and anticyclonic circulation during summer. Residual currents transport oceanic water northward (5 cm s^{-1}) along the Sonora coast. At El Golfo de Santa Clara currents turns west and current velocity decreases to 2 cm s^{-1}. South of Montague Island a bifurcation transports water southward and eastward (<1 cm s^{-1}) to the Baja California coast (Alvarez-Sánchez et al. 1993). Water mass formation occurs throughout the year when the Upper Gulf is in the inverse-estuarine mode. Owing to high evaporation rates in the Upper Gulf, the heavy, high salinity waters are advected by gravitational currents southward to depths of 30–50 m in summer and to more than 200 m in winter (Lavín et al. 1998).

Temperature difference between surface and bottom water is less than 0.5 °C, with higher temperatures at the bottom during winter and higher surface temperatures in the summer. Vertical temperature gradients may be reversed during summer, with up to 4 °C lower temperatures at the surface than at 5 m depth, owing to the strong tidal turbulence. The horizontal surface temperature gradient reverses at the beginning of spring and fall, due to the annual cycle of atmospheric temperature. Surface temperature increases from the southeast to the northwest in summer, while the opposite occurs in winter. Surface water temperatures range from 8.25 °C in December to 32.58 °C in August west of Montague Island, while in deeper waters of the southeast region seasonal temperatures range from 17 to 30.75 °C (Alvarez-Borrego et al. 1973).

The SE to NW transport along the Sonoran coast adds sediment from the ocean into the estuarine basin, while N to S transport along the Baja California coast removes sediments from the estuarine basin into the northern Gulf. The residual sediment transport pattern suggests a counter-clockwise path of materials exchange between the Colorado River delta and Upper Gulf of California (Carriquiry and Sánchez 1999). A net cross-basin sediment transport occurs in a NE-SW direction. The Sonoran sedimentary province is characterized by a heavy mineral suite of garnet and zircon (GZ), whose sediment supply originates from the sandy sediments of the Sonora Mesa deposits and Sonora Gran Desierto. The Baja Californian sedimentary province is characterized by a hornblende-epidote-piroxene (HEP) suite, whose sediments originate from earlier supplies of the Colorado River. The intense reworking of the Colorado River delta has dispersed HEP-rich sediments throughout the Upper Gulf and GZ-rich sediments are invading areas previously dominated by HPE sediments (Carriquiry and Sánchez 1999). The grain size of surface sediments at the Sonora (3.7-2.8 phi) and Baja California coast (6.8-4.7 phi) increases northward. However, while the organic matter (δ^{13} C and δ^{15} N) in surface sediments at the Sonora coast is mainly phytoplankton-derived, owing to intrusion of oceanic waters, in sediments off Baja the contribution of phytoplankton is low and bacterial biomass higher, probably due to anaerobic bottom waters (Aguíñiga-García 1999).

In general, turbidity increases towards the coasts, thus Sechii disk depths are about 2 m in deeper waters and between 0.5 and 1 m off Baja California and Sonora, respectively. Due to sediment resuspension Sechii depth around Montague Island is only a few centimeters. Seston values range from 16 mg l^{-1} in October to 132 mg l^{-1} in May with 23-100 % as particulate organic matter (García-de-Ballesteros and Larroque 1976). Surface concentrations of dissolved oxygen vary between 1.33 ml l^{-1} north of Montague Island in October to super-saturation (>130 %) in southern and central regions of the Upper Gulf (Alvarez-Borrego et al. 1973). Surface nutrient concentrations tend to increase from the central and northern Gulf of California (south of 31°N; 0.7–1.0 µM PO_4, 0–4.0 µM NO_3, 6.1–18 µM SiO_2; Alvarez-Borrego et al. 1978) towards the Upper Gulf adjacent to Montague Island (0.3–3.1 µM PO_4, 3.3–18.3 µM NO_3; Hernández-Ayón et al. 1993). Throughout the year, nutrient concentrations are highest in the estuary (2–15 µM NO_2, 40–53 µM NO_3, 5–11.5 µM PO_4, 60–92 µM SiO_2) but without any clear seasonal pattern. High concentrations in the estuary are possibly due to application of fertilizers in upstream agricultural areas, groundwater input, and sediment resuspension mainly during spring tides (Hernández-Ayón et al. 1993).

22.4 Biological Communities

Despite the ecological impact caused by the construction of dams, life in the estuary is abundant, even during the long periods without surface freshwater input. At most locations of the Gulf photosynthetic pigment concentrations vary seasonally, with maxima during strong upwelling events in "winter" (November to May) and minima under "summer" conditions between June and October. Owing to strong tidal mixing, seasonal variations of pigment concentrations in the Upper Gulf are less evident and concentrations are generally higher than 4 mg m^{-3} (Santamaría-Del-Angel et al. 1994). Despite the patchy distribution of phytoplankton, chlorophyll-*a* concentrations may reach values between 5 and 20 mg m^{-3} in turbid surface waters off El Golfo de Santa Clara and off San Felipe. Zooplankton biomass in the Upper Gulf of California generally increases west or northwestward and varies between 25–150 mg m^{-3} in August and 1–4 mg m^{-3} in October, though biomass values in the channels around Montague Island are consistently higher (up to 154 mg m^{-3}). In general, calanoid copepods are the most abundant taxonomic group, both numerically and in terms of biomass (Farfán and Alvarez-Borrego 1992).

Of the 73 fish species collected from soft bottoms (Guevara-Escamilla et al. 1973), 29 species are panamic, 31 occur in California waters and 13 species might be endemic to the area. The juveniles of most fish species occur in areas of high turbidity north of a line between El Golfo de Santa Clara and San Felipe, possibly owing to protection and food. Some species, like the endemic totoaba (*Totoaba macdonaldi*), supported an important fishery in the Upper Gulf since 1929; however, the drastic decline of stocks at the end of the 1960s placed the species on the endangered species list (Guevara-Escamilla et al. 1973). Despite a general scarcity of ecological studies, the Mexican Federal Government recognized the importance of the Upper Gulf of California as a reproduction and nursery area for fish and closed the area adjacent to the Colorado River mouth for fishing in 1955.

Among the invertebrates, including Cnidaria, Mollusca, Arthropoda, and Echinodermata, 17 of the most abundant species were reported around Montague Island and 48 from the oceanic region. Shrimp trawling before the decree of the Reserve of the Biosphere in 1993 impacted much of the fish and invertebrate fauna. As for fish, the area north of El Golfo de Santa Clara and San Felipe appears to be important for juveniles of the blue shrimp (*Penaeus stylirostris*; Félix-Pico and Mathews 1975). The landing of shrimp catches in San Felipe (>90 % *P. stylirostris*) and freshwater flow of previous years from the Colorado River into the Upper Gulf correlate significantly. While landings are approximately 200 tons year^{-1} in the absence of freshwater flow, they increase to more than 600 tons year^{-1} in years following freshwater input (S. Galindo-Bect and M. Page, pers. comm.).

Marine mammals are common in the Upper Gulf of California and adjacent areas. A colony of the abundant sea lion *Zalophus californiensis* exists off San Felipe and seven species of cetacean have been recorded after a period (1986–1987) of freshwater release (Silber et al. 1994). Bottlenose dolphins (*Tursiops truncatus*) are among the most frequently encountered groups (1,416 ind.) and are the only marine mammals in the Colorado River and in shallow waters of less than 2.5 m depth. The Upper Gulf is part of the habitat of an endangered endemic species, the small dolphin commonly known as "vaquita" (*Phocoena sinus*; 110), which occurs in small groups. The common dolphin *Delphinus delphis* (14239) is numerically the dominant cetacean, while fin whales, i.e., *Balaenoptera physalus* (215), Bryde's whales, i.e., *Balaenoptera edeni* (7), killer whales, i.e., *Orcinus orca* (17), and gray whales, i.e., *Eschrichtius robustus* (3) are less common.

Prior to changes in freshwater flux, the Colorado River delta was important for water birds and today it still offers habitat for 36 species (>100,000 ind.) in winter, though the habitat is less important during the summer season (~500 ind. with 15 species; Mellink et al. 1997). During the winter, Charadriiformes dominate the avifauna which rest and feed in the Colorado River delta. The vast majority (80 %) of shorebirds (163,744) in January are sandpipers (mainly *Calidris mauri*) together with more than 9000 American avocets and nearly 8000 willets (Morrison et al. 1992). Smaller shorebirds occur mainly on soft mudflats at the southern end of Montague and Pelicano islands while medium and larger birds visit harder mudflats at the western margin of the delta. The delta was one of the first Mexican sites to be incorporated into the Western Hemisphere Shorebird Reserve Network because of its importance for shorebirds. Montague

Island and adjacent areas are also important for seabirds. The brown pelican (*Pelecanus occidentalis*) is the most common species during early winter and double-crested cormorants (*Phalacrocorax auritus*) are abundant during both winter and spring. The delta supports only moderate numbers of American white pelicans (*Pelecanus erythrorhynchos*). Gulls account for 23% of the seabirds, the most common species are the ring-billed gull (*Larus delawarensis*), while terns and skimmers represent 16% of seabirds. The black skimmer (*Rhynchops niger*) and Caspian tern (*Sterna caspia*) are most abundant. During late spring and summer they are less obvious because they concentrate in breeding grounds but during courtship and pair-forming activities in early spring they are more conspicuous.

22.5 Human Impacts

Apart from San Felipe (pop. ~15,000) and El Golfo de Santa Clara (pop. ~600), which are sustained by fishing and recreational activities, there are no towns at the shores of the Upper Gulf. However, the Mexicali valley at the extreme NE of the Baja California peninsula, north of the Upper Gulf, is one of the most important agricultural regions in Mexico. The possible transport of pesticide from a total area of 328,000 ha, of which 186,000 ha are under irrigation, into the marine environment of the Upper Gulf has been a concern since the early 1970s. In 1972–1973, the average total DDT in the clam *Chione californiensis* was 0.067 ppm in specimens from Montague Island and 0.145 ppm in specimens from shores 30 km north of San Felipe, opDDD being the most abundant metabolite (Guardado-Puentes et al. 1973). However, about 10 years later the concentrations in *Chione californiensis* from beaches of El Golfo de Santa Clara had decreased (0.005–0.011 ppm) and low levels (0.008–0.009 ppm) were found in specimens of the mussel *Modiolus capax* 20 km south of San Felipe. Current levels are at least two orders of magnitude below the tolerance limits for human consumption (US Food and Drug Administration) and well below the action levels proposed by the US National Academy of Sciences as a danger to the environment (Gutierrez-Galindo et al. 1988).

References

Aguíñiga-García S (1999) Geoquímica de la cuenca estuarina del Río Colorado: δ^{13} C, δ^{15} N y biomarcadores lipídicos en sedimentos superficiales. PhD thesis, Universidade Autonoma Baja California, Ensenada, Mexico

Alvarez-Borrego S, Galindo-Bect LA, Flores-Baez BP (1973) Hidrología. Estudio Químico sobre la Contaminación por Insecticidas en la Desembocadura del Río Colorado, Tomo I, Reporte Final a la Dirección de Acuacultura de la Secretaría de Recursos Hidráulicos, Univ Aut Baja Calif, Ensenada

Alvarez-Borrego S, Rivera JA, Gaxiola-Castro G, Acosta-Ruiz MJ, Schwartzlose RA (1978) Nutrientes en el Golfo de California. Cienc Mar Baja Calif Mex 5:53–71

Alvarez-Sánchez LG, Godínez V, Lavín MF, Sánchez S (1993) Patrones de turbidez y corrientes en la Bahía de San Felipe al NW del Golfo de California. Comunicaciones Académicas CICESE CTOFT-9304, Ensenada

Argote ML, Amador A, Lavín MF (1995) Tidal dissipation and stratification in the Gulf of California. J Geophys Res 100:16103–16118

Argote ML, Lavín MF, Amador A (1998) Barotropic eulerian residual circulation in the Gulf of California due to the M_2 tide and wind stress. Atmósfera 11:173–197

Badan-Dangon A, Dorman CE, Merrifield MA, Winant C (1991) The lower atmosphere over the Gulf of California. J Geophys Res 96:16877–16896

Carriquiry JD, Sánchez A (1999) Sedimentation in the Colorado River Delta and Upper Gulf of California after nearly a century of discharge loss. Mar Geol 158:125–145

Carriquiry JD, Cupul AL, Castro PG (1992) Anomalía en el balance sedimentario en el delta del Río Colorado? GEOS Unión Geofís Mex 12:15

Cupul AL (1994) Flujos de sedimentos en suspensión y nutrientes en la cuenca estuarina del Río Colorado. MSc Thesis, Universidade Autonoma Baja California, Ensenada, Mexico

Farfán C, Alvarez-Borrego S (1992) Zooplankton biomass of the northernmost Gulf of California. Cienc Mar Baja Calif Mex 18:17–36

Félix-Pico E, Mathews CP (1975) Estudios preliminares sobre la ecología del camarón en la zona cercana a la desembocadura del Río Colorado. Cienc Mar Baja Calif Mex 2:68–85

García-de-Ballesteros G, Larroque M (1976) Elementos sobre la distribución de turbidez en el Alto Golfo de California. CalCOFI Rep 18:81–106

Guardado-Puentes J, Núñez-Esquer O, Flores-Muñoz G, Nishikawa-Kinomura KA (1973) Contaminación por pesticidas organoclorados. En: Estudio químico sobre la contaminación por insecticidas en la desembocadura del Río Colorado, Tomo I, Reporte Final a la Dirección de Acuacultura de la Secretaría de Recursos Hidráulicos, Univ Aut Baja Calif, Ensenada

Guevara-Escamilla S, Huerta-Díaz MA, Félix-Pico E, Farfán C, Mathews C (1973) Biología. En: Estudio Químico sobre la Contaminación por Insecticidas en la Desembocadura del Río Colorado, Tomo II, Reporte Final a la Dirección de Acuacultura de la Secretaría de Recursos Hidráulicos, Univ Aut Baja Calif, Ensenada

Gutierrez-Galindo EA, Flores-Muñoz G, Villaescusa-Celaya J (1988) Chlorinated hydrocarbons in molluscs of the Mexicali valley and Upper Gulf of California. Cienc Mar Baja Calif Mex 14:91–113

Gutierrez G, González JI (1989) Predicciones de marea de 1990: estaciones mareográficas del CICESE. Informe Técnico OC-89-01, CICESE, Ensenada

Hernández-Ayón M, Galindo-Bect S, Flores-Báez BP, Alvarez-Borrego S (1993) Nutrient concentrations are high in the turbid waters of the Colorado River delta. Estuar Coast Shelf Sci 37:593-602

Lavín MF, Organista S (1988) Surface heat flux in the northern Gulf of California. J Geophys Res 93:14033-14038

Lavín MF, Sánchez S (1999) On how the Colorado River affected the hydrography of the Upper Gulf of California. Continent Shelf Res 19:1545-1560

Lavín MF, Godínez VM, Alvarez LG (1998) Inverse estuarine features of the Upper Gulf of California. Estuar Coast Shelf Sci 47:769-795

León-Portilla M (1972) Descubrimiento en 1540 y primeras noticias de la isla de Cedros. Calafia 2:8-10

Meckel LD (1975) Holocene sand bodies in the Colorado delta area, northern Gulf of California. In: Broussard MC (ed) Deltas, models for exploration. Houston Geol Soc, Houston, pp 239-265

Mellink E, Palacios E, Gonzalez S (1997) Non-breeding waterbirds of the delta of the Rio Colorado, Mexico. J Field Ornithol 68:113-123

Morrison RIG, Ross RK, Torres MM (1992) Aerial surveys of Neartic shorebirds wintering in Mexico: some preliminary results. Progress note, Canadian Wildlife Service, Canadian Ministry of the Environment, Ottawa

Mosiño P, García E (1974) The climate of Mexico. In: Bryson RA, Hare FK (eds) World survey of climatology, vol II: climates of North America. Elsevier, New York

Roden GI (1964) Oceanographic aspects of the Gulf of California. In: Andel TjH van, Shor GG Jr (eds) Marine geology of the Gulf of California: a symposium. Am Assoc Petrol Geol Mem 3:30-58

Santamaría-Del-Angel E, Alvarez-Borrego S, Muller-Karger FE (1994) Gulf of California biogeographic regions based on coastal zone color scanner imagery. J Geophys Res 99:7411-7421

Silber GK, Newcomer MW, Silber PC, Pérez-Cortéz H, Ellis GM (1994) Cetaceans of the northern Gulf of California: distribution, occurrence, and relative abundance. Mar Mammal Sci 10:283-298

van Andel TJH, Shor GG Jr (1964) Marine geology of the Gulf of California: a symposium. Am Assoc Petrol Geol Mem 3:1-2

A Summary of Natural and Human-Induced Variables in Coastal Marine Ecosystems of Latin America

B. KJERFVE, U. SEELIGER, and L. DRUDE DE LACERDA

1 Climate and Hydrology

Coastal marine ecosystems along the coast of South and Central America from 56°S to 33°N latitude are largely governed by interactions of broad-scale oceanographic conditions with climate and hydrology. In consequence, a comprehensive understanding of variations in structure and processes of these ecosystems requires an overview of climatic, hydrological, and oceanographic variations as well as an assessment of the modifying impact of human activities.

In the far south of Latin America the climate is chilly subarctic marine but is tropical for most of the continent. Rainfall varies greatly between the extremes at Calama in the Atacama Desert on the northern coast of Chile, where rain has never been measured, to an annual precipitation of 12,717 mm in the upper Atrato Basin in Colombia (Eslava 1992). Freshwater runoff into the Pacific and Atlantic oceans was determined by the Miocene uplift of the Cordilleras (Andes), which separated the continent into a narrow, tectonically active Pacific coast and a vast, stable shield with Precambrian rocks and shallow sedimentary deposits to the east (see chapter by Kellogg and Mohriak). As a result, more than 90 % of the continental drainage flows into the Atlantic Ocean and the Caribbean Sea, including the drainages of the Amazon (Brazil), Orinoco (Venezuela), and Paraná (Argentina-Brazil-Paraguay-Bolivia), which are among the largest rivers in the world. Several other very large rivers also drain to the east or north, such as the Magdalena (Colombia), which empties into the Caribbean, and the Uruguay (Uruguay), Tocantins (Brazil), and São Francisco (Brazil) rivers which discharge into the Atlantic Ocean (Fig. 1; Table 1). The drainage basins east of the Andes are relatively large but small in numbers, whereas the Pacific coast is characterized by numerous small river basins. However, in the high rainfall area of Pacific Colombia the discharges are

Table 1. Statistics for the ten rivers in South America with the greatest water discharges. For locations see Fig. 1

River	Length (km)	Area (km^2)	Discharge (m^3 s^{-1})	Sediment load (10^6 t year^{-1})	Sediment yield (t km^{-2} year^{-1})	Source
Amazon (Brazil)	6,520	7,500,000	150,000	1182.60	158	J.P.M. Syvitski
Orinoco (Venezuela)	2,500	990,000	34,500	152.32	154	J.P.M. Syvitski
Paraná (Arg/Bra/Par/Bol)[a]	3,740	2,302,000	13,940	108.40	47	P.J. Depetris
Tocantins (Brazil)	2,700	760,000	11,000	–	–	J.P.M. Syvitski
Magdalena (Colombia)	1,612	257,438	7,200	144.00	560	J.D. Restrepo
Uruguay (Uruguay)	1,600	189,000	4,670	2.70	10	P.J. Depetris
Atrato (Colombia)	710	35,700	4,500	22.21	622	J.D. Restrepo
São Francisco (Brazil)	2,900	640,000	3,040	5.75	9	J.P.M. Syvitski
San Juan (Colombia)	352	14,300	2,550	16.42	1150	J.D. Restrepo
Guaibá (Brazil)	600	97,990	2,500	–	–	B. Kjerfve

[a] According to Goniadzki (1999), the Paraná River has experienced significantly increased discharge (mean >30,000 m^3 s^{-1}) during the past 20 years

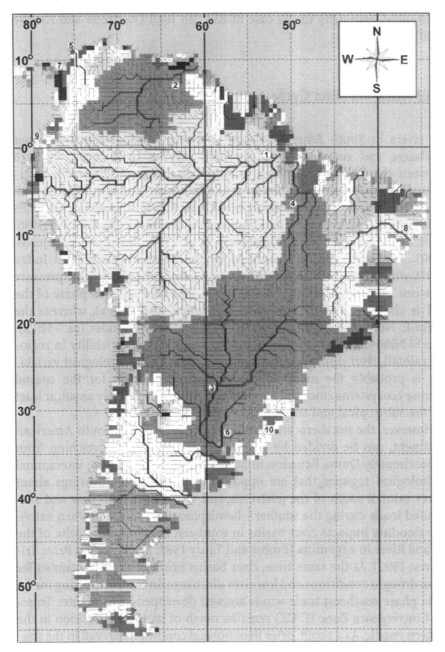

Fig. 1. Drainage basins map with a resolution of 30 min, showing boundaries and simulated topological network for all river basins of South America. *Numbers* refer to rivers in Table 1. Data were produced as part of the Global Hydrological Archive and Analysis System at the University of New Hampshire. (After C. Vörösmarty at http://www.r-hydronet.sr.unh.edu/grids/networks/sa.html

substantial, especially in the case of Rio San Juan (Restrepo and Kjerfve 2000).

2 El Niño–La Niña Cycle

All rivers in South America display a strong seasonal signal of water discharge and sediment load, typically varying by a factor of 5–10 between the low and the high monthly discharge. Additionally, interannual variability of water discharge and sediment load is associated with the El Niño–Southern Oscillation (ENSO) or El Niño–La Niña cycle with a factor of 2–4 between low and high annual discharge (Richey et al. 1986, 1989; Depetris et al. 1996; Vörösmarty et al. 1996; Restrepo and Kjerfve 2000). This cycle can be quantified by the Southern Oscillation Index (SOI), which is defined as the difference in atmospheric sea level pressure between Tahiti and Darwin (Glantz 1997). The cold La Niña phase of the SOI is characterized by a positive SOI index (ca. +5 hPa), whereas the warm El Niño phase is characterized by a negative SOI index (ca. -5 hPa). The El Niño–La Niña cycle gives rise to a significant variability in regional rainfall, river discharge, and sediment load. This hydrological variability is probably the most important forcing function for the coastal marine ecosystems since the temperature signal is relatively small, at least for the subtropical and tropical portions of the continent.

However, the northern and southern portions of the South American continent, can be divided by the subtropical jetstream, stretching from approximately Quito, Ecuador, to São Paulo, Brazil, and have interannual hydrological regimes that are opposite in phase. El Niño brings about heavy rainfall south of the jetstream, and rivers have high flow and suspended loads during the southern-hemisphere late summer, when extensive flooding impacts river basins in southern Brazil and the delta of the Paraná River in Argentina (Probst and Tardy 1989; Mechoso and Perez-Iribarren 1992). At the same time, river basins north of the jetstream suffer from drought conditions and low river discharge. In contrast, during the La Niña phase southeast trade winds are well developed and the Inter Tropical Convergence Zone (ITCZ) remains north of its typical position in the eastern Pacific. As a result, drier than normal conditions prevail in the southern portion of the South American continent but intense rainfall occurs in the northern parts and in portions of Central America (Ropelewski and Halpert 1987). Rivers in Colombia and Venezuela, in particular, experience catastrophic floods during La Niña seasons, often having drastic social and economic impacts.

3 Ocean Currents

The prevailing oceanographic conditions vary greatly along the coast of South and Central America. The Pacific coast is characterized by a north-flowing, cold-water, eastern boundary current, the Peru or Humboldt Current, with surface water temperatures of 11–17 °C, and a volume transport of approximately $20 \times 10^6 \, m^3 \, s^{-1}$. The current is not necessarily well defined and varies in magnitude between the inner shelf and seaward of the narrow shelf. During La Niña conditions, wind dynamics produce a strong offshore Ekman flux and upwelling just seaward of the shelf from as deep as 2,000 m in a band between 40°S and 4°S where approximately 20 % of the global industrial fish are harvested (see Chap. 16 by Tarazana and Arntz). Every 2–5 years, when equatorial winds weaken and the El Niño phase of the ENSO cycle peaks (i.e., 1982–1983, 1986–1987, 1991–1992, and 1997–1998), a catastrophic breakdown of the coastal marine upwelling system occurs with an enormous impact on coastal marine Pacific ecosystems as well as on ecosystems elsewhere. Coastal upwelling is also very much a feature of the Pacific coast of Mexico, where northwesterly winds and the southward flowing California Currents produce upwelling locally along the Pacific coast of Baja California (see Chap. 21 by Ibarra-Obando et al.) along with an arid coastal climate (Fig. 2).

The Atlantic coast of South America, with the exception of a section from 6°S (Rio Grande do Norte State) to 16°S (Abrolhos banks; see Chapter by Kellogg and Mohriak and Chap. 6 by Zelinda et al.), is characterized by relatively wide continental shelves. A western boundary current, the Brazil Current, flows southward from just south of the equator to approximately 35°S latitude and is part of the subtropical anticyclonic gyre in the South Atlantic. The Brazil current has a volume transport of $20–40 \times 10^6 \, m^3 \, s^{-1}$ and is characterized by coastal surface water temperatures between 26 and 30 °C (Castro and Miranda 1998). A number of coastal capes from Bahia to Rio de Janeiro, Brazil, introduce intense local coastal upwelling downstream of the capes. Upwelling is particularly intense and persistent atCabo Frio (see Chap. 7 by Valentin) where surface temperatures measure 12–18 °C even during the austral summer. The Brazil current converges with the northward flowing cold-water Malvinas Current between the mouth of Rio de la Plata and Lagoa dos Patos to form the Subtropical Convergence in the southwest Atlantic (see Chap. 11 by Odebrecht and Castello). The predominant flow on either side of the convergence zone continues towards the east (Fig. 2).

The Caribbean is characterized by a poorly organized surface flow from the east towards the west-northwest, the so-called Caribbean Current,

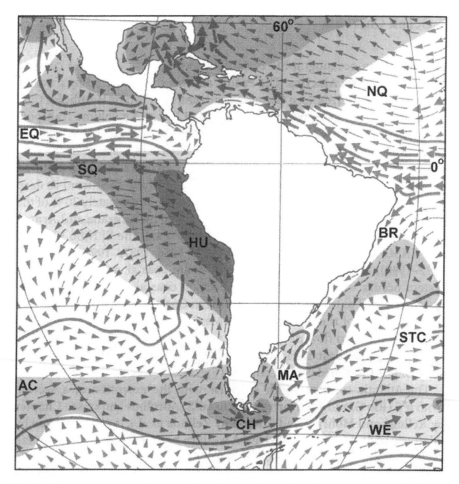

Fig. 2. Mean ocean currents around South America and surface temperature distribution deviations from the normal distribution on a globe covered by water. *EQ* Equatorial Current, *HU* Humboldt Current, *BR* Brazil Current, *MA* Malvinas Current, *NQ* North Equatorial Current, *CH* Cape Horn Current, *WE* Wendell Current, *SQ* South Equatorial Current, *STC* Subtropical Convergence, *AC* Antarctic Circumpolar Current. (Modified from Dietrich et al. 1980)

which is the continuation of the Guyana Current flowing towards the northwest along the coast of northern Brazil and the Guyanas. The Caribbean Current consists of a number of dynamic westward propagating anticyclonic and cyclonic eddies, which develop as the flow crosses the Lesser Antilles (Murphy et al. 1999). The anticyclonic eddies propagate along the northern half of the Caribbean basin, whereas cyclonic eddies are confined to the north coast of Soth America; the southwestern Caribbean gulf bordered by Colombia, Panama, and Costa Rica; and the Gulf of Honduras (see Chap. 1 by Heyman and Kjerfve).

4 Tides, Waves, and Relative Sea Level

Tidal processes break down vertical stratification in the water column, disperse sediment, nutrients, and pollutants and generate residual tidal circulation in estuaries and bays. Thus, it is important to understand local manifestation of tidal processes in managing coastal systems. Tidal conditions vary along the coasts of the South American continent. Tides along the Pacific coast are mostly mixed semidiurnal. Spring tides range from 1.5 m in the south of the continent, 5 m in the Gulf of Panama, to more than 7 m at the mouth of the Colorado River in the inner Gulf of California (see Chap. 22 by Alvarez-Borrego). Part of the Pacific coast of Mexico until the entrance to the Gulf of California experiences mixed, mainly diurnal tides with a tidal range of less than 1 m (see Chap. 20 by Flores Verdugo et al.). South of the equator, the predominant semidiurnal (M_2) tidal phase progresses from north to south along the Pacific coast of South America, consistent with the propagation of a Kelvin wave. The progression of the tidal phase from the Gulf of Panama to the Gulf of California cannot easily be attributed to a large-scale amphidromic system, whereas along the Pacific coast of Baja California the semidiurnal phase progresses northward, also consistent with the propagation of a Kelvin wave (Fig. 3). Along the Atlantic coast south of the mouth of the Amazon River, tides are mostly semidiurnal with a northward progression of the dominant M_2 phase. Macrotidal conditions occur on the broad Patagonian shelf, including the entrance to the Strait of Magellan, with peak spring tides exceeding 11 m at Rio Gallegos, Argentina. The coast between the mouth of the Rio de la Plata estuary and Cabo Frio, Brazil, experiences mostly mixed semidiurnal or diurnal tides (range <1 m). Macrotides with spring ranges exceeding 6 m also occur in inshore waters along the coast of northern Brazil (Maranhão and Pará) and the mouth of the Amazon River. North of the equator the spring tidal range decreases from 6 to 2 m in the delta of the Orinoco (see Chap. 4 by Conde) and the semidiurnal tide occurs largely in phase. Within the Caribbean, the tide is microtidal with an average range of 0.2 m. The tide is mixed and mainly diurnal in the eastern portion of the Caribbean basin, then becomes diurnal and further to the west mixed and mainly diurnal, and in the western portion of the basin mixed and mainly semidiurnal (Kjerfve 1981; Fig. 3).

Waves, and the longshore currents they produce, are the main cause of coastal erosion, alongshore transport, and sediment deposition. Storm

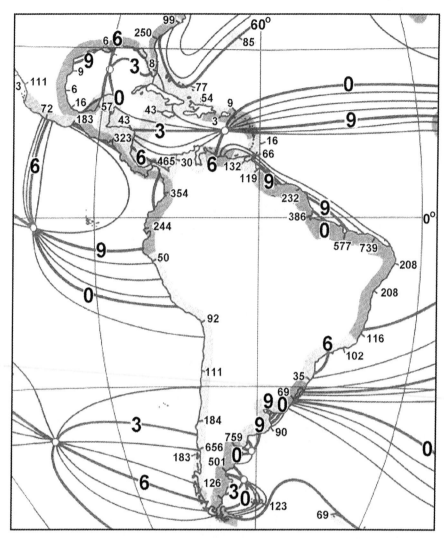

Fig. 3. Cotidal phase of the semidiurnal M_2 tide around South America, where the *large numbers* refer to the number of lunar hours the local high tide lags the passage of the moon across the Greenwich meridian (epoch). The *small numbers* along the coast are the mean semidiurnal spring tidal ranges, $2\times(M_2+S_2)$, in cm. (Modified from Dietrich et al. 1980

wave environments predominantly exist along the Pacific and Atlantic shores of the southern cone of South America. Long-period swells from the southwest, often with wave heights greater than 4 m and periods longer than 10 s, occur along the entire west coast of South and Central America (Davies 1980). The Pacific coast also experiences occasional, very destructive tsunamis in response to earthquakes around the rim of the Pacific basin. The Atlantic coast between the Rio de la Plata estuary and Recife, Brazil, is dominated by east coast swells from the south or southeast, whereas the coast north of Recife mainly experiences swells from the north and northeast. Tropical cyclones (hurricanes in the Caribbean and cyclones in the western Pacific) generate high and destructive waves several times annually to Caribbean shores and the Pacific coast of Mexico.

Because of global warming during the past century, sea level has been rising eustatically approximately 1.2 mm year^{-1}. However, the relative sea level depends not only on the change in global sea level but also on regional and local changes in land elevation and oceanographic, atmospheric, and hydrological conditions (Stewart et al. 1990). Relative sea levels along both east and west coasts of South America have in the past few decades been rising ever so slowly. Relative sea level change along the west coast has varied locally depending on uplift and slumping associated with geotectonics of the Cordilleras and plate subduction below the Pacific margin (see chapter by Kellogg and Mohriak). The east coast, on the other hand, has experienced a relative sea level fall of 3–7 m in the past 5,500 years (Isla 1989; Martin and Dominguez 1994; Gómez and Perillo 1995; Angulo and Lessa 1997; Lanfredi et al. 1998).

5 Human Impacts

The exponential increase of the human population in the past century is seriously impacting the global climate and coastal marine ecosystems. In 1996, Latin America and the Caribbean had a total population of 486 million, or 8.4 % of the global population compared to 12 % of the global land area. Although the population density for the continent is less than the global average, the doubling time for the population is 39 years (Population Reference Bureau 1996). The exploitation of habitats/resources is a function of human population size and density, generally being many times greater at the coast, and the population's level of socio-economic development. As a consequence, some of the highly productive coastal and marine ecosystems in South and Central America are subjected to high levels of anthropogenic impact.

Large coastal bays and gulfs, which for most parts have formed as a result of tectonic activity, are common features of South America's coasts. The Gulf of Guayaquil in Ecuador (see Chap. 17 by Twilley et al.) is the largest inland coastal marine system on the Pacific coast. Both Bahía de Buenaventura, Colombia (see Chap. 18 by Cantera and Blanco), and Baía de

Paranaguá, Brazil (see Chap. 10 by Lana et al.), are gulf-like, large drowned river valley estuaries on the west and east coast of the continent, respectively. Río de la Plata is a gigantic coastal plain estuary bordered by Buenos Aires and Montevideo (see Chap. 13 by Mianzan et al.). The Maracaibo system in Venezuela is characterized by anoxic conditions in a large area of the hypolimnion (see Chap. 3 by Rodríguez). Baía de Guanabara in Rio de Janeiro, Brazil, is eutrophic and polluted as a result of largely untreated drainage from a population of 8 million in the catchment (see Chap. 8 by Kjerfve et al.).

Freshwater runoff and sediment loads into the coastal areas have a profound effect on the character of coastal ecosystems. However, human activities within the river basins not only alter water and sediment supplies but also the inputs of nutrients and pollutants, which impact the quality of these environments. Most Latin American coastal marine systems have been impacted by the damming of rivers, closing of tidal inlets, or diversion of freshwater and deforestation within the watershed, which has changed the seasonal cycle and the amount of freshwater runoff and increased sediment loading of rivers due to soil erosion. Subsequent changes of salinity levels and cycles and high siltation have adversely affected shallow nearshore habitats and coastal wetlands, which play an important role as nursery areas. The damming of the Colorado River has produced a hypersaline environment in the inner Gulf of California (see Chap. 22 by Alvarez-Borrego). Humans have profoundly altered the Ciénaga Grande coastal lagoon during the past four decades with a general degradation of the ecosystem, primarily by eliminating coastal inlets when constructing a coastal highway (see Chap. 2 by Polania et al.). The diversion of freshwater and sediment into the Cananéia estuary has significantly modified most aspects of the estuarine environment (see Chap. 9 by Tundisi and Matsumura-Tundisi). The levels of nutrient flux from terrestrial ecosystems to the aquatic environment are a function of drainage and land use. Although the overall productivity of coastal and inland seas is normally increased by nutrient runoff, excess nutrients may negatively influence fisheries of semi-enclosed water bodies (see Chap. 4 by Conde and Chap. 2 by Polania et al.). Apart from the serious impact of many land-bound human activities on marine systems, oil and gas extraction in nearshore waters and semi-enclosed seas will also have to be reconciled with sustainable development of marine resources (see Chap. 3 by Rodríguez).

Furthermore, high density aquaculture in coastal water bodies with reduced water exchange can provoke excessive nutrient loading and local eutrophication with negative feedback on the aquaculture activities themselves. The extensive conversion of mangrove forests to shrimp pond culture may reduce natural shrimp populations which use mangrove systems as nursery grounds, and compromise the role the coastal wetlands play as buffer zones (see Chap. 17 by Twilley et al.). Hence, to avoid adverse impacts, aquaculture, and other coastal activities, should be planned within a framework of Integrated Coastal Area Management (ICAM). Mangrove clearance and wetland drainage also pose potential and actual conflicts with the interest of artisanal coastal fishermen and the rights of indigenous people (see Chap. 20 by Flores Verdugo et al.) which should be ensured as part of management plans for the coastal environment and its living resources. Other estuarine systems, such as Bahia Blanca, Argentina (see Chap. 14 by Perillo et al.), the Itamaracá estuary, Brazil (see Chap. 5 by Medeiros et al.), and the gigantic almost freshwater Patos Lagoon, Brazil (see Chap. 12 by Seeliger), have been impacted only slightly by human activities.

The main source of nutrients, either runoff, upwelling or tidal mixing, determines the productivity of the shelf ecosystems. Along most of the Atlantic coast of South America, shelf resources with high commercial value are fully exploited or overfished (see Chap. 11 by Odebrecht and Castello). Although overexploitation has contributed to the decline and collapse of populations at the Pacific coast of South America, synchronous natural variations in the abundance of stocks associated with ENSO phenomena also are important (see Chap. 16 by Tarazona and Arntz), particularly in upwelling systems. Less well known than the reefs of the Caribbean, the extensive eastern Pacific coral reefs (see Chap. 19 by Glynn) and the Abrolhos reefs (see Chap. 6 by Leão and Kikuchi) are characterized by many habitats and species that have evolved over long periods of time within a stable environment. In general, reef communities lack the ability to adjust to unusual stress, and thus are extremely sensitive to environmental change and are negatively affected by increased nutrients and turbidity. Some of the coral reef systems have been irreversibly degraded by human activities and their communities have been replaced with less complex systems based on algae growing on dead coral heads. For their restoration and preservation, damaging influences must be removed and access to reefs and their resources restricted.

In conclusion, effective measures to protect coastal marine ecosystems and their habitats along both the Pacific and Atlantic coasts of Latin America must consider natural short- and long-term variations of ecosystem forcing functions as well as human activities, especially in watersheds discharging into coastal seas. Therefore, integrated scientific research efforts related to system processes and long-term monitoring of ecosystem variables and resources need implementation and/or expansion. Coastal marine ecosystems and resources that are not yet degraded need to be brought within effective management regimes. The establishment of coastal marine parks or protected areas (see Chap. 1 by Heyman and Kjerfve and Chap. 19 by Glynn) may contribute to revenues from non-exploitive use of resources, as well as preserving important stocks.

References

Angulo RJ, Lessa GC (1997) The Brazilian sea-level curves: a critical review with emphasis on the curves from the Paranaguá and Cananéia regions. Mar Geol 140:141–166

Castro BM, Miranda LB (1998) Physical oceanography of the western Atlantic continental shelf located between 4°N and 34°S. In: Robinson AR (ed) The sea, vol 11. Wiley, New York, pp 209–251

Davies JL (1980) Geographical variation in coastal development, 2nd edn. Longman, London

Depetris PJ, Kempe S, Latif M, Mook WG (1996) ENSO-controlled flooding in the Paraná River (1904–1991). Naturwissenschaften 83:127–129

Dietrich G, Kalle K, Krauss W, Siedler G. (1980) General oceanography: an introduction, 2nd edn. Wiley, New York

Eslava J (1992) La precipitación en la región del Pacífico colombiano (Lloró: el sitio más lluvioso del mundo). Rev Zenit 3:47–71

Glantz MH (1997) Currents of change, El Niño's impact on climate and society. Cambridge University Press, Cambridge

Gómez EA, Perillo GME (1995) Sediment outcrops underneath shoreface-connected sand ridges, outer Bahía Blanca estuary, Argentina. Quat S Am Antarctica Peninsula 9:27–42

Goniadzki D (1999) Hydrological warning and information system for the Plata basin. Instituto Nacional del Agua y del Ambiente, Buenos Aires, Argentina

Isla FI (1989) Holocene sea-level fluctuation in the southern hemisphere. Quat Sci Rev 8:359–368

Kjerfve B (1981) Tides of the Caribbean Sea. J Geophys Res 86(C5):4243–4247

Lanfredi NW, Pousa JL, D'Onofrio EE (1998) Sea-level rise and related potential hazards on the Argentine coast. J Coast Res 14:47–60

Martin L, Dominguez JML (1994) Geological history of coastal lagoons. In: Kjerfve B (ed) Coastal lagoon processes. Elsevier, Amsterdam, pp 41–68

Mechoso CR, Perez-Iribarren G (1992) Streamflow in the Southeastern South America and the Southern Oscillation. J Climate 5:1535–1539

Murphy SJ, Hurlburt HE, O'Brien JJ (1999) The connectivity of eddy variability in the Caribbean Sea, the Gulf of Mexico, and the Atlantic Ocean. Journal of Geophysical Research 104:1431–1453

Population Reference Bureau (1996) World population data sheet. Demographic data and estimates for the countries and regions of the world. Washington, DC

Probst JL, Tardy Y (1989) Global runoff fluctuations during the last 80 years in relation to world temperature change. Am J Sci 289:267–285

Restrepo JD, Kjerfve B (2000) Water and sediment discharges from the western slopes of the Colombian Andes with focus on Rio San Juan. Journal of Geology 108:17–33

Ropelewski CF, Halpert MS (1987) Global and regional scales precipitation associated with El Niño-Southern Oscillation. Mon Weath Rev 115:1606–1626

Richey JE, Meade RH, Salati E, Devol AH, Nordin CF, dos Santos U (1986) Water discharge and suspended sediment concentrations in the Amazon River: 1982-1984. Water Resour Res 22:756-764

Richey JE, Nobre C, Deser C (1989) Amazon River discharge and climate variability: 1903 to 1985. Science 246:101-103

Stewart RW, Kjerfve B, Milliman J, Dwivedi SN (1990) Relative sea level change: a critical evaluation. UNESCO Rep Mar Sci 54:1-22

Vörösmarty CJ, Willmott CJ, Choudhury BJ, Schloss AL, Stearns TK, Robeson SM, Dorman TJ (1996) Analyzing the discharge regime of a large tropical river through remote sensing, ground-based climatic data, and modeling. Water Resour Res 32: 3137-3150

Subject Index

Abrolhos Bank   83
Abrolhos Depression   83
Acanthaster planci   284, 293
advective transport   74
algal blooms   138, 160
Amazon River   62
anadromous   192
anchovy   236
Andean margin   6, 7
anoxic conditions   50, 58
anthropic activities   276
anthropogenic   297
anthropogenic forcing   326
anthropogenic impact   349
anthropogenic threats   299
anticyclonic dominance   111
aquaculture   127
    mammals   66
aquatic vascular plant   64
Argentina   159
artisanal fisheries   160
artisanal fishermen   181
assemblages   156
Atlantic coastline   1
autochthonous carbon   175

backwaters   119
bacteria   152
bacterial populations   126
bacterioplankton   233
Bahia de Amatique   17
Bahia San Blas   205
Baixada Fluminense   108
Baja California   282, 285, 332
Baleia Point   85
banana industry   26
bank reefs   90
barrier island   268
beach types   223
Belize barrier reef   17, 25

benthic community   101, 170
benthic fauna   275
benthic organisms   235
benthos   211, 253
bioconcentration   115
biodiversity   25, 30, 142, 157
bioeroders   291
bioeroding mollusks   88
bioerosion   270, 284
biogeochemical process   323
biological change   321
bird fauna   66
birds   38
bivalves   102
bloom   41
boring organisms   269
Brazil   159
Brazil Current   85, 98, 147
Brazilian margin   3
breakup of Gondwana   1
breeding sites   173
Buenaventura Bay   265

calcification rates   294
calcium carbonate accumulation   295
California Current (CC) System   316
California Currents   282
carbonate platform   5
Caribbean deformed belt   12
Caribbean margin   10, 11
Caribbean sea   33
Catatumbo River   47
Cayman Trench   17
Central American seaway   285
cephalopd   155
Cetacea   55
Chilean coast   219
Chiloé Island   219
chlorophyll   76, 115, 255

Subject Index

clam fisheries   225
cliff erosion   269
Clipperton Atoll (largest E Pacific coral reef)   288
Coastal Arc   90
coastal current   229
coastal geomorphology   315
coastal lagoon   22, 33, 307, 322
coastal upwelling   10, 230, 345
coastal zone management   161
Cocos Island   291
Cocos Plate   307
cofferdam   67
Coiba Island   288
cold fronts   132
coliform   78
Colombia   265, 288
Colorado River   205, 331
commercial shipping docks   114
commercial species   160
cone-shaped hypolimnion   49, 51
consumer organisms   176
contamination   214
continental drainage flow   341
continental shelf   6, 345
coral   25, 281
coral bleaching   284, 298
coral fauna   86
coral pinnacles "chapeirões"   90
coralline algae   88
corallivores   292
Coriolis force   231
Costa Rica Coastal Current   282
cownose ray   22
crown-of-thorns sea star   284
crustacean guards   293
current   345
cyclonic circulation   334
cyclonic circulation gyre   346
cyclonic eddies   21, 152

Dagua River   266, 267
damselfish   292
Dean's parameter   224
decomposition   176, 267
degradation   40
demersal   156
demersal fishes   103
demersial fisheries   237
desert landscape   328
detritivores   272

detritivorous   194
detritus   172, 256
diatoms   238
dinoflagellate blooms   291
dissolved oxygen   122, 249, 267
diurnal variations   120
diversion of freshwater   350
domestic and industrial waste   114
dredging   56
drift-algae   172
drowned valley   265
dry sub-humid   307

eastern Brazil coastal region   85
eastern tropical Pacific   281
ebb deltas   206
ecological functions   122
Economic Exclusive Zone (EEZ)   159
economic impact   260
ecosystem management   259
ecotone   196
ecotourism   28
Ecuador   245, 282, 288
Ekman transport   231
El Niño   94, 151, 180, 238, 247, 267, 284, 288, 340, 344
El Salvador   287
elasmobranchs   155, 192
encrusting community   172
endemic   54, 65
endemism   317
energy flow   140
Engraulididae   37
ENSO   298, 321, 344,
environmental impact   199
environmental quality   258
epilimnion   49
epiphyte colonization   175
Equatorial Under Current   282
estuarine circulation   107
estuarine habitats   169
estuarine plume   149
estuarine stocks   181
estuarine zonation   133
estuary   185
euhaline   35
euryhaline forms   52
eutrophication   36, 181, 258, 260, 309
exotic species   303

Subject Index

faunal assemblages 271
fertilization 151
filamentosous bacteria 236
filter feeding 194
fine sediments 120
fish 320
fish assemblages 65
fish fauna 126
fisheries 40, 42, 56, 67, 68, 113, 127, 142, 199, 200, 278, 328
fishery policy 161
fishery products 27
food chain 176, 158, 238
food web 274, 276
foredeep basins 12
forested freshwater wetland 310
freshwater discharge 111
freshwater flow rates 168
freshwater input 135
freshwater runoff 110, 149
freshwater tributaries 207
fringing reefs 92
front 187, 188, 189, 199

Galapagos Islands 284, 288, 291
geomorphology 205
Gladden Spit 19, 26
Glen Canyon dam 333
Gorgona Island 288
gradients 36
gravel bottom 194
gravitational circulation 112
grazer 178, 235
grazing pressure 99
groundwater 328
groundwater infiltration 308
growth history of fringing reef 296
guano birds 238
Guatemala 27, 287
Guayas River 245
Gulf of California 331
Gulf of Guayaquil 245, 282
Gulf of Papagayo 287
Gulf of Venezuela 47, 48
Guyana Current 63

herbivore populations 295
herbivory 274
heterotrophic system 326
Holocene 62, 109
Hoover dam 333

Huatulco coast 284
human impact 141, 258
human population 327
Humbolt Current 247
humic substances 120, 124
humpback whale 89
hurricane 19, 308
hydrocorals 87, 284
hydrodynamic 121, 168
hydrological balance 73
hydrological restoration 44
hypersaline system 312
hyper-synchronous-type estuary 208
hypoxia 249

ichthyofauna 54, 170, 213
ichthyoplankton 191, 193
illegal exploitation 194
indicators 241
industrial fisheries 160
industrial wastes 214
Integrated Coastal Area Management 351
inter-annual variability 174
intertidal reef flat 91
Inter-Tropical Convergence Zone 265
introduced species 189, 190
invading species 241
invertebrates 340
Isla La Plata 288

jewfish 23

kelp forest 318

La Niña 242, 344, 345
lagoon system 122
landscape ecology 30
largest known coral reef 288
latitudinal regions 315
litter fall 255
litter production 274, 310
little ice age 287
littoral communities 51

Machalilla (southern-most E Pacific coral reef) 288
macrobenthic 125, 153
macroflora 101
macroinfauna 220, 222, 223, 224
Magdalena fan 13

Magdalena river   34
Malpelo Island   288
Malvinas Current   147
management   242
management strategies   182
manatee   77, 79
mangrove   22, 38, 52, 64, 74, 75, 76, 77, 78, 79, 119, 137, 251, 309
mangrove communities   113, 297
mangrove crab   255
mangrove forests   123
mangrove litter   122, 275
mangrove timber   67
mangrove types   271
Maria Cleofas Island   287
Marias Islands   287
marine mammal   337
marine parks   351
marine protected areas   299, 300, 301, 302
marine reserves   28
marine terraces   9
marine tourism   94
   marine turtles   89, 311
marsh physiography   172
marshes   192
mass balance   325
Maya Mountains   22
meanders   151
Mediterranean-type climate   316
meiobenthos   139
mercury   78
meroplankton   340
Mesodesma donacium   225
meso-scale gyres   20
metal   40, 277
microbenthos   139
micronekton   319
microplankton   158
Micropogonias furnieri   196
microtidal   347
micro-tide   21
migration   38, 170, 242
milky waters or white tides   233
Millepora   87, 92, 284
mixed layer depth   248
mixing processes   125
mixohaline   35
model   259
Montague Island   335
mud flats   272

muddy bottom   191, 194
Mussismilia   86

N:P ratio   136
nanoplankton   157
natural disturbances   297
Nayarit   285
negative estuary   323
nekton community   273
nesting site   42
Nicaragua   287
nitrogen   249
nursery   325
nursery ground   42, 195, 170
nutrients   51, 63, 158
nutrient cycling   295
nutrient distribution   135
nutrient dynamics   126
nutrient loading   350

ocean currents   220
oceanic fish   241
oceanographic conditions   341
octocorals   87
oil prospecting   67
ontogenic vertical migration   101
organic matter   125, 256
Orinoco   61
Outer Arc   93
overexploitation   161, 225
   over-fishing   299
overflow events   44
oxygen concentration   232
oyster   40

Pacific Andean coastline   1
Palythoa   88
Panama Canal   302
Panama Current   247
Paranaguá Bay   131
Parcel dos Abrolhos   93
particulate carbon   255
particulate organic matter   127
Patos Lagoon   149, 167
Pavona   285
pelagic fish   103
pelagic food web   100
Peru (or Humboldt) currents   229
Peru Coastal Current   282
Peru Oceanic Current   229
Peru Subsurface Countercurrent   230

Subject Index

Peruvian hake   237
pesticide   338
pH   249
photosynthesis   198
phytoplankton   53, 64, 76, 78, 124, 152, 212, 234, 253, 267, 273
phytoplankton community   319, 324
picoplankton   157
pigment concentration   336
pilchard   236
plankton community   169
plant cover   136
plant production   141
Pleistocene regression   83
pluvial regime   61
Pocillopora   285
polar fronts   148
pollution   312
porites   285
Port Honduras   23
post-larvae   254, 257
precambrian formation   62
predators   235
preservation   278
primary producers   21
primary production   39, 55, 76, 78, 99, 149, 174, 233, 273, 325
pristine   322
productivity   37, 254
protozoans   234
protozooplankton   152
provincial boundary   317
Puerto Barrios   26
pulses   35
Punta Sal   28

recruitment   37, 102
redox potential   125
reef walls   92
residual circulation   74
restoration   43
retention   196
Revillagigedo Islands   288
Rhizophora   251
rhodolith communities   297
rift   2
rifting   5
Rio de Janeiro   107
Río de La Plata   149, 185
river discharge   266, 344
river runoff   94

rivers in South America   342
rocky shores   269
Ruppia maritima beds   171

saline intrusion   327
salinity   50, 210
salinity front   185, 188
salinity preferences   198
salinity-energy gradient   134
salinization   180
salt flux   75
salt marshes   137
salt tectonics   2, 5
salt wedge   185, 187
sand dynamics   224
sand waves   110
sandy beach   268
Sanozama   10
Sauce Chico River   207
Scianidae   192
sea birds   103
sea level   93, 232, 349
sea stars   102
sea surface temperatures   232
seabirds   89, 275
seagrass   77, 79
sea-level canal   302
seasonal variability   222
seawater temperature   219
Secchi depth   335
secondary production   39, 179
sediment loads   350
sediment runoff   207
sediment transport   335
sedimentation rates   109
sewage   57, 277
shelf ecosystems   351
shell middens   311
Shore Bird Reserve Network   337
shrimp   254, 257
shrimp ponds   252, 258
Sierra Nevada de Santa Marta   34
silicate   250
siltation stress   288
soft bottom   171, 237, 272
soil salinity   124
Soldado Island   268
soluble reactive phosphate   250
South Atlantic central water (SACW)   97
South Pacific Ocean   230
southeastern Brazil   131

Southwest Atlantic   147, 167
squids   103
stationary fronts   132
stratification   35, 74, 112, 133, 210, 248, 267
subduction of oceanic ridges   8
sublittoral community   52
subsidence of coastline   7
subtropical convergence   147, 167
supralitoral level   191
suspended matter   168
sustainable management   278
sustainable yield   68
symbionts   292

Tablazo Bay   47, 48
tectonic activity   349
teleost   155
temperature   209
Tertiary depression   108
Tethys Sea   285
thermohaline circulation   333
tidal creeks   270
tidal currents   49
tidal energy   208
tidal flooding   65
tidal forcing   107, 134
tidal prism   112
tidal processes   347
tidal transport   125
tides   73, 74, 76, 149
timber exploitation   277
top carnivores   320
total allowable catch   161
total phosphorus   250
total suspended solid   75, 249
totoaba   336
toxic dinoflagellates   193
trawling   161
tributary channels   61
trophic dynamics   320
trophic flow   238
trophic interactions   76, 78, 157
trophic level   89, 178

trophic link   179
trophic relationships   295
trophic structure   140
tropical air masses   119
tsunamis   348
turbulence   74
turtles   22

uplift of coastal areas   7, 9
uplift of coral reefs   11
upper littoral   268
upwelling   97, 151, 187, 284, 285, 287, 288
Uruguay   159
Utria   288
UV radiation   160

vascular plants   324
vertical motions of coastline   1
vertical stratification   187
vertical zonation   269
Vitória-Trindade Ridge   4
volcanism   298
vortices   151

Warao   66
water circulation   122
water quality   259
water temperature   50
water turnover time   247
wave transport   74
West Indian manatee   24
whale sharks   26
wind   71, 74, 209, 231
wind stress   188
World Heritage Site   25

zonation   223
zooplankton   53, 76, 125, 153, 213, 234, 253, 273, 319
zooplankton aggregations   198
zooplankton biomass   336
zooplankton community   54
zooplankton grazing   138

Ecological Studies
Volumes published since 1994

Volume 105
Microbial Ecology of Lake Plußsee (1994)
J. Overbeck and R.J. Chrost (Eds.)

Volume 106
Minimum Animal Populations (1994)
H. Remmert (Ed.)

Volume 107
The Role of Fire in Mediterranean-Type Ecosystems (1994)
J.M. Moreno and W.C. Oechel (Eds.)

Volume 108
Ecology and Biogeography of Mediterranean Ecosystems in Chile, California and Australia (1995)
M.T.K. Arroyo, P.H. Zedler, and M.D. Fox (Eds.)

Volume 109
Mediterranean-Type Ecosystems. The Function of Biodiversity (1995)
G.W. Davis and D.M. Richardson (Eds.)

Volume 110
Tropical Montane Cloud Forests (1995)
L.S. Hamilton, J.O. Juvik, and F.N. Scatena (Eds.)

Volume 111
Peatland Forestry. Ecology and Principles (1995)
E. Paavilainen and J. Päivänen

Volume 112
Tropical Forests: Management and Ecology (1995)
A.E. Lugo and C. Lowe (Eds.)

Volume 113
Arctic and Alpine Biodiversity. Patterns, Causes and Ecosystem Consequences (1995)
F.S. Chapin III and C. Körner (Eds.)

Volume 114
Crassulacean Acid Metabolism. Biochemistry, Ecophysiology and Evolution (1996)
K. Winter and J.A.C. Smith (Eds.)

Volume 115
Islands. Biological Diversity and Ecosystem Function (1995)
P.M. Vitousek, L.L. Loope, and H. Adsersen (Eds.)

Volume 116
High Latitude Rainforests and Associated Ecosystems of the West Coast of the Americas: Climate, Hydrology, Ecology and Conservation (1996)
R.G. Lawford,, P. Alaback, and E. Fuentes (Eds.)

Volume 117
Global Change and Mediterranean-Type Ecosystems (1995)
J. Moreno and W.C. Oechel (Eds.)

Volume 118
Impact of Air Pollutants on Southern Pine Forests (1996)
S. Fox and R.A. Mickler (Eds.)

Volume 119
Freshwater Ecosystems of Alaska. Ecological Syntheses (1997)
A.M. Milner and M.W. Oswood (Eds.)

Volume 120
Landscape Function and Disturbance in Arctic Tundra (1996)
J.F. Reynolds and J.D. Tenhunen (Eds.)

Volume 121
Biodiversity and Savanna Ecosystem Processes. A Global Perspective (1996)
O.T. Solbrig, E. Medina, and J.F. Silva (Eds.)

Volume 122
Biodiversity and Ecosystem Processes in Tropical Forests (1996)
G.H. Orians, R. Dirzo, and J.H. Cushman (Eds.)

Volume 123
Marine Benthic Vegetation. Recent Changes and the Effects of Eutrophication (1996)
W. Schramm and P.H. Nienhuis (Eds.)

Volume 124
Global Change and Arctic Terrestrial Ecosystems (1997)
W.C. Oechel et al. (Eds.)

Volume 125
Ecology and Conservation of Great Plains
Vertebrates (1997)
F.L. Knopf and F.B. Samson (Eds.)

Volume 126
The Central Amazon Floodplain: Ecology
of a Pulsing System (1997)
W.J. Junk (Ed.)

Volume 127
Forest Decline and Ozone: A Comparison of
Controlled Chamber and Field Experiments
(1997)
H. Sandermann, A.R. Wellburn,
and R.L. Heath (Eds.)

Volume 128
The Productivity and Sustainability of
Southern Forest Ecosystems in a
Changing Environment (1998)
R.A. Mickler and S. Fox (Eds.)

Volume 129
Pelagic Nutrient Cycles: Herbivores
as Sources and Sinks (1997)
T. Andersen

Volume 130
Vertical Food Web Interactions:
Evolutionary Patterns and Driving Forces
(1997)
K. Dettner, G. Bauer, and W. Völkl (Eds.)

Volume 131
The Structuring Role of Submerged
Macrophytes in Lakes (1998)
E. Jeppesen et al. (Eds.)

Volume 132
Vegetation of the Tropical Pacific Islands
(1998)
D. Mueller-Dombois and F.R. Fosberg

Volume 133
Aquatic Humic Substances: Ecology
and Biogeochemistry (1998)
D.O. Hessen and L.J. Tranvik (Eds.)

Volume 134
Oxidant Air Pollution Impacts in the Montane
Forests of Southern California (1999)
P.R. Miller and J.R. McBride (Eds.)

Volume 135
Predation in Vertebrate Communities:
The Białowieża Primeval Forest as a Case
Study (1998)
B. Jędrzejewska and W. Jędrzejewski

Volume 136
Landscape Disturbance and Biodiversity in
Mediterranean-Type Ecosystems (1998)
P.W. Rundel, G. Montenegro,
and F.M. Jaksic (Eds.)

Volume 137
Ecology of Mediterranean Evergreen Oak
Forests (1999)
F. Rodà et al. (Eds.)

Volume 138
Fire, Climate Change and Carbon Cycling
in the North American Boreal Forest (2000)
E.S. Kasischke and B. Stocks (Eds.)

Volume 139
Responses of Northern U.S. Forests to
Environmental Change (2000)
R. Mickler, R.A. Birdsey, and J. Hom (Eds.)

Volume 140
Rainforest Ecosystems of East Kalimantan: El
Niño, Drought, Fire and Human Impacts
(2000)
E. Guhardja et al. (Eds.)

Volume 141
Activity Patterns in Small Mammals:
An Ecological Approach (2000)
S. Halle and N.C. Stenseth (Eds.)

Volume 142
Carbon and Nitrogen Cycling
in European Forest Ecosystems (2000)
E.-D. Schulze (Ed.)

Volume 143
Global Climate Change and Human Impacts
on Forest Ecosystems: Postglacial
Development, Present Situation and Future
Trends in Central Europe (2001)
J. Puhe and B. Ulrich

Volume 144
Coastal Marine Ecosystems of Latin America
(2001)
U. Seeliger and B. Kjerfve (Eds.)

Printed by Publishers' Graphics LLC